Wireless Information and Power Transfer

Wireless Information and Power Transfer: Theory and Practice

Edited by

Derrick Wing Kwan Ng
The University of New South Wales, Australia

Trung Q. Duong
Queen's University Belfast, United Kingdom

Caijun Zhong
Zhejiang University, China

Robert Schober
University of Erlangen-Nuremberg, Germany

This edition first published 2019

Registered Offices
John Wiley & Sons, Inc., 111 River Street, Hoboken, NJ 07030, USA
John Wiley & Sons Ltd, The Atrium, Southern Gate, Chichester, West Sussex, PO19 8SQ, UK

Editorial Office
The Atrium, Southern Gate, Chichester, West Sussex, PO19 8SQ, UK

For details of our global editorial offices, customer services, and more information about Wiley products visit us at www.wiley.com.

Wiley also publishes its books in a variety of electronic formats and by print-on-demand. Some content that appears in standard print versions of this book may not be available in other formats.

Library of Congress Cataloging-in-Publication Data
Names: Ng, Derrick Wing Kwan, editor. | Duong, Trung Q., editor. | Zhong, Caijun, (Professor of electrical engineering), editor. | Schober, Robert, editor.
Title: Wireless information and power transfer : theory and practice / edited by Derrick Wing Kwan Ng, The University of New South Wales, Sydney, Australia, Trung Q Duong, Queen's University Belfast, Belfast, UK, Caijun Zhong, Zhejiang University, Zhejiang, People's Republic of China, Robert Schober, University of Erlangen-Nuremberg, Erlangen, Germany.
Description: First edition. | Hoboken, NJ : John Wiley & Sons, Inc., [2019] | Includes bibliographical references and index. |
Identifiers: LCCN 2018032339 (print) | LCCN 2018033401 (ebook) | ISBN 9781119476849 (Adobe PDF) | ISBN 9781119476832 (ePub) | ISBN 9781119476795 (hardcover)
Subjects: LCSH: Wireless power transmission. | Wireless communication systems.
Classification: LCC TK3088 (ebook) | LCC TK3088 .W57 2018 (print) | DDC 621.381/044–dc23
LC record available at https://lccn.loc.gov/2018032339

Cover Design: Wiley
Cover Image: ©ktsimage/Getty Images

Set in 10/12pt Warnock by SPi Global, Pondicherry, India

Printed in Singapore by C.O.S. Printers Pte Ltd

10 9 8 7 6 5 4 3 2 1

To our loves

Contents

List of Contributors

Panos N. Alevizos
Technical University of Crete
Greece

Aggelos Bletsas
Technical University of Crete
Greece

He Chen
The University of Sydney
Australia

Wen Chen
Shanghai Jiao Tong University
China

Bruno Clerckx
Imperial College London
United Kingdom

Shuguang Cui
University of California
USA
and
Shenzhen Research Institute
of Big Data
China

Panagiotis D. Diamantoulakis
Friedrich-Alexander University
(FAU)
Germany

Trung Q. Duong
Queens University Belfast
United Kingdom

Maged Elkashlan
Queen Mary University of London
United Kingdom

Yifan Gu
The University of Sydney
Australia

Zoran Hadzi-Velkov
Ss. Cyril and Methodius University
Macedonia

Robert W. Heath Jr.
The University of Texas at Austin
USA

George K. Karagiannidis
Aristotle University of Thessaloniki
Greece

Talha Ahmed Khan
The University of Texas at Austin
USA

Ioannis Krikidis
University of Cyprus
Cyprus

Dong In Kim
Sungkyunkwan University (SKKU)
Korea

Kang Yoon Lee
Sungkyunkwan University (SKKU)
Korea

Yonghui Li
The University of Sydney
Australia

Yuanwei Liu
Queen Mary University of London
United Kingdom

Christos Masouros
University College London
United Kingdom

Jong Ho Moon
Sungkyunkwan University (SKKU)
Korea

Derrick Wing Kwan Ng
The University of New South Wales
Australia

Koralia N. Pappi
Intracom S. A. Telecom Solutions
Greece

and

Aristotle University of Thessaloniki
Greece

Jong Jin Park
Sungkyunkwan University (SKKU)
Korea

Slavche Pejoski
Ss. Cyril and Methodius University
Macedonia

Constantinos Psomas
University of Cyprus
Cyprus

Robert Schober
University of Erlangen-Nuremberg
Germany

Yuqing Su
The University of New South Wales
Australia

Stelios Timotheou
University of Cyprus
Cyprus

Morteza Varasteh
Imperial College London
United Kingdom

Feng Wang
Guangdong University of Technology
China

Xin Wang
Fudan University
China

Qingqing Wu
National University of Singapore
Singapore

Jie Xu
Guangdong University of Technology
China

Guangchi Zhang
Guangdong University of Technology
China

Rui Zhang
National University of Singapore
Singapore

Gan Zheng
Loughborough University
United Kingdom

Caijun Zhong
Zhejiang University
China

Nikola Zlatanov
Monash University
Australia

Preface

The goal of this book is to provide readers with a comprehensive insight into the theory, models, techniques, implementation, and application of wireless information and power transfer (WIPT) in energy-constrained wireless communication networks. Written by leading experts on the subject, this book includes 15 chapters, which cover various aspects of WIPT systems, including system modeling, physical layer techniques, resource allocation, and performance analysis. Chapter 1 serves as an introductory chapter and provides an overview of the key research problems regarding WIPT systems. The other 14 chapters aim to tackle specific research challenges for WIPT system design. All the chapters can be read independently.

We would like to express our gratitude to all the authors for their excellent contributions and timeliness in completing their respective chapters. In addition, we would like to thank the publisher team. Finally, we would like to thank the Australian Research Council (DE170100137) and the National Natural Science Foundation of China (61671406) for their financial support.

Derrick Wing Kwan Ng
Trung Q. Duong
Caijun Zhong
Robert Schober

1

The Era of Wireless Information and Power Transfer

Derrick Wing Kwan Ng[1], Trung Q. Duong[2], Caijun Zhong[3], and Robert Schober[4]*

[1] *School of Electrical Engineering and Telecommunications, The University of New South Wales, Australia*
[2] *School of Electronics, Electrical Engineering and Computer Science, Queens University Belfast, United Kingdom*
[3] *Institute of Information and Communication Engineering, Zhejiang University, China*
[4] *Institute for Digital Communications, Friedrich-Alexander-University Erlangen-Nuremberg (FAU), Germany*

1.1 Introduction

In recent decades, the rapid development of wireless communication technologies has triggered a massive growth in the number of wireless communication devices for various practical applications, including e-health, autonomous control, logistics and transportation, environmental monitoring, energy management, safety management, etc. It is expected that in the era of the Internet of Things (IoT), there will be 50 billion wireless communication devices connected together worldwide via the Internet with a connection density of 1 million devices per km^2 [1]. In particular, small wireless sensor modules will be unobtrusively and invisibly integrated into clothing, walls, and vehicles at locations which are inaccessible for wired/manual recharging. However, battery-powered wireless communication devices have limited energy storage capacity and their frequent replacement can be costly, cumbersome, or even impossible (e.g., biomedical implants), which creates a serious performance bottleneck for realizing reliable and ubiquitous wireless communication networks. A promising approach to prolong the lifetime of traditional wireless communication systems is to let the wireless communication devices harvest energy from the environment [2–4]. For example, solar, wind, and

*Corresponding author: w.k.ng@unsw.edu.au

Wireless Information and Power Transfer: Theory and Practice, First Edition.
Edited by Derrick Wing Kwan Ng, Trung Q. Duong, Caijun Zhong, and Robert Schober.

geothermal are the major renewable energy sources for generating electricity. Unfortunately, these conventional natural energy sources are usually climate and location dependent, which may be problematic for mobile devices. Also, the intermittent and uncontrollable nature of natural energy sources makes the use of energy harvesting in wireless communication systems, where providing a continuous and stable quality of service (QoS) is of paramount importance, challenging.

Wireless power transfer (WPT) offers a viable solution for facilitating efficient and sustainable communication networks serving energy-limited communication devices [5–8]. Specifically, in practical systems, wireless devices communicate with each other via electromagnetic (EM) waves in the radio frequency (RF) band. Indeed, RF signals carry both information and energy simultaneously. Thus, the RF energy of propagating signals radiated by transmitters can be recycled at receivers for prolonging the lifetime of networks and supporting the energy consumption required for information transmission. This technology eliminates the need for power cords and any physical contact for manual recharging. Moreover, the broadcast nature of wireless channels facilitates one-to-many wireless charging, which is crucial for wireless networks with large numbers of energy-limited devices. On the other hand, compared to natural renewable energy sources generating intermittent energy, RF-based energy harvesting enables a stable and controllable wireless energy supply for energy-limited communication receivers. More importantly, WPT technology enables simultaneous wireless information and power transfer (WIPT). It is expected that WIPT will serve as a building block for realizing self-sustained communication networks and as the key to unlock the potential of IoT networks. However, despite the conveniences introduced by WIPT technology, the integration of WIPT technology into communication networks also introduces many challenges. For instance, the WPT efficiency is usually low. In practice, wireless power has to be transferred via a carrier signal with a high carrier frequency such that antennas of reasonable size can be used for harvesting energy at handheld devices. However, the associated path loss severely attenuates the signal such that only a small amount of power can be harvested at the receiver. For example, for a communication distance of 10 m in free space, the attenuation of a wireless signal can be up to 50 dB for a carrier frequency of 915 MHz. Moreover, traditional communication networks were optimized for pure data communications. Therefore, it is expected that existing network protocols, resource allocation algorithms, and receiver structures will not be able to meet the unique challenges incurred by the nature of WPT. This book addresses these challenges and provides a comprehensive reference for various solutions for realizing efficient WIPT in practice. In the following sections, we will provide some background information on WPT and discuss exciting research directions. The specific details will then be covered in the subsequent chapters.

1.2 Background

The concept of WPT was first proposed by Nikola Tesla in 1899. The initial efforts on WPT focused on high-power-consumption applications. This raised serious public health concerns about strong electromagnetic radiation which prevented the further development of WPT in the late twentieth century. As a result, this area developed slowly until recent advances in silicon technology and multiple-antenna technology made WPT attractive once again. In fact, the use of WPT avoids the potentially high costs in planning, installing, displacing, and maintaining power cables in buildings and infrastructure. Hence, it is expected that innovative WPT networks are the key enabler of the IoT to connect all devices together via wireless powered sensors for the development of smart cities. The continued study of WPT in both industry and academia will produce frontier technologies by developing novel and cost-effective designs to enable breakthroughs in WPT in the information and communication technology industry sector. For example, it is estimated that the development of IoT for logistics and transportation has a total potential economic impact of 1.9 trillion per year in the next decade [9]. In order to seize the rising business opportunities, recently different companies, e.g., Samsung Electronics and Huawei Technologies, have begun to launch various research and study groups to facilitate the development and standardization of WPT for powering small wireless communication devices.

1.2.1 RF-Based Wireless Power Transfer

The existing WPT technologies can be categorized into three classes: inductive coupling, magnetic resonant coupling, and RF-based WPT. The first two technologies rely on near-field EM waves, which do not provide any mobility to energy-limited wireless communication devices due to the limited wireless charging distances (a few meters) and the required alignment of the EM-field with the energy harvesting circuits. In contrast, RF-based WIPT exploits the far-field properties of EM waves, which enable concurrent wireless charging and data communication over long distances (hundreds of meters). Moreover, RF energy is omnipresent and can be harvested from the signals in the environment transmitted by Wi-Fi access points, TV base station towers, cellular communication base stations, etc. Also, RF-based WIPT utilizes the RF spectrum and the radiation is regulated by the government to ensure safety. More importantly, RF signals can serve as a dual-purpose carriers for conveying both information and power simultaneously.

Nowadays, prototype RF-based energy harvesting circuits are able to harvest microwatts to milliwatts of power over the range of 10 m for a transmit power of 1 W (typical transmit power of a Wi-Fi router) and carrier frequencies of less than 1 GHz [10]. The harvested energy is sufficient to power not only wireless sensors (e.g., fire alarm sensors), but also digital clocks mounted on the wall, which reduces the inconvenience of battery replacement. Although WIPT is critical to the design and implementation of sustainable communication networks, existing system models and resource allocation algorithms have only been proposed and optimized for pure information transfer. In practice, network designers need to strike a balance in the non-trivial trade-off between information and power transfer, leading to significantly different resource allocation algorithms, system models, and interference management schemes, compared to conventional wireless data communications. The introduction of RF-based WIPT imposes new challenges for the design of communication networks since traditional techniques used for the design of data communications cannot solve the fundamental problems in WIPT networks. Hence, there is an emerging need for the development of novel design theories, hardware circuit architectures, and signal processing techniques to unlock the potential of WIPT networks.

1.2.2 Receiver Structure for WIPT

RF-based energy harvesting technology enables the possibility of simultaneous WIPT (SWIPT), wireless-powered communication (WPC), and wireless-powered backscatter communication (WPBC), e.g., [11–15]. Specifically, in SWIPT networks, cf. Figure 1.1a, a transmitter broadcasts an information-carrying signal to provide information and energy delivery service simultaneously. In wireless-powered communication networks (WPCNs), cf. Figure 1.1b, wireless-powered devices first harvest energy, either from a dedicated power station or from ambient RF signals, and then exploit the harvested energy to transmit information signals. In WPBC, cf. Figure 1.1c, energy is transferred in the downlink and information is transferred in the uplink, where backscatter modulation at a tag is used to reflect and modulate the incoming RF signal for communication with a reader (e.g., access point). Since no oscillators are needed at the tags to generate carrier signals, backscatter communications generally entail orders-of-magnitude lower power consumption than conventional radio communications.

In practice, existing RF-based energy harvesting circuits harvest the energy of the received signal directly in the RF domain. In fact, the energy harvesting process destroys the modulated information (e.g., phase-embedded information) in the signal. In addition, conventional information decoding is performed in the digital baseband and the frequency down-converted signals cannot be used for energy harvesting. In other words, information decoding and energy harvesting cannot be performed on the same received signal. As a result, various types of practical energy harvesting receivers have been proposed to enable SWIPT/WPCN. In particular, for SWIPT, the receiver should separate the energy harvesting and information decoding processes.

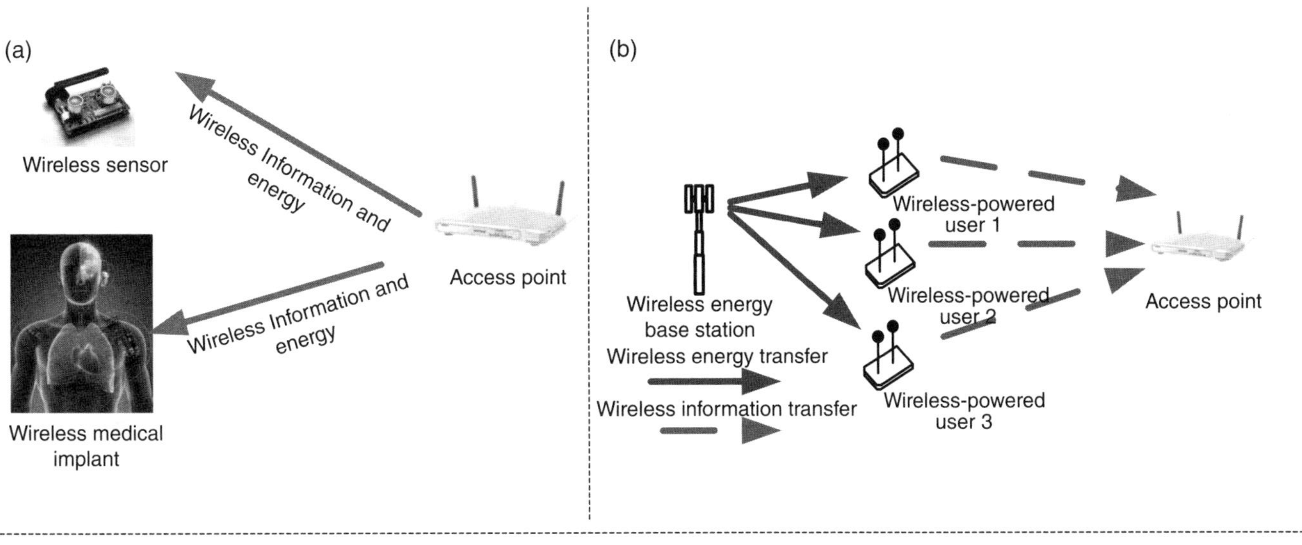

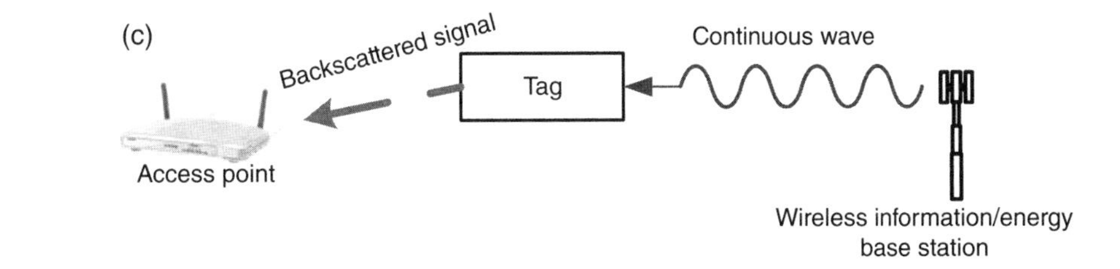

Figure 1.1 Three commonly adopted WIPT network architectures: (a) a SWIPT network, (b) a WPCN, and (c) a WPBC network.

A practical solution is to split the received RF power into two distinct parts, one for energy harvesting and the other one for information decoding. In the following, we discuss two receiver architectures commonly adopted in the literature to achieve this signal splitting.

Time Switching (TS) Receiver: For TS receivers, the transmission is divided into two orthogonal time slots, one for transferring wireless energy and the other one for conveying information, cf. Figure 1.2a. The receiver switches between the co-located energy harvesting circuit and information decoding circuit for harvesting energy and decoding information in successive time slots [16]. In practice, by taking into account the channel statistics and QoS requirements for power transfer, the time durations and the switching sequence for wireless information transfer and energy transfer can be optimized to achieve different system design objectives. Although the TS receiver structure allows for a simple transceiver hardware implementation, it requires accurate time synchronization and information/energy scheduling and the associated control signaling overhead can be demanding, especially in multi-user systems.

Power Splitting (PS) Receiver: A PS receiver splits the signal received at an antenna into two streams at different power levels using a PS unit, cf. Figure 1.2b. In particular, one stream is conveyed to the RF-based energy harvesting circuit for energy harvesting, and the other one is down-converted to baseband for information decoding [16]. Obviously, the PS process incurs a higher receiver complexity compared to the TS process. In addition, optimization of the ratio of the two power streams is generally needed to achieve a balance between the information decoding and energy harvesting performance. Furthermore, insertion loss, additional noise, and circuit related interference may be introduced by the PS process [17]. However, PS receivers enable the possibility of SWIPT, as the received signal can be concurrently exploited for both information decoding and energy harvesting. Therefore, the PS receiver is more suitable for applications with critical information/energy or delay constraints [5] than the TS receiver.

In this book we present the current research trends in system modeling, physical layer design, and resource allocation algorithm design to overcome the challenges in implementing WIPT networks, which is needed to bridge the gap between theory and practice. In the following, we first provide an overview of some of these exciting research problems which will then be discussed in detail in the subsequent chapters.

1.3 Energy Harvesting Model and Waveform Design

To enable RF-based energy harvesting at a wireless communication receiver, a rectenna is usually deployed for converting electromagnetic energy into direct current (DC) electricity. In practice, various rectifier technologies

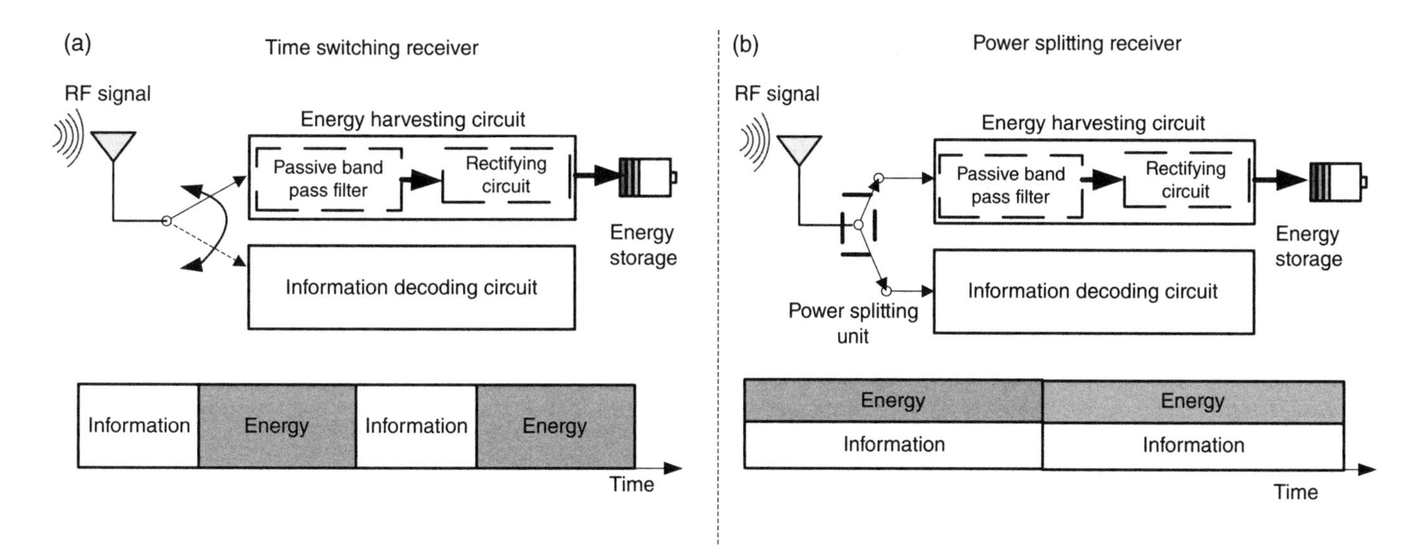

Figure 1.2 Block diagram of two receiver structures for WIPT: (a) time switching receiver and (b) power splitting receiver.

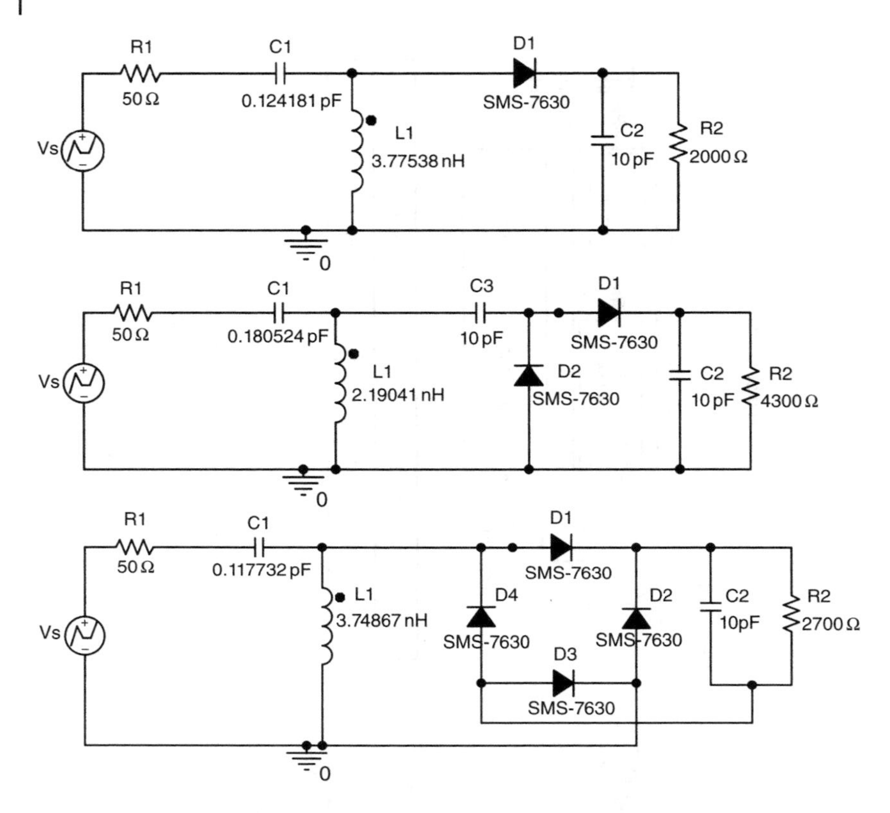

Figure 1.3 Examples of single series, voltage doubler, and diode bridge rectifiers designed for an average RF input power of −20 dBm at 5.18 GHz. v_s is the voltage source of the antenna [24]. R1 models the antenna impedance. C1 and L1 form the matching network. D1, D2, D3, and D4 refer to the Schottky diodes. C2 and R2 form the low-pass filter with R2 being the output load.

(including the popular Schottky diodes, complementary metal oxdide semiconductor (CMOS), backward tunnel diodes, etc.) and topologies (with single and multiple diode rectifiers) exist [18]. Examples of single series, voltage doubler, and diode bridge rectifiers consisting of one, two, and four Schottky diodes, respectively, are shown in Figure 1.3. In general, an accurate energy harvesting model can be obtained by deriving mathematical equations to describe the input–output characteristic of an energy harvesting circuit based on its schematic, e.g., Figure 1.3. However, this usually leads to complicated expressions which are intractable for performance analysis and resource allocation algorithm design. As a result, a linear model is often assumed for characterizing the harvested power after the rectifying circuit. In particular, a constant energy harvesting efficiency is assumed to model the capability of the RF-to-DC conversion circuit, e.g., [19, 20]. In other words, with this model, the power conversion efficiency is independent of the input power level of the energy harvesting circuit. Although the linear model is tractable, it has been verified by experimental data that RF-based energy harvesting circuits usually result in nonlinear end-to-end wireless power transfer [21]. More importantly, employing the linear energy harvesting model for resource allocation algorithm design may lead to unsatisfactory suboptimal performance, e.g., [18, 22, 23].

On the other hand, the performance of wireless energy harvesting is directly related to waveform design. For example, experiments have shown that signals with high peak-to-average power ratio (PAPR), such as multisine and chaotic signals, tend to yield higher DC powers for a given average incident RF power compared to constant envelope signals [25, 26]. This is because compared to a constant-envelope signal, for instance, a pulsed high-PAPR signal having the same average power charges the capacitor used in the rectifier to a higher peak amplitude, leading to a higher output DC voltage during the discharge time of the capacitor. On the other hand, for information transfer, Gaussian signals are optimal while constant-envelope signals are desirable from an implementation point of view. Hence, from a waveform design perspective, there is a non-trivial tradeoff between the performance of wireless information transfer and wireless power transfer. Further details on nonlinear energy harvesting models and waveform design can be found in Chapters 1, 3, and 11.

1.4 Efficiency and Interference Management in WIPT Systems

Different from conventional wireless communication systems, where data rate and energy efficiency are the most important system performance metrics [4, 27, 28], in WIPT systems, the wireless energy transfer efficiency and fairness in resource allocation are equally important QoS metrics. Thus, the design of resource allocation algorithms should take into account the emerging need for energy transfer efficiency. In fact, the introduction of the energy harvesting capability to energy-limited receivers introduces a paradigm shift for system design. For instance, in conventional communication networks, co-channel interference is regarded as one of the major factors that limits the system performance, especially in multi-user systems. Hence, most interference management techniques designed for pure information communication systems aim to suppress or avoid interference via sophisticated resource allocation. However, in WIPT systems, the receivers may embrace strong interference since it can act as a vital source of energy. In fact, injecting artificial interference into the communication network may be beneficial for overall system performance, especially when the receivers do not have enough energy to support their normal operations, since in this case keeping the receivers "alive" via energy harvesting is more important than information decoding. Moreover, by exploiting the extra degrees of freedom offered by multiple antennas, constructive interference can be created, which improves the performance of both information and energy transfer. Hence, there is a non-trivial tradeoff between interference, total harvested power, and energy consumption in WIPT networks. Resource allocation fairness issues in WIPT networks is covered in Chapter 7. A thorough study of the spectral and energy efficiency of WIPT networks is provided in Chapter 13, while the use of multiple antennas for improving the spectral efficiency of WIPT networks is investigated in Chapter 5. In addition, a detailed study on creating constructive interference for improving the performance of WIPT networks is presented in Chapter 10.

1.5 Security in SWIPT Systems

In SWIPT systems, one can increase the energy of the information-carrying signal to facilitate energy harvesting at energy-limited receivers. However, this method will also increase the susceptibility to eavesdropping due to the broadcast nature of wireless channels. Moreover, the escalating number of wireless energy harvesting devices also poses a security threat in future wireless communication networks due to the enormous amount of data transmitted over wireless channels [29–34]. Nowadays, cryptographic encryption algorithms operating in the application layer are adopted to ensure wireless communication security. Unfortunately, these traditional security methods may not be applicable in future wireless networks with large numbers of transceivers, since encryption algorithms usually require secure secret key distribution and centralized secrete key management via an authenticated third party. As an alternative or a complementary technology to the existing encryption algorithms, physical layer (PHY) security has been proposed for guaranteeing secure communication. The principle of PHY layer security is to exploit the unique characteristics of wireless channels, such as fading, noise, and interference, to protect the communication between legitimate devices from eavesdropping. In particular, it has been shown that in a wire-tap channel, a source and a destination can exchange perfectly secure information if the source–destination channel has better conditions compared to the source–eavesdropper channel [35]. Hence, multiple-antenna technology has been proposed to ensure secure communication, e.g., [36–40]. Specifically, by exploiting the extra spatial degrees of freedom offered by multiple antennas, an artificial noise/interference signal can be injected into the communication channel deliberately to impair the received signals at the potential eavesdroppers. Thereby, communication secrecy can be guaranteed at the expense of allocating a large portion of the transmit power to artificial noise generation. On the other hand, the transmitted artificial noise can be harvested by the receivers and the recycled energy can be used to extend the lifetime of energy-constrained portable devices. The dual use of artificial noise for securing communication and facilitating energy harvesting is an important and unique property of secure SWIPT systems. This issue is discussed in detail in Chapter 11.

1.6 Cooperative WIPT Systems

Cooperative techniques for improving the performance of communication systems have drawn significant interest over the past decade. The basic idea of cooperative communication is that multiple single-antenna terminals of a multi-user system share their antennas to form a virtual multiple-input multiple-output (MIMO) communication system. Roughly speaking, there are three types of cooperation, namely, user cooperation, base station cooperation (distributed antennas), and relaying. In particular, cooperative relaying offers a low-cost implementation for achieving coverage extensions, diversity gains, and throughput gains. Nevertheless, in practical systems, relays may be equipped with limited energy supply and there is no strong incentive for the relays to cooperate. In contrast, with WIPT technology, operators can not only share the spectrum with users, but also energy and time. For instance, base stations run by the operators can broadcast wireless energy to charge up energy-limited cooperative devices as a token for future cooperation. This approach creates more incentives for both parties to cooperate and therefore improves the systems' overall spectrum efficiency without requiring external energy sources, e.g., [15, 41]. On the other hand, the distributed antenna system architecture of cooperative networks can be used to reduce the distance between transmitters and receivers. Furthermore, it inherently provides spatial diversity for combating path loss and shadowing, which can be exploited to facilitate efficient WIPT, e.g., [32]. Hence, cooperative techniques can be seamlessly integrated with WIPT for realizing sustainable communication networks. In Chapter 15, a macroscopic approach for characterizing the performance of large-scale wireless networks is discussed. In Chapter 12, the performance of a typical three-node cooperative WPCN is analyzed and evaluated.

1.7 WIPT for 5G Applications

The fifth-generation (5G) wireless networks will be a revolutionary design compared to existing communication systems. It is anticipated that 5G networks will connect at least 100 billion devices worldwide with approximately

Table 1.1 Requirements for 5G wireless communication systems [42]

Figure of merit	5G requirement	Comparison with 4G
Peak data rate	10 Gb/s	100 times higher
Guaranteed data rate	50 Mb/s	–
Mobile data volume	10 Tb/s/km^2	1000 times higher
End-to-end latency	Less than 1 ms	25 times lower
Number of devices	1 M/km^2	1000 times higher
Total number of human-oriented terminals	$\geq$ 20 billion	–
Total number of IoT terminals	$\geq$ 1 trillion	–
Reliability	99.999 %	99.99%
Energy consumption	–	90% less
Mobility	500 km/h	–

7.6 billion mobile subscribers due to the tremendous popularity of smartphones, laptops, sensors, etc., and provide an individual user experience of up to 10 Gb/s. Some of the requirements for 5G wireless networks are listed in Table 1.1. In order to fulfill the requirements of 5G communication networks, various disruptive techniques, such as non-orthogonal multiple access (NOMA), millimetre wave (mmWave) communications, and mobile-edge computing, have been proposed in the literature. On top of this, advanced signal processing algorithms have been developed to improve the energy efficiency of communication systems. Nevertheless, despite the potential system throughput improvements and the reductions in consumed energy brought by these techniques, energy-limited communication devices with short life span still create a system performance bottleneck. Therefore, combing WIPT technology with 5G-enabling techniques has become an emerging research topic. In Chapters 6, 8, and 14, the application of WIPT in systems employing NOMA, mmWave, and mobile-edge computing is studied in detail.

1.8 Conclusion

This book is aimed at graduate students, researchers, and engineers in the field who are interested in WIPT communication networks and their applications. In particular, the first few chapters provide an introduction to WIPT and some basic hardware designs enabling WIPT. The middle part of the book is at a more advanced level and provides a further understanding and knowledge of WIPT systems. In the last part, we will study the fundamental problems in WIPT networks, including communication security in WIPT systems, energy transfer efficiency, and interference management in WIPT systems. In addition, various practical applications of WIPT and corresponding case studies are included in the last part of the book, which is aimed at practitioners.

Acknowledgement

This work is supported by the Australian Research Councils Discovery Early Career Researcher Award funding scheme under Grant DE170100137.

Bibliography

1 M. Zorzi, A. Gluhak, S. Lange, and A. Bassi (2010) From today's INTRAnet of Things to a future INTERnet of Things: A wireless- and mobility-related view. *IEEE Wireless Commun.*, **17** (6): 44–51.
2 V. Chamola and B. Sikdar (2016) Solar powered cellular base stations: current scenario, issues and proposed solutions. *IEEE Commun. Mag.*, **54** (5): 108–114.
3 D.W.K. Ng, E.S. Lo, and R. Schober (2013) Energy-efficient resource allocation in OFDMA systems with hybrid energy harvesting base station. *IEEE Trans. Wireless Commun.*, **12** (7): 3412–3427.
4 Q. Wu, G.Y. Li, W. Chen, D.W.K. Ng, and R. Schober (2017) An overview of sustainable green 5G networks. *IEEE Wireless Commun.*, **24** (4): 72–80.
5 I. Krikidis, S. Timotheou, S. Nikolaou, G. Zheng, D.W.K. Ng, and R. Schober (2014) Simultaneous wireless information and power transfer in modern communication systems. *IEEE Commun. Mag.*, **52** (11): 104–110.
6 X. Chen, Z. Zhang, H.H. Chen, and H. Zhang (2015) Enhancing wireless information and power transfer by exploiting multi-antenna techniques. *IEEE Commun. Mag.*, **53** (4): 133–141.
7 S. Bi, Y. Zeng, and R. Zhang (2016) Wireless powered communication networks: An overview. *IEEE Wireless Commun.*, **23** (2): 10–18.
8 Z. Ding, C. Zhong, D.W.K. Ng, M. Peng, H.A. Suraweera, R. Schober, and H.V. Poor (2015) Application of smart antenna technologies in simultaneous wireless information and power transfer *IEEE Commun. Mag.*, **53** (4): 86–93.
9 DHL and Cisco. Internet Of Things in logistics, Technical Report. [Online]. Available: https://www.dpdhl.com/content/dam/dpdhl/presse/pdf/2015/DHLTrendReport_Internet_of_things.pdf.
10 P. Corporation (2011) RF energy harvesting and wireless power for low-power applications. [Online]. Available: http://www.mouser.com/pdfdocs/Powercast-Overview-2011-01-25.pdf.

11 Q. Wu, M. Tao, D.W.K. Ng, W. Chen, and R. Schober (2016) Energy-efficient resource allocation for wireless powered communication networks. *IEEE Trans. Wireless Commun.*, **15** (3): 2312–2327.

12 H. Ju and R. Zhang (2014) Throughput maximization in wireless powered communication networks. *IEEE Trans. Wireless Commun.*, **13** (1): 418–428.

13 C. Boyer and S. Roy (2014) Backscatter communication and RFID: Coding, energy, and MIMO analysis. *IEEE Trans. Commun.*, **62** (3): 770–785.

14 G. Yang, C.K. Ho, and Y.L. Guan (2015) Multi-antenna wireless energy transfer for backscatter communication systems. *IEEE J. Select. Areas Commun.*, **33** (12): 2974–2987.

15 H. Chen, Y. Li, J.L. Rebelatto, B.F. Ucha-Filho, and B. Vucetic (2015) Harvest-then-cooperate: Wireless-powered cooperative communications. *IEEE Trans. Signal Process.*, **63** (7): 1700–1711.

16 R. Zhang and C.K. Ho (2013) MIMO broadcasting for simultaneous wireless information and power transfer. *IEEE Trans. Wireless Commun.*, **12**: 1989–2001.

17 D.W.K. Ng, E.S. Lo, and R. Schober (2013) Wireless information and power transfer: Energy efficiency optimization in OFDMA systems. *IEEE Trans. Wireless Commun.*, **12**: 6352–6370.

18 B. Clerckx, R. Zhang, R. Schober, D.W.K. Ng, D.I. Kim, and H.V. Poor (2018) Fundamentals of wireless information and power transfer: From RF energy harvester models to signal and system designs. *IEEE JSAC* special issue on wireless transmission of information and power, accepted.

19 J. Zhu, Y. Li, N. Wang, and W. Xu (2017) Wireless information and power transfer in secure massive MIMO downlink with phase noise. *IEEE Wireless Commun. Lett.*, **6** (3): 298–301.

20 Y. Liu (2016) Wireless information and power transfer for multirelay-assisted cooperative communication. *IEEE Commun. Lett.*, **20** (4): 784–787.

21 D. Wang and R. Negra (2013) Design of a dual-band rectifier for wireless power transmission. *2013 IEEE Wireless Power Transfer (WPT)*, pp. 127–130.

22 E. Boshkovska, D.W.K. Ng, N. Zlatanov, and R. Schober (2015) Practical non-linear energy harvesting model and resource allocation for SWIPT systems. *IEEE Commun. Lett.*, **19** (12): 2082–2085.

23 E. Boshkovska, D.W.K. Ng, L. Dai, and R. Schober (2017) Power-efficient and secure WPCNs with hardware impairments and non-linear EH circuit. *IEEE Trans. Commun.*, submitted.

24 B. Clerckx and E. Bayguzina (2017) Low-complexity adaptive multisine waveform design for wireless power transfer. *IEEE Antennas and Wireless Propag. Lett.*, **16**: 2207–2210.

25 A. Boaventura, A. Collado, N.B. Carvalho, and A. Georgiadis (2013) Optimum behavior: Wireless power transmission system design through

behavioral models and efficient synthesis techniques. *IEEE Microwave Mag.*, **14** (2): 26–35.

26 R. Morsi, V. Jamali, D.W.K. Ng, and R. Schober (2018) On the capacity of SWIPT systems with a nonlinear energy harvesting circuit. *Proceedings of the IEEE ICC.*

27 C.Y. Wong, R.S. Cheng, K.B. Lataief, and R.D. Murch (1999) Multiuser OFDM with adaptive subcarrier, bit, and power allocation. *IEEE J. Select. Areas Commun.*, **17** (10): 1747–1758.

28 D.W.K. Ng, E.S. Lo, and R. Schober (2012) Energy-efficient resource allocation in multi-cell OFDMA systems with limited backhaul capacity. *IEEE Trans. Wireless Commun.*, **11** (10): 3618–3631.

29 X. Chen, D.W.K. Ng, and H.H. Chen (2016) Secrecy wireless information and power transfer: Challenges and opportunities. *IEEE Wireless Commun.*, **23** (2): 54–61.

30 M. Liu and Y. Liu (2017) Power allocation for secure SWIPT systems with wireless-powered cooperative jamming. *IEEE Commun. Lett.*, **21** (6): 1353–1356.

31 Y. Wu, X. Chen, C. Yuen, and C. Zhong (2016) Robust resource allocation for secrecy wireless powered communication networks. *IEEE Commun. Lett.*, **20** (12): 2430–2433.

32 D.W.K. Ng and R. Schober (2015) Secure and green SWIPT in distributed antenna networks with limited backhaul capacity. *IEEE Trans. Wireless Commun.*, **14**: 5082–5097.

33 X. Chen, J. Chen, and T. Liu (2016) Secure transmission in wireless powered massive MIMO relaying systems: Performance analysis and optimization. *IEEE Trans. Veh. Technol.*, **65** (10): 8025–8035.

34 W. Liu, X. Zhou, S. Durrani, and P. Popovski (2016) Secure communication with a wireless-powered friendly jammer. *IEEE Trans. Wireless Commun.*, **15** (1): 401–415.

35 A. D. Wyner (1975) The wire-tap channel. *Bell System Technical Journal*, **54** (8): 1355–1387.

36 D.W.K. Ng, E.S. Lo, and R. Schober (2014) Robust beamforming for secure communication in systems with wireless information and power transfer. *IEEE Trans. Wireless Commun.*, **13**: 4599–4615.

37 X. Chen, X. Wang, and X. Chen (2013) Energy-efficient optimization for wireless information and power transfer in large-scale mimo systems employing energy beamforming. *IEEE Wireless Commun. Lett.*, **2**: 667–670.

38 D.W.K. Ng, E.S. Lo, and R. Schober (2016) Multi-objective resource allocation for secure communication in cognitive radio networks with wireless information and power transfer. *IEEE Trans. Veh. Technol.*, **65** (5): 3166–3184.

39 L. Liu, R. Zhang, and K.C. Chua (2014) Secrecy wireless information and power transfer with MISO beamforming. *IEEE Trans. Signal Process.*, **62**: 1850–1863.

40 T.A. Le, Q.T. Vien, H.X. Nguyen, D.W.K. Ng, and R. Schober (2017) Robust chance-constrained optimization for power-efficient and secure SWIPT systems. *IEEE Trans. Green Commun. Network.*, **1** (3): 333–346.

41 Q. Wu, W. Chen, D.W.K. Ng, J. Li, and R. Schober (2016) User-centric energy efficiency maximization for wireless powered communications. *IEEE Trans. Wireless Commun.*, **15** (10): 6898–6912.

42 5G Infrastructure Association. The 5G infrastructure public private partnership: The next generation of communication networks and services. Technical Report. [Online]. Available: https://5g-ppp.eu/wp-content/uploads/2015/02/5G-Vision-Brochure-v1.pdf.

2

Fundamentals of Signal Design for WPT and SWIPT

Bruno Clerckx and Morteza Varasteh*

Department of Electrical and Electronic Engineering, Imperial College London, United Kingdom

2.1 Introduction

Standardized wireless communication systems based on radio frequency (RF) radiation are one of the important technologies in our modern societies. However, demand growth for using wireless-based applications has made the spectral efficiency a bottleneck for this technology. Moreover, the emergence of low-power autonomous wireless devices, such as ubiquitous sensing through wireless sensor networks (WSN) or an Internet-of-Things (IoT), has made the powering of such devices another challenge for the next generation 5G mobile networks. One solution for this powering problem is wireless power transfer (WPT). Indeed for very short ranges, WPT via inductive power is a reality with available products and standards, such as the Wireless Power Consortium, Power Matters Alliance, Alliance for Wireless Power, and Rezence. For long distances (1 m and beyond), though, far-field or radiative WPT[1] is a promising technology that has attracted growing attention, motivated by the increase in the electrical efficiency of computer technology [1].

In general, one can either harvest ambient wireless energy (known as wireless energy harvesting, WEH) or transmit dedicated signals (simply denoted as WPT throughout this chapter). In the case of WEH, the aim is to harvest the ambient energy of the RF signals exclusively designed for communication purposes. On the other hand, in the case of WPT, the aim is to design signals and the entire end-to-end architecture exclusively for the purpose of delivering wireless energy.

Recent results in the research literature reveal that the future of wireless technology goes beyond communication systems. Indeed, wireless technology

*Corresponding author: Bruno Clerckx; b.clerckx@imperial.ac.uk
1 Throughout this chapter, WPT refers to radiative WPT.

Wireless Information and Power Transfer: Theory and Practice, First Edition.
Edited by Derrick Wing Kwan Ng, Trung Q. Duong, Caijun Zhong, and Robert Schober.

has the potential to disrupt WPT as it has disrupted mobile communications in the last 40 years. So far, wireless network designs have focused on communication-centric transmissions and are becoming a highly matured technology evolving towards its next generation, 5G. On the other hand, wireless power is much less mature as there is currently not a single standard on far-field WPT.

WPT creates a lot of potential opportunities. In fact, energy can be transferred in a controlled and an efficient way without the need for a central energy supply. WPT provides the possibility of eliminating wires and reducing the size of batteries, which in turn results in smaller manufactured devices. It can also be a substantial solution to tackle the growing concern of the production, maintenance or disposal of batteries. WPT is also predictable and reliable in contrast to the ambient energy harvesting technologies, such as thermal, solar or vibration. These advantages are indeed very relevant to future wireless networks, where autonomous low-power and energy-limited devices are expected to become increasingly popular.

RF signals potentially can be used to transmit information and transfer energy simultaneously, however, so far these two opportunities, i.e. Wireless Information Transmission (WIT) and WPT, have evolved separately. As a result, the energy of the RF signals used for communication purposes has not been exploited. Moreover, since current wireless network design is communication-centric, delivering wireless power may require the design of an alternative wireless power network, which is not very efficient from an infrastructure perspective. Re-thinking wireless network design in a unified manner is much more promising. WIT and WPT would be designed jointly to enable simultaneous wireless information and power transfer (SWIPT) in order to make the optimum use of the available energy resources, RF spectrum, and network infrastructures.

The main structural requirements and design challenges of the envisioned network are as follows:

1) The range of the delivered power varies from 5 to 100 m in indoor/outdoor environments.
2) The efficiency of the delivered power can be enhanced up to a fraction of a percent or a few percent.
3) Non-line-of-sight (NLOS) as well as line-of-sight (LOS) users are supported in order to broaden the applications of the network.
4) All mobile users, or at least users at low speeds, are supported.
5) The service is available and accessible within the network coverage area.
6) The network and the RF spectrum can be used for WIT and WPT simultaneously.
7) The safety and health issues of the RF systems are resolved and comply with the regulations.

8) The energy consumption of devices that are powered via RF signaling are confined.

In the following section we provide an overview of a WPT architecture and the latest advances in the field of signal design for WPT and SWIPT. We also discuss how these emerging signal strategies tackle the above challenges.

2.2 WPT Architecture

In contrast with earlier works on WPT that focused on high-power transmission, the latest research has focused on low-power applications. Recently, there has been a fast-growing need to build reliable WPT systems in order to provide power to devices with low power consumption, such as RFID tags, wireless sensors, and consumer electronics [2, 3]. As a result, there is an interest in WPT for low to medium power delivery (on the order of μW to a few W) for moderate distances (on the order of a few meters to hundreds of meters) [4, 5].

In Figure 2.1, a general structure of a WPT system is illustrated, which consists of an RF transmitter and an energy harvester composed of a rectenna[2] and a power management unit (PMU). Note that a rectifier is required in order to convert the RF signal to a DC signal, which is either fed to the devices in order to supply their power consumption, or stored as power in a battery for later use. The end-to-end power transfer efficiency can be decomposed as follows:

$$e = \frac{P_{\text{dc,ST}}}{P^t_{\text{dc}}} = \underbrace{\frac{P^t_{\text{rf}}}{P^t_{\text{dc}}}}_{e_1} \cdot \underbrace{\frac{P^r_{\text{rf}}}{P^t_{\text{rf}}}}_{e_2} \cdot \underbrace{\frac{P^r_{\text{dc}}}{P^r_{\text{rf}}}}_{e_3} \cdot \underbrace{\frac{P_{\text{dc,ST}}}{P^r_{\text{dc}}}}_{e_4}, \tag{2.1}$$

where P^t_{dc}, P^t_{rf}, P^r_{rf}, P^r_{dc}, and $P_{\text{dc,ST}}$ are the different power levels illustrated in Figure 2.1. Note that in WPT, the power transmitter can be optimized and, accordingly, can bring high control and reliability on the delivered power at the user end. In the following, we briefly review the techniques that can be used in order to enhance the different gains in (2.1):

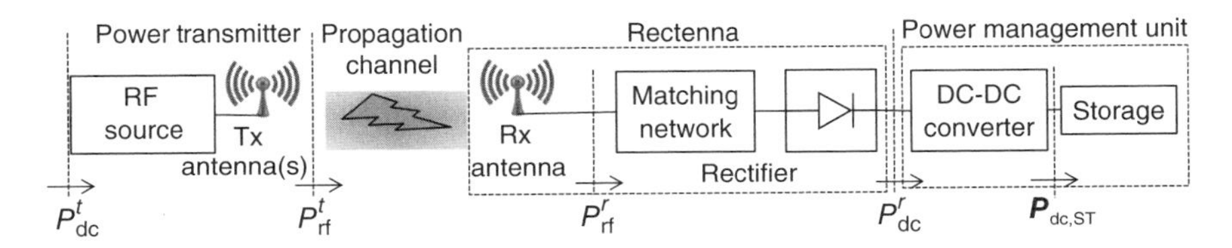

Figure 2.1 Block diagram of a conventional far-field WPT architecture.

2 The rectenna is modeled as an antenna concatenated with a rectifier.

- DC-to-RF conversion efficiency (e_1): This gain can be improved by using efficient power amplifiers (PA) and transmitting signals subject to peak-to-average power ratio (PAPR) constraint.
- The RF-to-RF conversion efficiency (e_2): This gain is a bottleneck of the system, and typically a highly directional transmission is required. Some common approaches in the RF literature are real-time reconfiguration of time-modulated arrays based on localization of the power receivers [6], phased-arrays [7] or retrodirective arrays [8].
- The RF-to-DC conversion efficiency (e_3): This gain depends on the efficiency of the rectenna. A rectenna is a device that captures electromagnetic waves, and rectifies and filters them via a low-pass filter. Due to its nonlinear behavior, resulting from the presence of nonlinear devices, such as diodes in the rectifier, the rectenna design and analysis is challenging. Indeed, its implementation is sensitive due to several losses resulting from threshold and reverse-breakdown voltages, different parasitic sources, impedance mismatching, and harmonic generation [9–11]. In WPT, the rectenna can be optimized for a specific frequency and input power. A single diode rectifier and multiple diode rectifier are preferable in the low power range (1–500 μW) and high power range (above 500 μW), respectively [12, 13]. Another factor impacting the gain e_3 is the output load variations. One alternative to tackle the sensitivity of the gain e_3 on output load is to utilize a resistance compression network [14]. Note that to enlarge the operating range of a rectenna versus input power variations, multiple rectifying devices can be utilized, where each one of them is optimized for a specific range of power levels [15]. This in turn can resolve the saturation effect problem in applications where some level of complexity is allowed at the rectenna. We also note that, in the low power regime, other than the rectenna's design, the gain e_3 is also dependent on the design of the input waveform (from both power and shape of the waveform point of view) [16–20] and the output filter [21, 22].
- The DC-to-DC conversion efficiency (e_4): This gain is improved by tracking the rectifier optimum load dynamically [23, 24]. Note that the rectenna's load is variable and the diode impedance varies drastically with power level and rectifier nonlinearity. Therefore, matching problem and joint optimization of the matching and the load is quite challenging [25].

Unfortunately, maximizing the whole gain e is not achieved by maximizing the different gains e_1, e_2, e_3, and e_4 independently of each other. This is due to the fact that the nonlinearity of the rectifier creates coupling among different gains, especially in the low input power range (1 μW to 1 mW). Note that since the gain e_3 is a function of the input waveform of the rectenna, it is therefore dependent on the design of the transmitted signal (shape and power level) and on the wireless channel state information (CSI). Accordingly, due to the aforementioned coupling effect among different gains, the gains e_2 and e_1 also depend on the transmitted signal and CSI. From the above discussions, it is observed that:

1) Most of the effort in the wireless power literature has been put into the design of the energy harvester.
2) The results obtained in the literature have focused on point-to-point links.
3) Although the effect of the rectenna's nonlinearity on the design of WPT systems has been recognized by the microwave research community, the focus so far has been mainly on decoupling the WPT design by independently optimizing the transmitter and the energy harvester.
4) In the design of WPT systems, multipath and fast fading effects, which are crucial in NLOS transmissions, have not been considered.
5) No WPT design so far has considered the application of feedback in the system in order to improve the system performance by reporting the CSI to the transmitter.
6) There exists no result or systematic approach towards systematic signal design for WPT.

In view of the above and in order to tackle the challenges of the envisioned network, we require:

1) A WPT structure that provides a closed loop and an adaptive communication link between the transmitter and the receiver in order to support channel feedback/training, energy feedback, charging control, etc.
2) A flexible transmitter capable of optimizing the transmitted signal jointly over time, frequency, and CSI. This along with a systematic approach for designing and optimizing a signal can indeed lead to maximizing the end-to-end power transfer efficiency.
3) An approach for designing a link and a system that takes advantage of multiple transmitters and/or receivers (such as co-located transmit antennas delivering power to multiple receivers).

2.3 WPT Signal and System Design

Noting that wireless power and communication systems share the same transmission medium, and inspired by communication techniques such as multiple-input multiple-output (MIMO), feedback, CSI acquisition, etc., we expect to make use of similar techniques in WPT. Unfortunately, the aforementioned techniques cannot be directly applied to WPT due to the differences in objective functions, sensitivity of the receivers, and the effect of interference and modeling of the receivers. Accordingly, we need a modified version of these techniques in order to design WPT systems.

In the following we review emerging and contemporary results on communications and signal design for WPT [26], which comprises (i) efficient transmit signal design in terms of beamforming, waveform, and power allocation, (ii) different CSI acquisition strategies, (iii) multi-user WPT transceiver strategies, and (iv) prototyping. As mentioned earlier, since nonlinearity of the rectenna creates coupling among different conversion efficiencies, it is crucial to consider the nonlinearity effect in analysing and optimizing the system design.

The first systematic signal design approach for adaptive closed-loop WPT was first proposed in [27, 28] for a point-to-point transmission, where maximization of the gains e_2 and e_3 are considered under PAPR constraint. The goal is to optimize the transmit signal as a function of CSI accounting for multisine waveforms, beamforming, and power allocation, jointly. The proposed approach resolves some of the limitations of WPT by exploiting the beamforming gain, the frequency-diversity gain, and the rectifier nonlinearity. The model considered for the rectifier is a Taylor expansion of the diode characteristic function. It is shown that the optimal phase of the proposed waveform is obtained explicitly, whereas the locally optimal magnitude of the waveform is obtained via solving a non-convex optimization problem utilizing a convex optimization technique known as the reversed geometric program (GP). Based on the multiple observations made in [27, 28] the following are observed:

1) It is shown that the adaptive signals optimized accounting for nonlinearity are more efficient than non-adaptive multisine signals (as also used in [16–19]).
2) In the low power regime, it is shown that the nonlinearity of the rectifier is an essential factor in signal design and can be exploited to boost the end-to-end power transfer efficiency.
3) The optimized waveform reveals a tradeoff between allocating power over multiple frequencies to exploit the rectifier nonlinearity and allocating power to a single frequency to exploit the channel frequency selectivity. As a result, optimal power allocation is performed over multiple frequencies but strong frequency-domain channels are allocated high powers.
4) It is shown that multipath and frequency-selective channels are beneficial to system performance if transmit WPT waveforms are adaptive to the CSI, whereas for non-adaptive waveforms multipath is detrimental.

In Figure 2.2 (top), the magnitude of the frequency response of a given realization of a wireless channel over a 10 MHz-bandwidth is illustrated. A multisine waveform with 16 sinewaves uniformly spread within 10 MHz is considered. Based on the assumption that the channel is known at the transmitter, the magnitudes of the optimized multisine waveforms on the 16 frequencies are illustrated in Figure 2.2 (bottom). In comparison with the non-adaptive waveform commonly used in the literature [16–20], the proposed adaptive waveform design allocates more power to frequencies with larger channel gains.

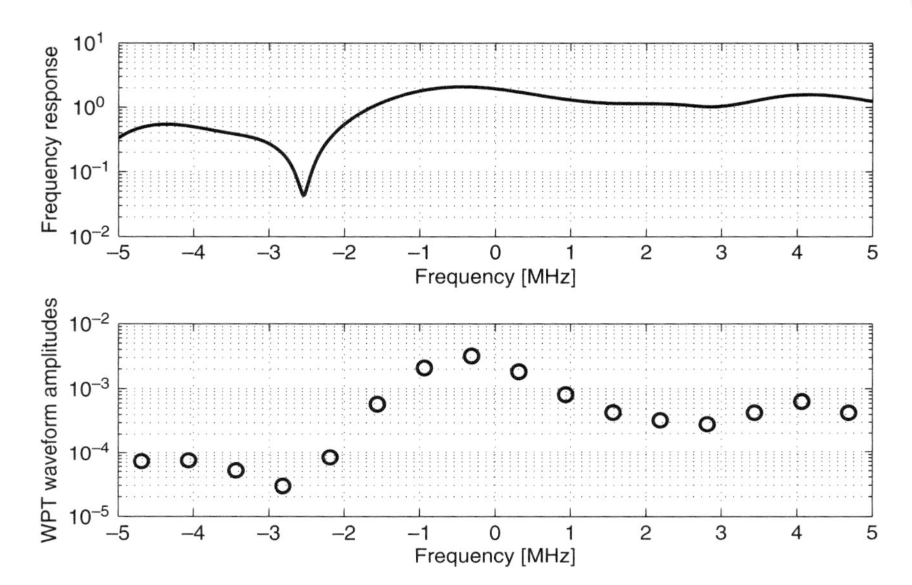

Figure 2.2 Frequency response of a wireless channel (top) and WPT waveform magnitudes ($N = 16$) (bottom) for 10 MHz bandwidth [27].

The advantages of the proposed adaptive waveform design compared to the non-adaptive waveform design proposed in [16–20] have been validated using ADS and PSpice simulations (with a single series rectifier, a WiFi-like environment at 5.18 GHz, and an average input power of about −12 dBm and −20 dBm in [28] and [27, 29], respectively). In Figure 2.3, the harvested DC power with respect to different numbers of sinewaves (for a single transmit antenna, a single series rectifier, under an average input power constraint of −12 dBm and multipath fading) is illustrated. It is observed that for four sinewaves the gain of harvested power for adaptive waveforms with respect to their non-adaptive counterpart is over 100%, and for eight sinewaves the gain is over 200%. Significant performance gains have also been validated in [27] at −20 dBm average input power (for various bandwidths and in the presence of multiple transmit antennas), where waveform and beamforming are jointly optimized in the system design. Moreover, it is shown in [29] that the systematic signal design approach of [27] in fact is applicable in a wide range of rectifier topologies with one and multiple diodes (providing significant gains), e.g., single series, voltage doubler, and diode bridge.

As an important result, the systematic approach for the signal design proposed in [27] shows that contrary to what is considered in [19, 20], PAPR maximization is not always the only and proper approach to design efficient WPT signals. Indeed, in WPT over a frequency flat channel with multisine waveforms as the channel input, high PAPR is a valid metric,

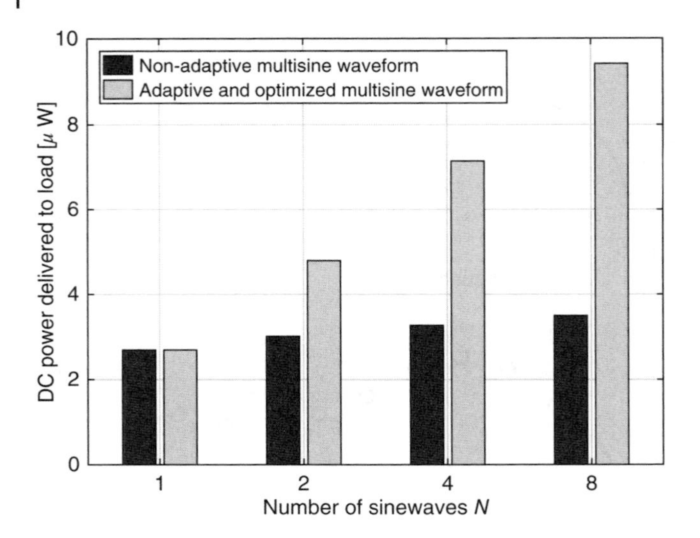

Figure 2.3 DC power vs. number of sinewaves N for adaptive and non-adaptive waveforms [27].

whereas for the same scenario in the presence of multipath, PAPR is not an appropriate metric any longer. This is illustrated in Figure 2.3, where the DC power obtained through adaptive waveform design is compared with the non-adaptive counterpart. It is observed that the adaptive multisine waveform leads to a higher DC power (despite exhibiting a significantly lower PAPR) compared to the non-adaptive multisine waveforms. Recall that in the adaptive multisine waveform the scheme allocates more of the available power budget to stronger frequency domain channels and accordingly it leads to lower PAPR waveforms than non-adaptive in-phase multisine waveforms with uniform power allocation.

The results in [27] also highlight the potential of a large-scale multisine multi-antenna closed-loop WPT architecture. In [30, 31], such a promising architecture was designed and shown to be an essential technique in order to improve the total gain e in (2.1), and also to increase the range of WPT for devices with low power consumption. The proposed technique provides highly efficient far-field charging transmitted signals optimized over many transmit antennas and a large number of frequency components, jointly. This therefore brings together the advantages of pencil beam and waveform design in order to exploit the large beamforming gain of the transmit antenna array and the nonlinearity of the rectifier specifically at long distances. When the dimension of the architecture gets large, the problem is more challenging, which accordingly calls for a reformulation of the optimization problem. The new design enables reduction of orders of magnitude complexity in signal

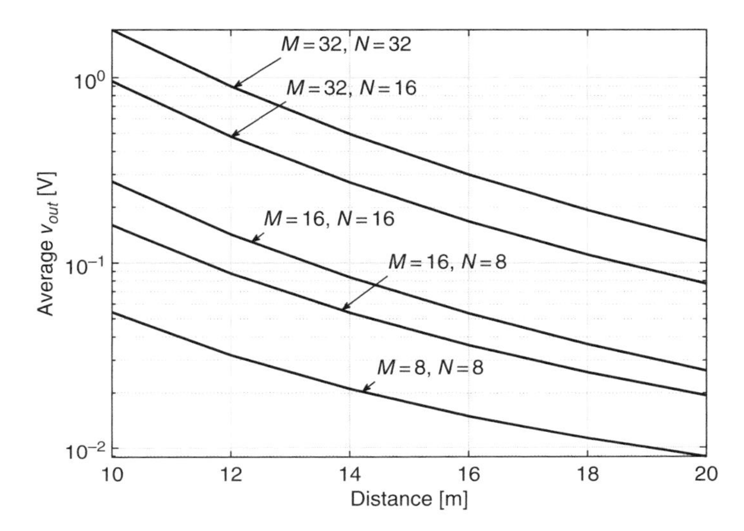

Figure 2.4 Rectifier average output voltage as a function of the Tx–Rx distance [30].

design compared to the reverse GP approach. In [29], another systematic and closed-form low-complexity adaptive waveform design approach is proposed. It is shown that the proposed scheme performs close to the optimal performance. In Figure 2.4 the output voltage of the rectifier versus the distance of the transmitter and the receiver is presented for several values of the number of sinewaves N and transmit antennas M in the multisine transmit waveform. It is observed that the range of the output voltage is expanded with both N and M. This is indeed due to the optimized channel-adaptive multisine waveforms that exploit a beamforming gain, a frequency-diversity gain, and the rectifier nonlinearity, jointly.

Although in recent progress on signal design for WPT many transmit antennas are considered, in the receiver side a single antenna and rectifier per terminal is assumed. Therefore, understanding how to extend the signal design to multiple antennas at the receiver is an interesting and practical problem. In this respect, in [32–34] the problem of combining RF, DC or mixed RF-DC signals is considered.

So far the focus of the discussions has been on waveform designs based on deterministic multisine signals, whereas it is also of significant interest to understand the performance of modulated random waveforms for WPT structures and compare their performance with deterministic waveforms. Indeed, since in the latter we allow the waveform to be random, we are interested in the statistical distribution of the channel input rather than its deterministic shape. As a result, it is interesting to find channel input distributions (or simply modulations) that are specifically designed for WPT in order to boost the end-to-end power transfer efficiency. This could also be extended to the design of unified and efficient signals which are good for transmission of information and power simultaneously. The amount of DC power is impacted by the randomness of a modulated signal. In [35], it is shown that for a single carrier transmission, utilizing a circularly symmetric complex Gaussian (CSCG) distribution as a channel input is advantageous compared to an unmodulated continuous waveform. The improvement in the output DC power for CSCG distributed channel inputs compared to unmodulated continuous waveforms is as follows. Due to the rectifier nonlinearity, it is shown that with a good approximation, the output DC power is dependent on the fourth-order moment as well as the second-order moment. Since the CSCG distributed channel inputs have higher fourth-order moment compared to unmodulated continuous waveforms, as a result the output DC power is also higher. In [36], it is further shown that more improvement in the gain can be obtained using asymmetric Gaussian inputs. Interestingly, it has recently appeared in [37] that input distributions departing from Gaussian but exhibiting a low probability of high amplitude signals can achieve even higher harvested energy. This opens the door to the design of novel energy modulation for WPT. Unlike single-carrier transmission, utilizing CSCG modulated waveforms for multicarrier transmissions is less efficient than for multisine [35]. This is due to the fact that for CSCG distributed channel inputs in multicarrier transmissions there is independent randomness across different carriers, which leads to random fluctuations that are destructive for WPT purposes. This contrasts with the periodic behavior of deterministic multisine waveforms that are more suitable to turn on and off the rectifier periodically. Despite the recent progresses on WPT signal design, finding more efficient input distribution, modulation, and waveform for WPT is an exciting and promising research area.

In order to enhance the end-to-end power transfer efficiency and the range of WPT, we require to jointly exploit the beamforming gain, the channel frequency-selectivity, and the rectifier nonlinearity via a systematic waveform design. This requires the CSI to be known perfectly at the transmitter. However, in practice acquiring the perfect knowledge of the CSI at the transmitter is not feasible and indeed is a challenging topic. In the literature, there are various techniques and strategies in order to acquire CSI, for example forward-link training with CSI feedback, reverse-link training via channel reciprocity, and power probing with limited feedback [26]. The first two are reminiscent of strategies used in modern communication systems [38]. The last one, power probing with limited feedback, is more suitable for WPT purposes and is more promising since it requires low communication and signal processing requirements in order to be implemented. Indeed, this technique relies on the measurements of the DC power output and on the small number of feedback bits for waveform refinement and waveform selection strategies [39]. In the waveform refinement strategy, the transmitter sends two waveforms in each stage consecutively. Upon reception of the waveforms, the receiver sends one bit as a feedback to report increase or decrease in the harvested power during the same stage. Multiple one-bit feedbacks enable the transmitter to successively refine the waveform precoders in a tree-structured codebook over multiple stages. In the waveform selection strategy, the transmitter transmits over multiple time slots with every time a different waveform precoder within a codebook and accordingly the receiver feeds back the index of the precoder yielding the largest harvested power.

A wireless power network consisting of a single transmitter and a single receiver is a special case of a general WPT network. Assume a scenario where there is one single transmitter and there are multiple receivers each equipped with one rectenna. In this multi-user setting, one waveform designed to be suitable for one receiver may be found inefficient or even destructive for the other receivers. Accordingly, the amount of harvested energy from one rectenna depends on the amount of energy harvested from the other rectennas in the system. Hence, there is a tradeoff between the energy harvested by the different rectennas. The energy region formulates this tradeoff by expressing the achievable harvested energy with respect to the set of all rectennas simultaneously. It is characterized as a weighted sum of harvested energy, where by changing the weights we can operate on a different point of the energy region boundary. Discussions related to designing WPT waveforms regarding multi-user settings can be found in [27, 30]. An energy region for a two-user scenario with a multisine waveform spanning 20 transmit antennas and 10 frequencies is illustrated in Figure 2.5. The important observation is that by optimizing the waveform jointly to deliver power to the two users simultaneously, a larger energy region is obtained compared to the region

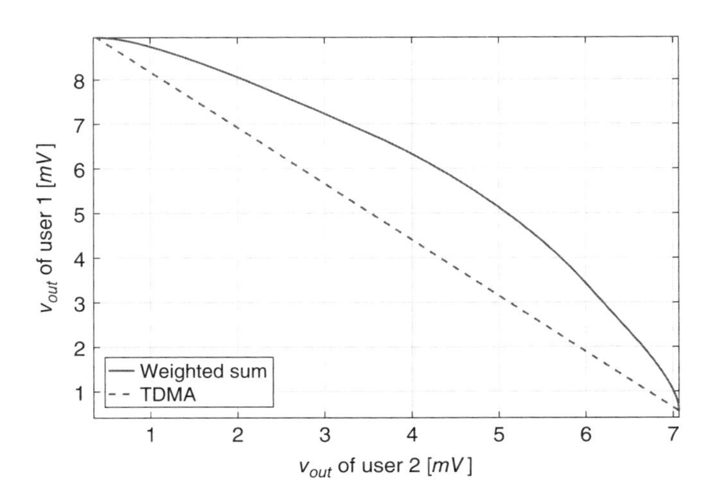

Figure 2.5 Two-user energy region with $M = 20$ and $N = 10$ [30].

obtained through time sharing technique, i.e. TDMA, where the transmitted waveform is optimized for only one user at each time.

In order to study the entire network consisting of many transmitters and receivers, a systematic architecture is required to be defined [26]. In the case when cooperation among transmitters is allowed, in order to serve multiple receivers efficiently, the transmit signal design must be optimized among all the transmitters jointly. As another scenario, if local coordination among the transmitters is possible, a different set of single receivers can be served by a different subset of transmitters. Alternatively, the simplest setup has each receiver served by a single transmitter. For each of these setups, different techniques for resource allocations, charging control, and different strategies (centralized or distributed) are required. Depending on the system model, the CSI sharing and acquisition techniques may differ. In [26], it is shown that in order to enhance the accessibility of wireless power and distribute energy more evenly in a coverage area, it is preferable to distribute the antennas in the area and allow cooperation among them. This is in contrast with the setup where the antennas are co-located. As a compliment of distributed antenna setup, note that the strong energy beams in the direction of the user can be easily avoided, which is crucial from a health and safety point of view.

In order to demonstrate the feasibility of the proposed signal and system designs for WPT, prototyping and experimentation are required. Hence, implementation of a closed-loop WPT structure with a real-time over-the-air transmission (based on a frame structure switching between a WPT phase and a CSI acquisition phase) is needed. Similarly to CSI acquisition in communication systems, here the channel is to be acquired at the millisecond level. Different parts of the system, such as rectenna, channel estimation module, and channel-adaptive waveform generator, have to be built. In [40], the first prototype of a closed-loop WPT system is reported, which is based on channel-adaptive waveform optimization and dynamic CSI acquisition. The reported prototyping and experimentation in [40] validates the theoretical results developed in [27, 29] and confirms the following observations:

1) The nonlinearity of the rectifier is a crucial factor to be considered in WPT waveform design, especially in low power regimes
2) Similar to the problems in communication systems, the wireless propagation medium has a significant impact on waveform design for WPT.
3) In order to boost the end-to-end performance in frequency-selective channels and NLOS scenarios, CSI acquisition and adaptive waveform design are essential.
4) Utilizing larger bandwidth improves WPT performance thanks to the exploitation of the channel frequency diversity gain.
5) PAPR is not an appropriate criterion to design WPT waveforms.

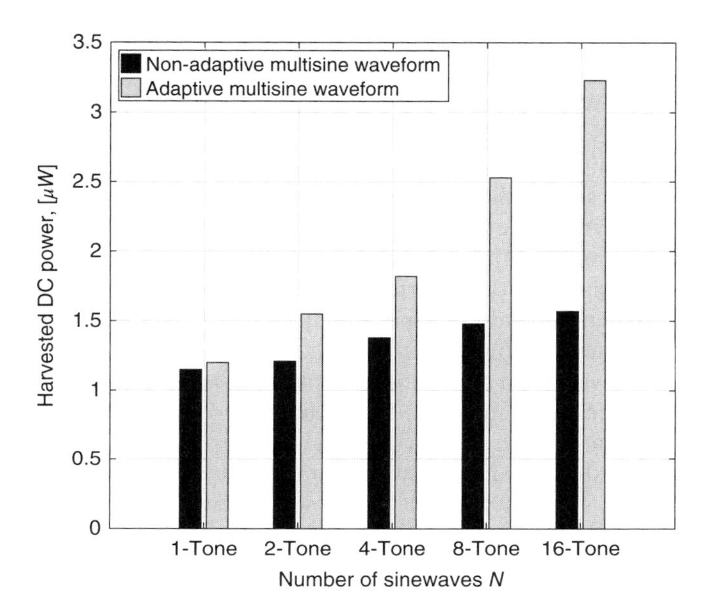

Figure 2.6 Harvested DC power in an indoor NLOS deployment as a function of N uniformly spread within a 10 MHz bandwidth [40].

In Figure 2.6 the harvested DC power through adaptive and non-adaptive multisine waveform (in a NLOS setup with a single antenna at the transmitter and receiver) is illustrated. It is observed that improvement in the DC output power is 105% more than for an open-loop WPT architecture with non-adaptive multisine waveform (with the same number of sinewaves) and 170% more than for a continuous wave.

2.4 SWIPT Signal and System Design

In the SWIPT paradigm, we require both WPT and WIT be integrated. Hence, it is of significant interest to find the fundamental limits regarding the ultimate transmission rate and harvested power under certain design constraints, and accordingly to characterize possible tradeoffs between conveying information and energy wirelessly [41–43], and to identify corresponding transmission strategies. This fundamental tradeoff is characterized using a rate–energy region that contains all pairs of rate and harvested energy levels that can be achieved. The goal is to identify the largest achievable rate–energy region and the corresponding strategy. In the literature, a linear model of the rectifier (ignoring its nonlinearity) is commonly used and the obtained results rely on the conventional capacity-achieving CSCG channel input distributions [41–43]. Utilizing the WPT waveform designs mentioned earlier [35, 36, 44], it is shown that the rectifier nonlinearity has a significant impact on the performance of a SWIPT system. Indeed, nonlinear rectifier modeling leads to input distributions that are not CSCG any longer. This is in contrast with the linear model for the rectifier, where CSCG distributed channel inputs are optimal. In [36], it is shown that for a single-carrier transmission over frequency flat channels, asymmetric Gaussian inputs perform better than their CSCG counterparts. Also in [35, 44], it is shown for multicarrier transmission that non-zero mean Gaussian inputs outperform CSCG inputs. Nevertheless, the optimal input distribution and transmit signal strategy for almost all SWIPT architectures remain unknown.

In [35], the problem of waveform and transceiver design for SWIPT with a nonlinear rectenna at the receiver, and accordingly characterizing the tradeoff region, is studied. In what follows, we briefly review the main results in [35]. First, the model of the rectenna nonlinearity (introduced in [27] for multicarrier unmodulated signal) is extended. The scaling laws of harvested energy related to single-carrier and multicarrier modulated and unmodulated waveforms are analytically derived as a function of the number of carriers and the propagation conditions. It is shown that by adapting to the linear model, both unmodulated and modulated waveforms perform equally for WPT purposes, whereas in the nonlinear model for the rectenna the modulated waveform is beneficial in single-carrier transmission, but detrimental in multicarrier transmission. Second, a novel SWIPT architecture is introduced where at the transmitter side the channel input signal is the superposition of multicarrier unmodulated and modulated waveforms, and at the receiver side there is a power-splitter equipped with an energy harvester and an information decoder. Accordingly, in order to characterize the maximum achievable rate–energy region, the superposed waveform and the power splitter are optimized jointly. As key observations, it is worth noting that (i) a multicarrier unmodulated waveform is useful to enlarge the rate–energy region of SWIPT, (ii) a combination of power splitting and time sharing is in general the best strategy, and (iii) a non-zero mean Gaussian distributed channel input outperforms the conventional capacity-achieving zero-mean Gaussian distributed channel input in multicarrier transmissions.

As an interesting observation in [35], it is noted that input distribution in every sub-band with a linear model-based design is CSCG. This leads to the classical result of water-filling, where the mean of the channel input is zero and the corresponding variance is frequency dependent. However, considering the superposition of multicarrier modulated and unmodulated waveforms, mentioned earlier, the channel input is not zero-mean any longer and its magnitude is Ricean distributed with a certain K-factor in each sub-band. The more the harvested power requirement, the more the power allocation to the multicarrier unmodulated waveform, and accordingly the larger the mean of the

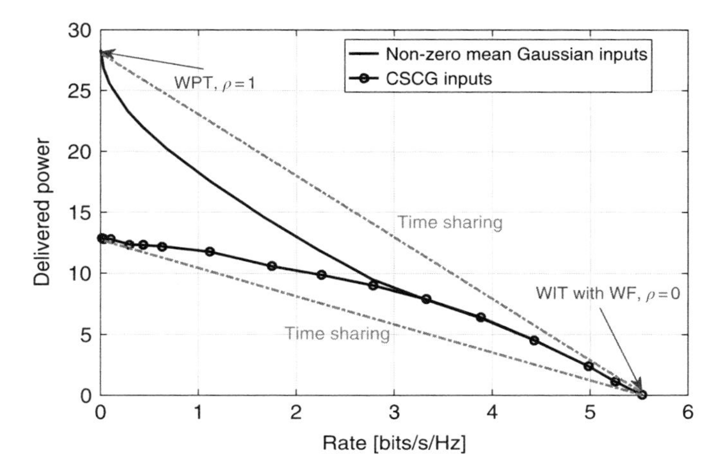

Figure 2.7 Rate (bits/s/Hz)–energy region of a multicarrier transmission with $N = 16$ sub-bands with CSCG and non-zero-mean Gaussian inputs [35]. (The rectenna's output in is in the form of current in amperes. However, since power is proportional to current, with abuse of notation, the vertical axis is referred to as power.)

channel input. It is also worth noting that the variance of the channel input decreases as the corresponding mean of the waveform increases. This overall leads to an input distribution whose K-factor increases as more emphasis is put on maximizing the delivery of energy. In Figure 2.7, the harvested power versus the information rate is illustrated for CSCG and non-zero-mean superposed inputs. It is observed that the tradeoff between the energy and information is a function of the power splitting ratio (denoted by ρ). When ρ is small (assigning most of the signal to information purposes) both of the schemes perform similarly, but as ρ increases, the non-zero-mean input outperforms CSCG channel inputs.

In [37], the capacity[3] problem of SWIPT over a point-to-point and complex additive white Gaussian noise (AWGN) channel is considered, taking into account the nonlinearities of the rectifier. The capacity is studied under channel input average power, channel input amplitude, and receiver delivered power constraints. The results obtained in [37] are listed as follows. First, leveraging the rectifier model introduced in [27], it is shown that the output DC power of the rectenna in the low power regime can be lower bounded by considering the even moments of the channel baseband equivalent output. Accordingly, the delivered power at the receiver is modeled as a linear combination of even moments of the channel baseband equivalent output, being larger than

3 From an information theoretic perspective, the capacity of a channel is defined as the maximum number of bits that can be transmitted in each channel use, such that the probability of information decoding error approaches zero, arbitrarily.

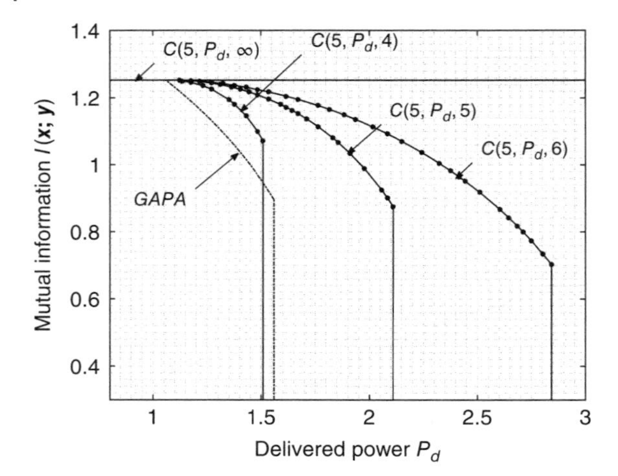

Figure 2.8 Mutual information $I(\mathbf{x}; \mathbf{y})$ corresponding to the complex Gaussian inputs (denoted by GAPA). Mutual information $I(\mathbf{x}; \mathbf{y})$ corresponding to different values of the minimum delivered power constraint P_d with amplitude constraints 4, 5, 6, and ∞. Average power constraint is 5[37]. (The rectenna's output in is in the form of current with units of amperes. However, since power is proportional to current, with abuse of notation, the vertical axis is referred as power.)

a threshold. Second, it is shown that – regardless of practical limitations – the capacity of an AWGN channel under a transmit average power constraint and receiver delivered power constraint is the same as the capacity of an AWGN channel.[4] That is, considering an energy harvester at the receiver with the aforementioned nonlinear model and given a fixed average power for the channel input – regardless of practical limitations – there is no loss in information transmission and any value for the delivered power is feasible. Third, it is shown that under transmit average power, amplitude, and receiver delivered power constraints, the optimal input distributions – jointly good for information and power – are discrete, with a finite number of mass points on their amplitude.[5] In Figure 2.8 the simulation result for the transmitted information in terms of mutual information $I(\boldsymbol{x}; \boldsymbol{y})$ [6] versus harvested power P_d is illustrated for an average power constraint $P_a = 5$. $C(\cdot, \cdot, \cdot)$ on each line represents the maximum achievable rate, where the first, second, and third arguments are the average power, amplitude, and delivered power constraints, respectively. The horizontal line related to $C(5, P_d, \infty)$ corresponds to the AWGN channel capacity under an average power constraint $P_a = 5$. $C(5, P_d, 4)$, and $C(5, P_d, 5)$ and $C(5, P_d, 6)$ correspond to the maximum achievable rates corresponding to the average power

4 Note that the capacity of a complex AWGN channel with SNR γ is given as $\log(1 + \gamma)$.
5 The phase of the channel input is uniformly distributed over $[-\pi, \pi)$.
6 $\boldsymbol{x}$ and $\boldsymbol{y}$ represent the channel baseband equivalent input and output, respectively.

of 5 and amplitude constraints of 4, 5, and 6, respectively. The rate–power region obtained from Gaussian inputs is denoted by the Gaussian asymmetric power allocation (GAPA). It is observed that the optimized input distributions (those corresponding to $C(5, P_d, 4)$, $C(5, P_d, 5)$, and $C(5, P_d, 6)$) yield significantly larger rate–power region compared to the region corresponding to GAPA. It is also observed that by increasing the amplitude constraint, the rate–power region tends to the rate–power region corresponding to the problem with no amplitude constraint at the receiver, i.e. $C(5, P_d, \infty)$.

2.5 Conclusions and Observations

In the following we bring together some of the important observations and discussions given in this chapter. First, taking into account the nonlinearity of the rectifier, there is huge potential to improve the systematic signal and system design and optimization approach towards efficient WPT and SWIPT. Second, the nonlinearity of the rectifier profoundly changes the design approach of WPT and SWIPT systems, i.e. (i) it has a different design method compared to the conventional wireless communications, whose channel is assumed to be linear, (ii) it leads to different channel input distributions, transceiver design, and use of the RF spectrum, and (iii) it is advantageous in the low power regime, which results in an increased rate–energy region, range, and sensitivity. Third, adaptive signal design leads to an architecture that provides the possibility of fixing the rectenna (e.g., with a fixed load), whereas the transmit signal is adaptive. This is in contrast with the traditional WPT design where the waveform is fixed and the rectenna is adaptive. Although both approaches are complementary, it can be impractical to have energy-constrained devices whose rectenna is dynamically adjusted with the channel condition. This is due to the fact that the wireless channel changes very quickly (on the order of 10 ms). In the adaptive waveform design approach, the transmitter is smarter and there is no need to run power-hungry optimizations at the receiver side. However, a challenging problem in this approach is to find proper techniques in order to acquire channel CSI appropriately. Ultimately, it is envisioned that an entire end-to-end optimization of the system should be conducted, likely resulting in an architecture where the transmit signals and the rectennas adapt themselves dynamically as a function of the CSI.

Bibliography

1 S. Hemour and K. Wu (2014) Radio-frequency rectifier for electromagnetic energy harvesting: Development path and future outlook. *Proceedings of the IEEE*, **102** (11): 1667–1691.

2 H.J. Visser and R.J.M. Vullers (2013) RF energy harvesting and transport for wireless sensor network applications: Principles and requirements. *Proceedings of the IEEE*, **101** (6): 1410–1423.
3 Z. Popovic, E.A. Falkenstein, D. Costinett, and R. Zane (2013) Low-power far-field wireless powering for wireless sensors. *Proceedings of the IEEE*, **101** (6): 1397–1409.
4 E. Falkenstein, M. Roberg, and Z. Popovic (2012) Low-power wireless power delivery. *IEEE Transactions on Microwave Theory and Techniques*, **60** (7): 2277–2286.
5 Z. Popovic (2013) Cut the cord: Low-power far-field wireless powering. *IEEE Microwave Mag.*, **14** (2): 55–62.
6 D. Masotti, A. Costanzo, M.D. Prete, and V. Rizzoli (2016) Time-modulation of linear arrays for real-time reconfigurable wireless power transmission. *IEEE Trans Microwave Theory Techniques*, **64** (2): 331–342.
7 T. Takahashi, T. Mizuno, M. Sawa, T. Sasaki, T. Takahashi, and N. Shinohara (2011) Development of phased array for high accurate microwave power transmission. In *2011 IEEE MTT-S International Microwave Workshop Series on Innovative Wireless Power Transmission: Technologies, Systems, and Applications*, May 2011, pp. 157–160.
8 R.Y. Miyamoto and T. Itoh (2002) Retrodirective arrays for wireless communications. *IEEE Microwave Mag.*, **3** (1): 71–79.
9 T.W. Yoo and K. Chang (1992) Theoretical and experimental development of 10 and 35 GHz rectennas. *IEEE Trans Microwave Theory Techniques*, **40** (6): 1259–1266.
10 B. Strassner and K. Chang (2013) Microwave power transmission: Historical milestones and system components. *Proceedings of the IEEE*, **101** (6): 1379–1396.
11 C.R. Valenta and G.D. Durgin (2014) Harvesting wireless power: Survey of energy-harvester conversion efficiency in far-field, wireless power transfer systems. *IEEE Microwave Mag.*, **15** (4): 108–120.
12 A. Costanzo and D. Masotti (2016) Smart solutions in smart spaces: Getting the most from far-field wireless power transfer. *IEEE Microwave Mag.*, **17** (5): 30–45.
13 A. Boaventura, A. Collado, N.B. Carvalho, and A. Georgiadis (2013) Optimum behaviour: Wireless power transmission system design through behavioural models and efficient synthesis techniques. *IEEE Microwave Mag.*, **14** (2): 26–35.
14 K. Niotaki, A. Collado, A. Georgiadis, S. Kim, and M.M. Tentzeris (2014) Solar and electromagnetic energy harvesting and wireless power transmission. *Proceedings of the IEEE*, **102** (11): 1712–1722.

15 H. Sun, Z. Zhong, and Y.X. Guo (2013) An adaptive reconfigurable rectifier for wireless power transmission. *IEEE Microwave and Wireless Components Lett.*, **23** (9): 492–494.
16 M.S. Trotter, J.D. Griffin, and G.D. Durgin (2009) Power-optimized waveforms for improving the range and reliability of RFID systems. In *2009 IEEE International Conference on RFID*, pp. 80–87.
17 A.S. Boaventura and N.B. Carvalho (2011) Maximizing DC power in energy harvesting circuits using multisine excitation. In *2011 IEEE MTT-S International Microwave Symposium*, pp. 1–4.
18 C.R. Valenta and G.D. Durgin (2013) Rectenna performance under power-optimized waveform excitation. In *2013 IEEE International Conference on RFID*, pp. 237–244.
19 C.R. Valenta, M.M. Morys, and G.D. Durgin (2015) Theoretical energy-conversion efficiency for energy-harvesting circuits under power-optimized waveform excitation. *IEEE Trans Microwave Theory Techniques*, **63** (5): 1758–1767.
20 A. Collado and A. Georgiadis (2014) Optimal waveforms for efficient wireless power transmission. *IEEE Microwave and Wireless Components Lett.*, **24** (5): 354–356.
21 A. Boaventura, N. Carvalho, and A. Georgiadis (2014) The impact of multi-sine tone separation on RF-DC efficiency. *Proceedings of Asia-Pacific Microwave Conference.*
22 N. Pan, A.S. Boaventura, M. Rajabi et al. (2015) Amplitude and frequency analysis of multi-sine wireless power transfer. 2015 *Integrated Nonlinear Microwave and Millimetre-wave Circuits Workshop (INMMiC)*, Taormina, pp. 1–3.
23 A. Dolgov, R. Zane, and Z. Popovic (2010) Power management system for online low power RF energy harvesting optimization. *IEEE Trans Circuits and Systems I: Regular Papers*, **57** (7): 1802–1811.
24 A. Costanzo, A. Romani, D. Masotti, N. Arbizzani, and V. Rizzoli (2012) RF and baseband co-design of switching receivers for multiband microwave energy harvesting. *Sens. Actuators A, Phys.*, **179**: 158–168.
25 F. Bolos, J. Blanco, A. Collado, and A. Georgiadis (2016) RF energy harvesting from multi-tone and digitally modulated signals. *IEEE Trans. on Microwave Theory and Techniques*, **64** (6): 1918–1927.
26 Y. Zeng, B. Clerckx, and R. Zhang (2017) Communications and signals design for wireless power transmission. *IEEE Trans. Commun.*, **65** (5): 2264–2290.
27 B. Clerckx and E. Bayguzina (2016) Waveform design for wireless power transfer. *IEEE Trans. Signal Process.*, **64** (23): 6313–6328.
28 B. Clerckx, E. Bayguzina, D. Yates, and P.D. Mitcheson (2015) Waveform optimization for wireless power transfer with nonlinear energy harvester

modelling. In *2015 International Symposium on Wireless Communication Systems (ISWCS)*, pp. 276–280.

29 B. Clerckx and E. Bayguzina (2017) Low-complexity adaptive multisine waveform design for wireless power transfer. *IEEE Antennas and Wireless Propagation Lett.*, **16**: 2207–2210.

30 Y. Huang and B. Clerckx (2017) Large-scale multiantenna multisine wireless power transfer. *IEEE Trans. Signal Process.*, **65** (21): 5812–5827.

31 Y. Huang and B. Clerckx (2016) Waveform optimization for large-scale multi-antenna multi-sine wireless power transfer. In *2016 IEEE 17th International Workshop on Signal Processing Advances in Wireless Communications (SPAWC)*, pp. 1–5.

32 Z. Popovic, S. Korhummel, S. Dunbar, R. Scheeler, A. Dolgov, R. Zane, E. Falkenstein, and J. Hagerty (2014) Scalable RF energy harvesting. *IEEE Trans. Microwave Theory and Techniques*, **62** (4): 1046–1056.

33 R.J. Gutmann and J.M. Borrego (1979) Power combining in an array of microwave power rectifiers. In *IEEE MTT-S International Microwave Symposium Digest*, pp. 453–455.

34 N. Shinohara and H. Matsumoto (1998) Dependence of DC output of a rectenna array on the method of interconnection of its array elements. *Electr. Eng. Jpn.*, **125** (1): 9–17.

35 B. Clerckx (2018) Wireless information and power transfer: Nonlinearity, waveform design, and rate–energy tradeoff. *IEEE Trans. Signal Process.*, **66** (4): 847–862.

36 M. Varasteh, B. Rassouli, and B. Clerckx (2017) Wireless information and power transfer over an AWGN channel: Nonlinearity and asymmetric gaussian signaling. *CoRR*, abs/1705.06350. [Online]. Available: http://arxiv.org/abs/1705.06350.

37 M. Varasteh, B. Rassouli, and B. Clerckx (2017) On capacity-achieving distributions for complex AWGN channels under nonlinear power constraints and their applications to swipt. CoRR, abs/1712.01226, 2017. [Online]. Available: http://arxiv.org/abs/1712.01226.

38 B. Clerckx and C. Oestges (2013) *MIMO wireless networks: Channels, techniques and standards for multi-antenna, multi-user and multi-cell systems*, Academic Press (Elsevier).

39 Y. Huang and B. Clerckx (2017) Waveform design for wireless power transfer with limited feedback. *IEEE Trans. Wireless Commun.*, **17** (1): 415–429.

40 J. Kim, B. Clerckx, and P.D. Mitcheson (2017) Prototyping and experimentation of a closed-loop wireless power transmission with channel acquisition and waveform optimization. In *2017 IEEE Wireless Power Transfer Conference (WPTC)*, pp. 1–4.

41 L.R. Varshney (2008) Transporting information and energy simultaneously. In *IEEE International Symposium on Information Theory*, pp. 1612–1616.

42 P. Grover and A. Sahai (2010) Shannon meets Tesla: Wireless information and power transfer. In *IEEE International Symposium on Information Theory*, pp. 2363–2367.

43 R. Zhang and C.K. Ho, "MIMO broadcasting for simultaneous wireless information and power transfer," *IEEE Transactions on Wireless Communications*, vol. 12, no. 5, pp. 1989–2001, May 2013.

44 B. Clerckx (2016) Waveform optimization for swipt with nonlinear energy harvester modelling. In *WSA 2016; 20th International ITG Workshop on Smart Antennas*, pp. 1–5.

3

Unified Design of Wireless Information and Power Transmission

Dong In Kim, Jong Jin Park, Jong Ho Moon, and Kang Yoon Lee*

School of Information and Communication Engineering, Sungkyunkwan University (SKKU), Korea

3.1 Introduction

With the growing interest in the Internet-of-Things (IoT), massive connection of low-power IoT/wearable devices is desired for machine-type communication (MTC), and battery lifetime is becoming a crucial issue for operating IoT networks perpetually [1, 2]. Hence, a fundamental paradigm shift should be launched to develop low-power transceivers, such as ambient backscatters [3] and LoRa backscatters using chirp spread spectrum (CSS) [4], enabling the vision of low-power smart devices with ubiquitous connectivity. Furthermore, smart devices may need to be self-powered for energy neutral operation. Self-powered smart devices, charged by ambient renewable resources instead of requiring battery replacement, can be deployed to enable low-power wide area networks (LP-WAN) for IoT [5]. One promising solution is simultaneous wireless information and power transfer (SWIPT), a technique that transfers both information and power in the air using a radio frequency (RF) signal. Existing SWIPT approaches aim to enhance the rate–energy tradeoff [6, 7] through optimum waveform design for SWIPT on the one hand, and via implementation of appropriate receiver structures on the other hand, at both circuit and system levels.

SWIPT uses a single tone (carrier) with amplitude and/or phase modulation for low-complexity devices. Recently, it was shown that using multi-tone waveforms can boost up the efficiency of wireless power transfer (WPT) due to the nonlinear rectification process [8, 9], but a simple modulation based on a single tone cannot be used with multi-tone waveforms. To tackle this difficulty, we have proposed peak-to-average power ratio (PAPR) based SWIPT [10], a

*Corresponding author: Dong In Kim; dikim@skku.ac.kr

Wireless Information and Power Transfer: Theory and Practice, First Edition.
Edited by Derrick Wing Kwan Ng, Trung Q. Duong, Caijun Zhong, and Robert Schober.

technique that uses multi-tone waveforms for energy transfer and their distinct levels of PAPR to convey information. Multi-tone (PAPR) based SWIPT yields higher WPT efficiency with increased operational range and low-power information decoding, but it suffers from a lower rate than single tone based SWIPT. Therefore, we should optimize the tradeoff between achievable rate and harvested energy as well as the operational range.

To empower the LP-WAN with ubiquitous connectivity for IoT, the operational range of self-powered smart devices needs to be enlarged, which depends largely on the end-to-end efficiency of WPT. In this chapter, we reveal that the efficiency in fact varies with both the RF input power at the rectifier and the specific RF signal waveform being used for SWIPT, which has motivated us to implement dual mode SWIPT. For this, we propose a new transceiver architecture for SWIPT with adaptive mode switching (MS) between single and multi-tone waveforms via an adaptive power management and information decoding (PM-&-ID) module to control the operational range for energy neutral operation. Furthermore, the WPT efficiency relies on the circuit design of the rectifier because of the nonlinear rectification process, and we propose a reconfigurable energy harvester with multiple energy harvesting (EH) circuits implemented in parallel before the battery.

The rest of this chapter is organized as follows. In Section 3.2 we present nonlinear EH models along with the measured RF-to-direct current (DC) power conversion efficiency (PCE), which provides useful insights for designing dual mode SWIPT. Section 3.3 proposes a new waveform and transceiver design for multi-tone (PAPR) based SWIPT with higher WPT efficiency and dual mode SWIPT for an increased operational range. In Section 3.4 we further propose the reconfigurable energy harvester for enlarging the linear region of the PCE over wide input power ranges.

Notations. The main notations used in this chapter are summarized in Table 3.1. We use boldface lower case letters to denote vectors. $(\cdot)^*$ represents the complex conjugate; $|\cdot|$ and $\|\cdot\|$ denote the absolute value of a complex scalar and the l_2-norm of a vector, respectively; the normal distribution is denoted by $\mathcal{N}(\mu, \sigma^2)$ with mean μ and variance σ^2; $\sim$ stands for "distributed as"; **Re** and **E** denote the real parts of a complex number and expectation, respectively; Pr$[X]$ represents the probability of event X.

3.2 Nonlinear EH Models

We consider a SWIPT system which uses single/multi-tone waveforms for signaling and is equipped with $K \geq 1$ transmit antennas and a single receive antenna. Most previous studies on SWIPT systems assumed an ideal linear energy harvester [11]–[13], in which the RF-to-DC PCE ζ is modeled as a fixed

Table 3.1 Nomenclature adopted in this chapter

Notation	Description
P_T/P_R	Transmit/receive power
P_{in}	Input power of EH circuit
P_{EH} ($P_{EH,x}$)	EH power ($x = L, NL$ for linear/nonlinear)
P_{sat}	Maximum EH power due to diode saturation
P_{th}	Input power threshold for dual mode operation
P_C	Circuit power consumption of the receiver
ζ	Power conversion efficiency (PCE)
R	Achievable data rate
B_T/B_C	Transmit/coherence bandwidth
W	Signal bandwidth
Q (N)	Number of total (selected) multi-tone waveforms
f_c	Carrier frequency
Δf	Minimum spacing of multi-tone waveforms
h	Complex-valued channel gain
$\eta(t)$	Noise signal
N_o	Noise power spectral density level
T (T_x)	Symbol time ($x = s, m$ for single/multi-tone)
K	Number of array antennas
M	Modulation index
ρ	Power-splitting ratio
p_{out}	Outage probability
p_b	Bit-error rate (BER)
p_{tag}	Target BER
N_{EH}	Number of EH circuits
α	Receive mode indicator variable
B	Number of fading blocks

constant independent of the input power P_{in}. Under the linear EH model, the harvested power can be described as $P_{EH,L} = \zeta P_{in}$ $(0 \leq \zeta \leq 1)$. However, considering the nonlinear rectification process (diode turn-on/reverse breakdown voltages, diode nonlinearity, and saturation effects reported in [8, 9, 14]), the harvested power P_{EH} cannot be simply predicted by the conventional linear model. In fact, the PCE of the EH circuit is generally a function of the input signal waveform (e.g., both the amplitude and phase in the case of a sinewave). Thus, the PCE is a function of not only the input power (e.g., the squared amplitude of the sinewave), but also the shape of the input signal (e.g., the phase of the sinewave).

To address the nonlinear characteristics and to improve accuracy, new nonlinear EH models were proposed in [9], [14]. On the one hand, the nonlinear EH model in [9] shows that for the single diode rectifier circuit, the output DC current of the diode (i.e., the output DC power of the rectifier for a given load resistance) with the fourth-order Taylor series truncation is given by

$$i_{out} = c_2\mathbf{E}\,[y^2(t)] + c_4\mathbf{E}\,[y^4(t)] \tag{3.1}$$

where $c_j = \frac{i_s}{j!n_f^j v_t^j}(\sqrt{R_{\text{ant}}})^j$ for $j = 2, 4$, i_s, n_f, v_t, and R_{ant} are the reverse bias saturation current, the ideality factor, the thermal voltage of the diode, and the resistance of the receive antenna, respectively. Also, in (3.1), $y(t)$ denotes the received RF signal, which is given by

$$y(t) = \mathbf{Re}\left[\sum_{n=1}^{N} h_n s_n e^{j(2\pi f_n t+\phi_n)}\right]. \tag{3.2}$$

Here, h_n, f_n, s_n, and ϕ_n denote the complex channel, the carrier frequency, the amplitude, and the phase for the nth sinewave. As can be seen from (3.1) and (3.2), the PCE is a function of the input signal waveform (i.e., s_n and ϕ_n values) because the output DC current i_{out} is a function of the received RF signal $y(t)$, which is actually a function of the input power and the shape of the input signal.

However, the nonlinear model of (3.1) is valid only for a small signal, resulting in huge inaccuracy outside the small-signal range. On the other hand, a nonlinear EH model based on logistic function was suggested in [14], where the fitted harvested power $P_{EH,NL}$ is expressed as

$$P_{EH,NL} = \frac{P_{sat}(1 - e^{-c_1 P_{in}})}{1 + e^{-c_1(P_{in}-c_0)}}. \tag{3.3}$$

Here, c_0 and c_1 are constants related to the EH circuit specification and the shape of the input signal. Given the EH circuit and input signal, we can obtain P_{sat}, c_0, and c_1 using a curve-fitting tool. Specifically, the nonlinear model of (3.3) tackles the nonlinearity of the PCE of the rectifier (including the turn-on and saturation nonlinearities) and it characterizes the nonlinearity in terms of the input power, whereas the model of (3.1) tackles the diode nonlinearity due to the nonlinear diode characteristic and it characterizes the nonlinearity in terms of the input signal waveform. However, the nonlinear model of (3.3) may be inaccurate in the low power region for wide input power ranges.

Figure 3.1 shows the *measured* RF-to-DC PCE of the single diode rectifier circuit over a wide input power range (−20–15 dBm) when single/multi-tone waveforms ($N = 1, 2, 4, 8, 16$) are used with a single transmit antenna ($K = 1$). For a given shape (i.e., fixed N) of the input signal, the PCE depends mainly on the input power (i.e., P_{in}). In this case, the PCE typically increases with the input power, and then remains nearly constant where the input power exceeds a

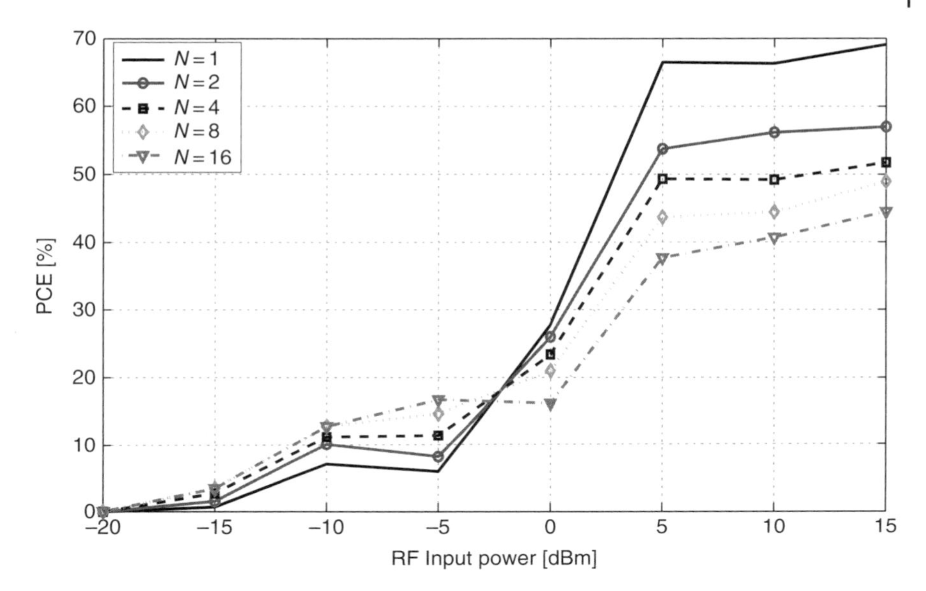

Figure 3.1 Measured RF-to-DC power conversion efficiency (PCE) of the single diode rectifier circuit for both single tone ($N = 1$) and multi-tone waveforms ($N = 2, 4, 8, 16$).

certain threshold. On the other hand, for a given input power the PCE depends mainly on the shape of the input signal (i.e., N).

Therefore, we may consider adaptive MS between single and multi-tone waveforms to achieve the maximum PCE. For this, we define the input power threshold P_{th} for adaptive MS, which depends only on the circuit specification. For example, P_{th} is −3 dBm for the single-stage rectifier in Figure 3.1.

3.3 Waveform and Transceiver Design

3.3.1 Multi-tone (PAPR) based SWIPT

Multi-tone Waveforms and Transmit/Receive PAPR. A new transceiver architecture for SWIPT using multi-tone waveforms and their distinct PAPR is shown in Figure 3.2. First, the information is encoded by selecting a subset of the multi-tone waveforms having a specific level of PAPR as follows. Suppose there are Q tones available for SWIPT with minimum spacing Δf, such that

$$\mathcal{F} = \{f_1, f_2, \dots, f_Q\} \tag{3.4}$$

where $f_i = f_1 + (i-1)\Delta f$ for $i = 1, \dots, Q$. The subset $\mathcal{F}_N \subset \mathcal{F}$ is defined as

$$\mathcal{F}_N = \{f_{(1)}, f_{(2)}, \dots, f_{(N)}\} \tag{3.5}$$

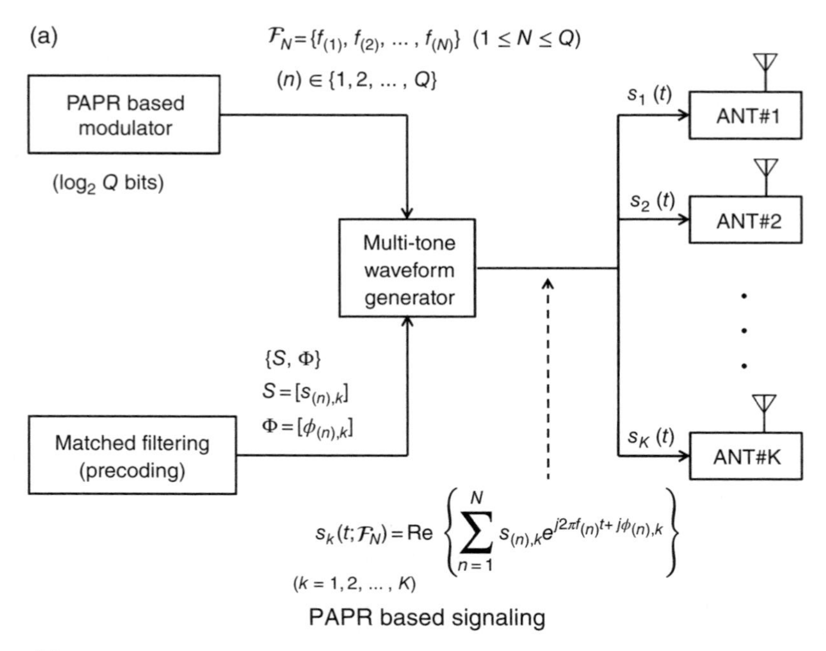

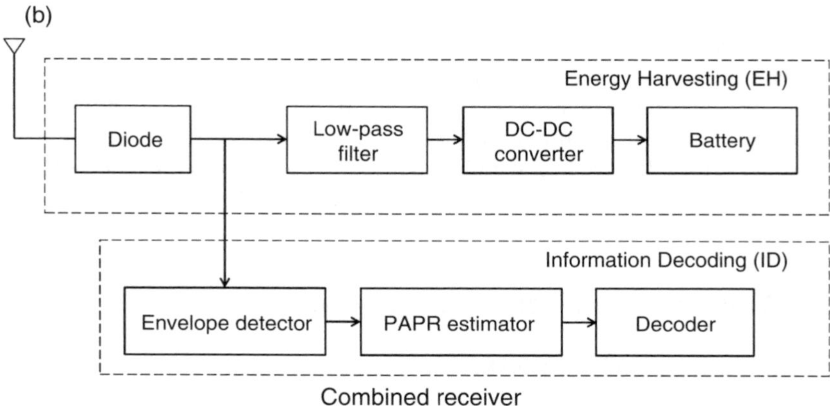

Figure 3.2 A new transceiver architecture for SWIPT using multi-tone waveforms for energy transfer and their distinct PAPR for information transmission. In the combined receiver (b), the time constant for envelope detection is shorter than that for low-pass filtering to yield the DC output.

Table 3.2 An example mapping when $Q = 4$

2 bits	N	$PAPR_{TX}$	$\mathcal{F}_N$
[00]	1	2	$\{f_{(1)}\}$
[01]	2	4	$\{f_{(1)}, f_{(2)}\}$
[10]	3	6	$\{f_{(1)}, f_{(2)}, f_{(3)}\}$
[11]	4	8	$\{f_1, f_2, f_3, f_4\}$

where $(n) \in \{1, 2, \dots, Q\}$ for $1 \leq n \leq N$ and $(n) \neq (n')$ if $n \neq n'$. The $\log_2 Q$-bit information can be one-to-one mapped to $\mathcal{F}_N$ in an increasing order of $N = |\mathcal{F}_N| = 1, \dots, Q$, as illustrated in Table 3.2. Note that all distinct subsets of same size N produce the same level of PAPR in (3.8) below, and $\mathcal{F}_N$ acts as the *best* subset whose N tones experience relatively better channel gains, especially for the frequency-selective (FS) channel.[1] Assume that an array of K antennas is used at the transmitter for enhancing WPT efficiency. Then, the multi-tone waveforms for the kth antenna, which are generated by subset $\mathcal{F}_N$, can be expressed as

$$s_k(t; \mathcal{F}_N) = \mathbf{Re} \left\{ \sum_{n=1}^{N} s_{(n),k} \exp[\, j2\pi f_{(n)} t + j\phi_{(n),k} \,] \right\} \tag{3.6}$$

for $k = 1, \dots, K$. Here, $s_{(n),k}$ and $\phi_{(n),k}$ are the magnitude and initial phase associated with the (n)th tone $f_{(n)}$ at the kth antenna. Note that a total transmit power of P_T is shared by the K antennas, subject to

$$\sum_{k=1}^{K} \mathbf{E}\,\{|s_k(t; \mathcal{F}_N)|^2\} \leq P_T. \tag{3.7}$$

Now, the $\log_2 Q$-bit information is conveyed via the distinct PAPR of $s_k(t; \mathcal{F}_N)$ $(k = 1, \dots, K)$ through the subset $\mathcal{F}_N$, which can be evaluated as

$$PAPR_{TX}^{k}(N) = \frac{\max_{t\in[0,T]} |s_k(t; \mathcal{F}_N)|^2}{\frac{1}{T}\int_T |s_k(t; \mathcal{F}_N)|^2 dt} \leq 2N. \tag{3.8}$$

Here, the upper bound on the *transmit* PAPR, namely $PAPR_{TX} = 2N$, can be achieved with uniform power allocation among N tones and K antennas and all initial phases aligned, such that

$$s_{(n),k} = \sqrt{\frac{2P_T}{NK}} \;\text{ and }\; \phi_{(n),k} = \phi \;\text{ for }\; \forall\; n, k. \tag{3.9}$$

1 This implies that knowledge of the channel state information (CSI) will allow selection of the *best* subset $\mathcal{F}_N$ of N tones in a given transmission time.

Thus, it is necessary to perform the matched-filtering (MF) through the channel to achieve amplitude matching and phase alignment at the receiver.

Given an array of K antennas is used for SWIPT, the received signal can be expressed as

$$r(t;\mathcal{F}_N) = \mathbf{Re}\left\{\sum_{k=1}^{K}\sum_{n=1}^{N} s_{(n),k}h_{(n),k}\exp[j2\pi f_{(n)}t + j\phi_{(n),k}]\right\} + \eta(t), \quad (3.10)$$

where the noise $\eta(t)$ is zero mean with variance $\sigma^2 = N_o W$. Here, N tones in $\mathcal{F}_N$ are assumed to be spaced over a *selective* sub-band of the coherence bandwidth B_c (i.e., W), which gives similar Rayleigh-faded channel gains for different tones but independent channel gains for different antennas. The composite channel gains $\{h_{(n),k}\}$ are assumed to be estimated by sending a short pilot signal to the transmitter before each frame transmission, assuming the channel reciprocity holds for SWIPT.

To achieve a high WPT efficiency and also the upper bound of the transmit PAPR, we assume that the transmitter performs the MF precoding for the amplitude matching and phase alignment at the receiver. Therefore, the complex envelope of the (n)th tone in the multi-tone waveforms for the kth antenna should be adjusted as

$$s_{(n),k}\exp(j\phi_{(n),k}) = \sqrt{2P_T}\,\frac{h^*_{(n),k}}{\sqrt{\sum_{k=1}^{K}\sum_{n=1}^{N}|h_{(n),k}|^2}}, \quad (3.11)$$

where the transmit power constraint in (3.7) has been applied. Using the channel vector $\mathbf{h}_{(n)} = [h_{(n),1}, h_{(n),2}, \ldots, h_{(n),K}]$, the received signal in (3.10) after the precoding can be simplified to

$$r(t;\mathcal{F}_N) = \sqrt{\frac{2P_T}{\sum_{n=1}^{N}\|\mathbf{h}_{(n)}\|^2}}\sum_{n=1}^{N}\|\mathbf{h}_{(n)}\|^2\cos(2\pi f_{(n)}t) + \eta(t). \quad (3.12)$$

At the receiver, the rectifier in Figure 3.2b performs the RF-to-DC conversion and follows the envelope of the received signal in (3.12), based upon which the *receive* PAPR is measured as

$$PAPR_{RX}(N) = \frac{\max_{t\in[0,T]}|r(t;\mathcal{F}_N)|^2}{\frac{1}{T}\int_T |r(t;\mathcal{F}_N)|^2 dt} \cong \frac{\left|\sqrt{2P_T\sum_{n=1}^{N}\|\mathbf{h}_{(n)}\|^2} + \eta\right|^2}{\frac{P_T\sum_{n=1}^{N}\|\mathbf{h}_{(n)}\|^4}{\sum_{n=1}^{N}\|\mathbf{h}_{(n)}\|^2} + N_o B_c}. \quad (3.13)$$

If the noise can be ignored, given the received power level is −10 dBm for energy signals, unlike the −60 dBm for information signals, the receive PAPR reduces to

$$PAPR_{RX}(N) \cong \frac{2\left(\sum_{n=1}^{N}\|\mathbf{h}_{(n)}\|^2\right)^2}{\sum_{n=1}^{N}\|\mathbf{h}_{(n)}\|^4}. \quad (3.14)$$

Moreover, if the channel appears frequency flat (FF), namely $\mathbf{h}_{(n)} = \mathbf{h}$ for $\forall n$, then the receive PAPR achieves the upper bound as

$$PAPR_{RX}(N) \cong \frac{2N \parallel \mathbf{h} \|^4}{\parallel \mathbf{h} \|^4} = 2N. \tag{3.15}$$

Remarks. It is interesting to see that the receive PAPR can achieve the upper bound $PAPR_{RX} = PAPR_{TX} = 2N$ regardless of the channel vector $\mathbf{h}_{(n)}$, which implies that the WPT can be maximized with the MF precoding at the transmitter, while achieving the upper bound on the transmit PAPR at the receiver. *In particular, the channel estimation is not required at the receiver for the PAPR based information decoding in an FF channel.* Even if the channel appears FS, we can identify a sub-band with high channel gain where the channel appears FF, thereby avoiding channel estimation.

Evaluation. To validate the PAPR based information transmission, the bit-error rate (BER) performance is evaluated as

$$p_b(N) = 1 - \Pr\left[2N - 1 \leq PAPR_{RX}(N) < 2N + 1\right] \tag{3.16}$$

for $2 \leq N \leq Q - 1$ and

$$p_b(N) = \begin{cases} \Pr\left[PAPR_{RX}(N) \geq 3\right] & \text{for } N = 1, \\ 1 - \Pr\left[PAPR_{RX}(N) \geq 2Q - 1\right] & \text{for } N = Q. \end{cases} \tag{3.17}$$

Figure 3.3 shows the BER performance versus the average received SNR when $K = 1$ for both FF ($B_T = 1$ MHz) and FS ($B_T = 10$ MHz) channels. In the FS channel, we select a sub-band of $B_c = 1$ MHz where the channel appears FF. Here, the numerical BER evaluations based on (3.13), (3.16) and (3.17) are validated by Monte Carlo simulations. Furthermore, the BER performance of PAPR based signaling ($Q = 16$) is compared with that of single tone signaling with $M = 16$-ary PAM. We observe that their BER performance is comparable in the SNR range of interest.

The multi-tone waveforms for SWIPT are generated with the minimum spacing $\Delta f = T^{-1}$. Hence, if we define $B_c \ll B_T$ as a sub-band with higher channel gain for the FS channel, the information rate can be evaluated as

$$R_b = \frac{1}{T} \log_2(1 + B_c T) \quad \text{bits/s} \tag{3.18}$$

for $Q = 1 + B_c T$. We observe that as Q increases, the harvested power offered by multi-tone waveforms is increased, referring to [9]

$$\text{DC output} = \frac{1}{Q} \sum_{N=1}^{Q} \left(c_2 P_R + 2c_4 \frac{2N^2 + 1}{2N} P_R^2 \right) \tag{3.19}$$

when $K = 1$. However, T in (3.18) is also increased with fixed B_c, resulting in reduced R_b. Hence, there exists a rate-energy tradeoff for varying Q by which

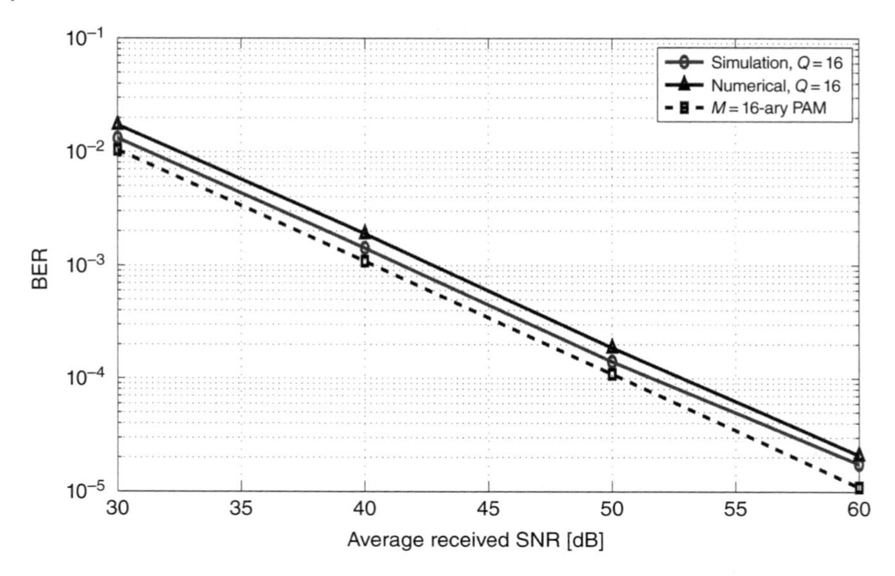

Figure 3.3 BER versus average received SNR for PAPR based signaling with $Q = 16$ tones and single tone signaling with $M = 16$-ary pulse amplitude modulation (PAM).

B_c can be properly selected, considering a specific channel gain offered by a sub-band within B_T in the FS channel.

In Figure 3.4, the DC output is normalized to one for a single tone ($Q = 1$), where the achievable data rate increases with increased $M = 2, 4, 8, 16, 32, 64$, subject to the target BER of 10^{-3}. Note that the data rate of the single tone with M-ary PAM is set to $R_{PAM} = B_c \log_2 M$ for $B_c = 1$ MHz, subject to the target BER for all M above. This is true when the energy signal yields the received SNR above 50 dB at the received power $P_R = -20$ dBm with the noise level of typically -130 dBm/Hz over the sub-band of 1 MHz. The proposed SWIPT using multi-tone waveforms with $Q = 2, 4, 8, 16, 32, 64$ produces much higher normalized DC output in (3.19), relative to the single tone ($Q = 1$), though the data rate in (3.18) decreases with increased Q. For low-rate IoT sensors, the operational range of RF energy transfer to a sensor node is most critical and can be greatly enlarged with multi-tone (PAPR) SWIPT.

3.3.2 Dual Mode SWIPT

Waveform Design and Transmitter Architecture. For dual mode SWIPT, an integrated transmitter generates two types of single and multi-tone waveforms, as shown in Figure 3.5. The transmitter selects a waveform according to the adaptive MS policy, which depends on the received power at the receiver.

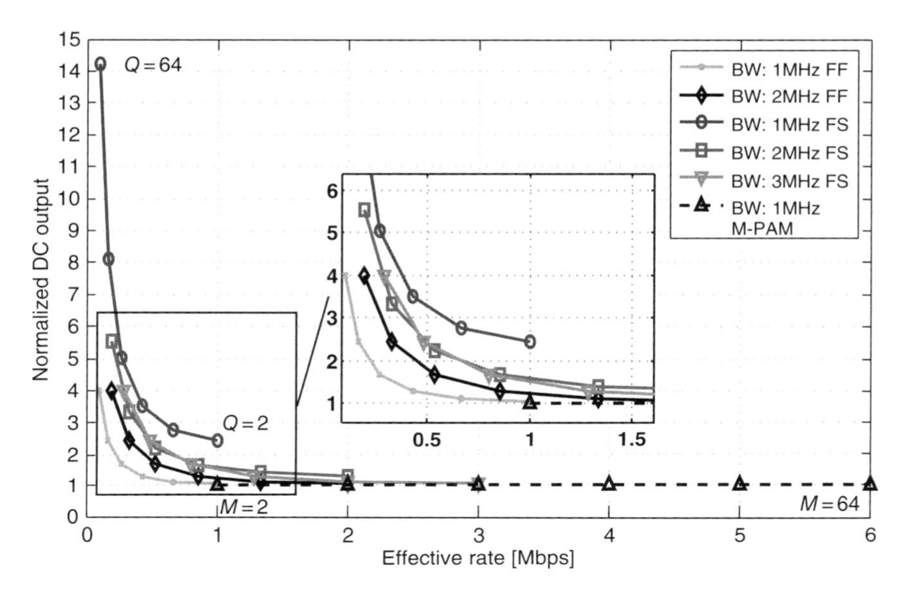

Figure 3.4 Rate–energy tradeoff for varying Q and M in multi-tone waveforms with PAPR and single tone waveform with M-ary PAM, respectively.

Single tone/multilevel PSK. The single tone waveform is optimized for high data rate, unlike the multi-tone waveforms. We consider single tone signaling of multiple energy levels and phase-shift keying (PSK) as

$$s_s(t) = \mathbf{Re}\,\{A\exp(j2\pi f_c t + j\theta)\}, \tag{3.20}$$

where $A \in \{d\sqrt{2}, 2d\sqrt{2}, \ldots, N_e d\sqrt{2}\}$ and $\theta \in \{\frac{2\pi}{N_p}, 2\frac{2\pi}{N_p}, \ldots, 2\pi\}$ are the amplitude and phase of the modulated symbol, and f_c is the center frequency of the signal. N_e denotes the number of energy levels (e.g., the inner/outer circles in Figure 3.5), where d is the constellation unit distance, and N_p is the number of signal (phase) points per level. The unit distance d can be obtained from the average signal power relationship as

$$P_T = \frac{N_p \sum_{n=1}^{N_e} (nd)^2}{M}, \tag{3.21}$$

where $M = N_e N_p$. The energy-level/phase information is decoded jointly at the receiver.

Multi-tone/PAPR modulation. As the distance from the access point (AP) is increased, the transmit power should be increased to compensate for the severe propagation loss. We may enhance the efficiency of WPT by using multi-tone waveforms, which effectively increases the operational range, thanks to the nonlinear rectification process. Furthermore, the PAPR based

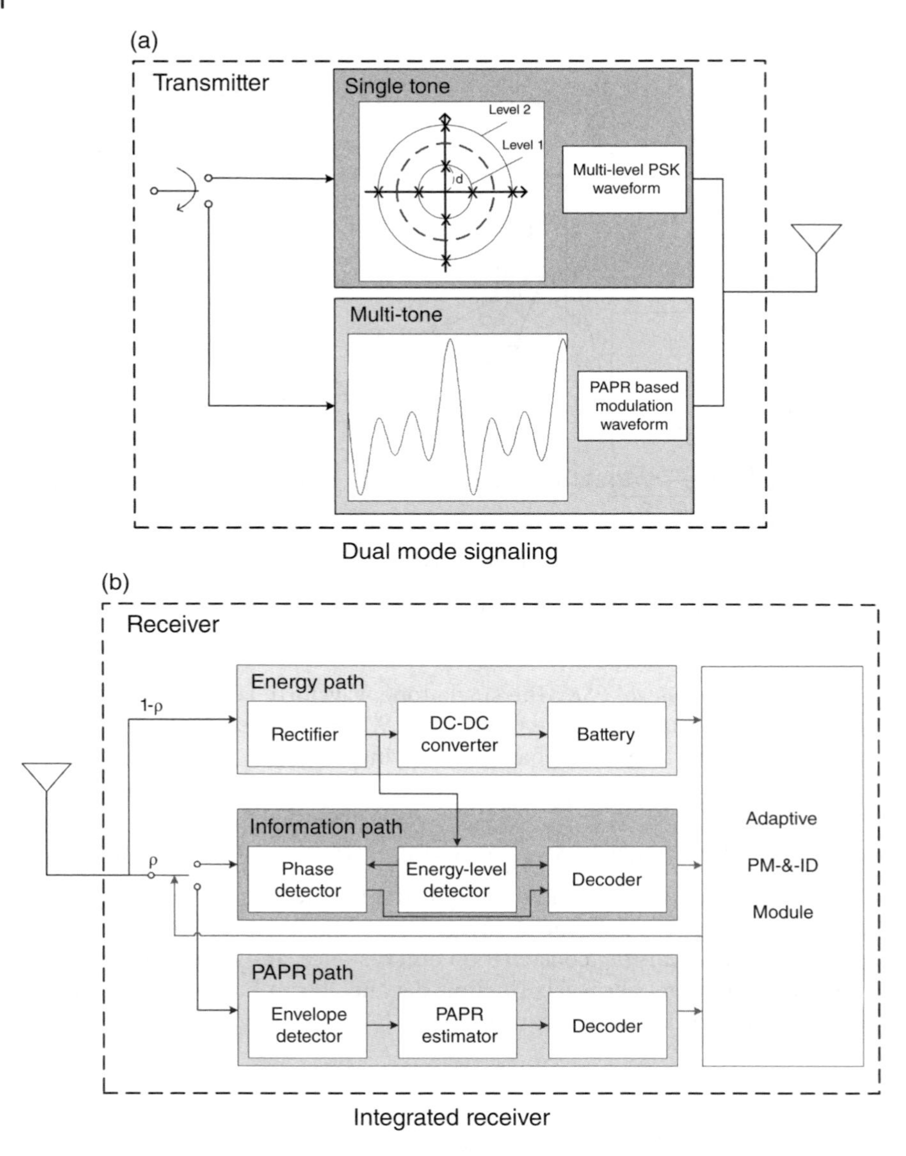

Figure 3.5 A new transceiver architecture for dual mode SWIPT.

information transmission facilitates low-power decoding via simple PAPR measurements [10]. For the maximum transmit PAPR with a single antenna $K = 1$ in (3.9), the PAPR modulated signal using a subset of N tones from Q available tones (i.e., $N \in \{1, \ldots, Q\}$) is given by

$$s_m(t) = \mathbf{Re}\left\{\sum_{n=1}^{N}\sqrt{\frac{2P_T}{N}}\exp[\,j2\pi f_{(n)}t + j\phi\,]\right\}. \tag{3.22}$$

To achieve a high WPT efficiency, we assume that the transmitter performs the MF precoding for amplitude matching and phase alignment at the receiver with the acquired CSI. After the MF precoding that leads to the maximal-ratio transmission (MRT), the transmitted signal can be expressed as

$$s(t) = \begin{cases} A\frac{h^*}{|h|}\cos(2\pi f_c t + \theta) & \text{for } s_s(t), \\ \sqrt{\frac{2P_T}{N}}\frac{h^*}{|h|}\sum_{n=1}^{N}\cos[2\pi f_{(n)}t] & \text{for } s_m(t). \end{cases} \tag{3.23}$$

Receiver Architecture. A new receiver architecture is shown in Figure 3.5, which consists of three (energy/information/PAPR) paths and an adaptive PM-&-ID module. To enable the receiver to perform both ID and EH from the same signal, the power splitting is performed in front of the three paths with power-splitting ratio ρ ($0 \leq \rho \leq 1$). We assume that infinitesimally small ρ is enough for reliable ID with the integrated receiver [12]. Then, the received signal in the *energy path* can be expressed as

$$y_{EH}(t) = \sqrt{1-\rho}hs(t) + \eta(t), \tag{3.24}$$

where $\eta(t) \sim \mathcal{N}(0, \sigma^2)$. The resulting DC power P_{EH} is evaluated as

$$P_{EH} = \zeta_{NL}(N, P_{in}) \times P_{in}, \tag{3.25}$$

where ζ_{NL} is a nonlinear PCE function given N and $P_{in} = \mathbf{E}\{|y_{EH}(t)|^2\} = (1-\rho)|h|^2P_T = (1-\rho)P_R \cong P_R$ (noise energy ignored), which is used for charging the battery and ID in the information path (see Figure 3.5). As the signal power in the energy path is sufficiently large to ignore the noise, the energy-level ID can be performed reliably with high SNR, unlike conventional SWIPT.

The received signal in the *information path* is of the form

$$y_{ID}(t) = \sqrt{\rho}hs(t) + \eta(t), \tag{3.26}$$

which is used to decode the phase information for a given energy level. Since channel estimation is required, decoding in the information path causes more circuit power consumption than PAPR based decoding. Note that the energy and information paths are jointly combined here for single tone SWIPT.

The *PAPR path* is used for PAPR based ID. For this, a PAPR estimator measures the PAPR of the received signal envelope by dividing the peak power with the root mean square (RMS) power. The received PAPR in FF channel is evaluated as

$$PAPR = \frac{\max_{t\in[0,T_m]} |y_{ID}(t)|^2}{\frac{1}{T_m}\int_{T_m} |y_{ID}(t)|^2 dt} \cong 2N. \tag{3.27}$$

PAPR based ID does not require power-hungry active devices, such as mixer and ADC, as well as channel estimation. Furthermore, the efficiency of WPT is enhanced in the low power region, and the PAPR path is suitable for low circuit power consumption while increasing the operational range with lower rate.

Adaptive Mode Switching. In conventional SWIPT, the achievable rate depends on the distance from the AP. With the proposed dual mode operation for SWIPT, however, the rate dependence can be mitigated well, leading to an increased operational range. The achievable rates with single and multi-tone modes are evaluated as

$$R_s = [1 - p_{out}(M)]\log_2 M, \tag{3.28}$$

$$R_m = \frac{1}{B_c T_m}[1 - p_{out}(Q)]\log_2 Q, \tag{3.29}$$

respectively, where $B_c = T_s^{-1}$. The outage probability can be defined as $p_{out} = \Pr[p_b > p_{tag}]$, i.e. the BER given M or Q is higher than the target BER p_{tag}. We assume that $p_{tag,s}$ for the single tone mode is lower than $p_{tag,m}$ for the multi-tone mode because the former requires tight QoS for high data rate. Also, the circuit power consumption should not exceed the harvested power to ensure the energy causality at the receiver. Note that the energy constraint of each mode is different because the single tone mode requires a more complex decoding procedure due to the channel estimation than the multi-tone mode with PAPR based ID.

We can formulate an adaptive MS optimization problem which chooses a proper mode with modulation index M and multi-tone Q to maximize the achievable rate, depending on the received power threshold P_{th} obtained from the nonlinear EH model and the QoS constraints given above. The optimization problem can be formulated as

$$\begin{aligned} \text{(P1)} : \max_{L} \quad & R_x \\ \text{s.t.} \quad & P_{EH} \geq P_{C,x}, \quad p_b(L) \leq p_{tag,x}, \end{aligned}$$

where $x \in \{s, m\}$, $L \in \{M, Q\}$, $P_{C,s}$ ($P_{C,m}$) is the circuit power consumption of the single tone (multi-tone) mode. We can obtain the optimal solution (M^*, Q^*) numerically by running the algorithm described below. Note that, if M^* or Q^* is less than 2, the transceiver is assumed to operate for EH only.

Algorithm 1 Algorithm for adaptive MS policy

Initialize M,Q and calculate P_{EH} from P_{in}
if $P_{in} \geq P_{th}$ and $P_{EH} \geq P_{C,s}$ then
  Choose maximum M such that BER constraints are met
  return M^*
else if $P_{in} < P_{th}$ and $P_{EH} = P_{C,m}$ then
  Choose minimum Q such that BER constraints are met
  return Q^*
else
  return $M^* = 1$ and $Q^* = 1$
end if

Evaluation. The performance of dual mode SWIPT and conventional SWIPT is compared. We assume $P_T = 40$ dBm and $B_c = 1$ MHz where f_c is 900 MHz. We consider a Rayleigh FF channel with path-loss exponent 2.5. Also, the noise power is assumed to be -130 dBm/Hz and ρ is set to 10^{-6}. Here, we have assumed that the single tone mode needs a strict BER constraint to support high data rates. Thus, the target BER of each mode is set to $p_{tag,s} = 0.01$ and $p_{tag,m} = 0.05$, accordingly. We set $P_{th} = 0.5$ mW, and the energy causality constraint is set to be $P_{C,s} = 0.2$ mW and $P_{C,m} = 0.01$ mW.

Figure 3.6 shows the achievable rate versus the distance from the AP subject to the QoS constraints. The rate is obtained by taking the average over different channel realizations. Compared to conventional SWIPT, dual mode SWIPT increases the operational range, where MS occurs at around 5 m for the given P_{th}. We see that dual mode SWIPT provides not only an increased operational range compared to conventional SWIPT with quadrature amplitude modulation (QAM), but also yields higher achievable rate than the multi-tone (PAPR) SWIPT when the receiver is close to the transmitter. This is because the adaptive MS policy allows the choice of optimum values for M and Q according to the received power for maximum rate.

3.4 Energy Harvesting Circuit Design

The nonlinear rectification process also characterizes the saturation effects where the efficiency of WPT rapidly decreases as the RF input power exceeds a certain threshold. To overcome this saturation nonlinearity, we propose a reconfigurable energy harvester to split the RF input power of the rectifier

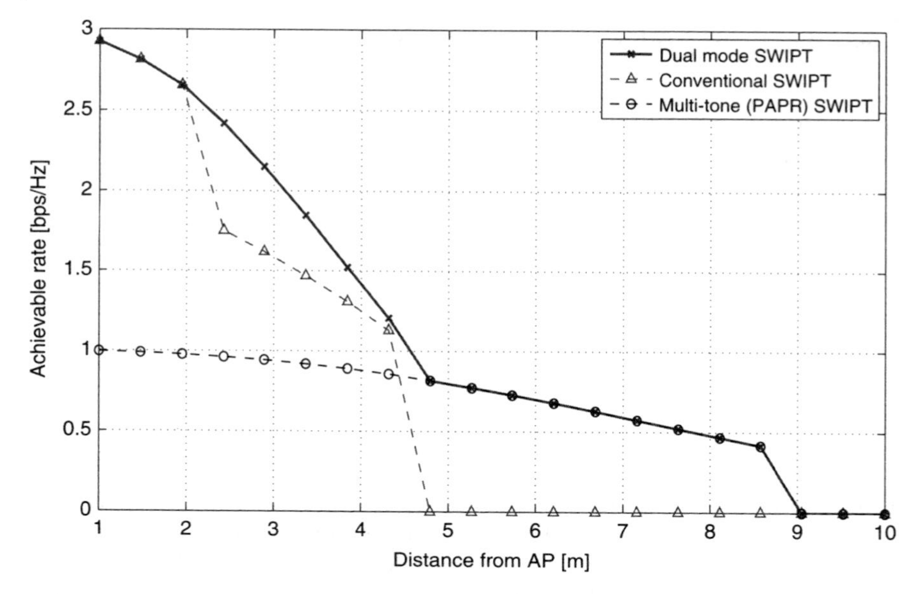

Figure 3.6 Achievable rate distribution subject to the QoS constraints.

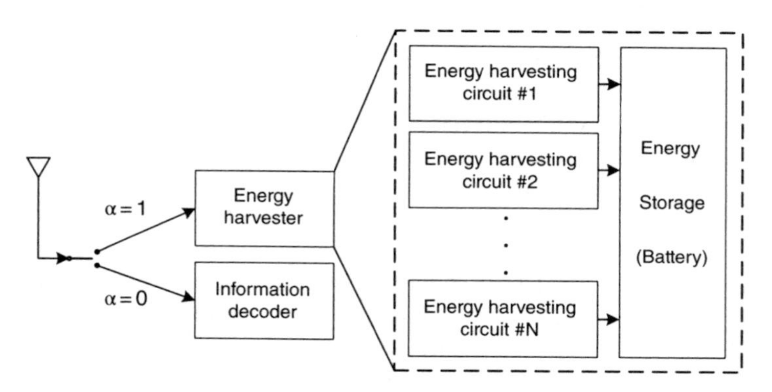

Figure 3.7 The proposed receiver structure for SWIPT.

among multiple EH circuits to ensure its reliable operation outside the saturation region.

Proposed Reconfigurable Energy Harvester. Figure 3.7 shows the proposed receiver structure with N_{EH} multiple EH circuits implemented in parallel and followed by a battery. The nonlinear energy harvester is reconfigured adaptively depending on the received power. We assume an identical nonlinear EH model for each EH circuit. Hence, the harvested power at the ith EH circuit, denoted by $P_{EH,i}$, is expressed as

$$P_{EH,i}(\alpha, P_R) = \alpha \frac{P_{sat}(1 - e^{-c_1 P_R})}{1 + e^{-c_1(P_R - c_0)}}. \tag{3.30}$$

Here, the mode switching indicator variable α is set to either ID ($\alpha = 0$) or EH ($\alpha = 1$), considering the channel state, as shown in Figure 3.7.

When $\alpha = 1$, all received signal is fed to the energy harvester and then split evenly among N_{EH} multiple EH circuits. We assume an ideal energy harvester such that there is no energy loss during power splitting and energy conversion from multiple EH circuits to the battery. Under the ideal EH assumption, the harvested power at the battery is linearly increased to

$$P_{EH,tot} = \sum_{i=1}^{N_{EH}} P_{EH,i}(\rho_i P_R), \tag{3.31}$$

where ρ_i is the power-splitting ratio for the ith EH circuit. As each EH circuit is modeled by the identical nonlinear function with $\rho_i = 1/N_{EH}$ and $P_{EH,i} = P_{EH,1}$ for $\forall i$, the total harvested power becomes

$$P_{EH,tot} = N_{EH} P_{EH,1}(P_R/N_{EH}). \tag{3.32}$$

To gain some insights into the effect of the power-splitting ratio $\boldsymbol{\rho} = [\rho_1, \ldots, \rho_{N_{EH}}]$, we consider a toy example of the energy harvester with two EH circuits in parallel. In this case, we can drop the circuit index of the power-splitting ratio and simplify the total harvested power to

$$P_{EH,tot}(\rho, P_R) = P_{EH,1}(\rho P_R) + P_{EH,1}((1 - \rho)P_R), \tag{3.33}$$

where $\rho \in [0, 1]$. $P_{EH,tot}$ varies as the power-splitting ratio ρ or the received power P_R changes. With fixed P_R, $P_{EH,tot}$ can be maximized by optimizing ρ.

Figure 3.8 illustrates $P_{EH,tot}$ as a function of ρ and P_R while Figure 3.9 shows the contours of $P_{EH,tot}$ as functions of ρ and P_R. The parameters of the practical EH circuits here are set to $P_{sat} = 3mW$, $c_0 = 0.0033$, and $c_1 = 1000$. We observe that the optimal power-splitting ratio ρ^* changes with the received power P_R. When P_R is low, only one EH circuit is activated while switching off the other by setting $\rho = 0$ or 1. For high P_R, it is preferable to activate both EH circuits by splitting the received power evenly ($\rho = 0.5$). This is a direct consequence of the nonlinear behavior of the EH circuit. The EH efficiency becomes very low for low P_R, and thus using a single EH circuit is better than using both. On the other hand, using a single EH circuit causes saturation for high P_R, and it is desirable to activate both EH circuits. When saturation occurs, splitting the received power evenly can reduce the excess power due to the saturation. Based on these observations from the above toy example, the total harvested power can be expressed as $P_{EH,tot} = \max\{P_{EH,1}(P_R),\ 2P_{EH,1}(P_R/2)\}$.

When the number of EH circuits increases, we can reconfigure the energy harvester in a similar manner according to the received power with even split. The total harvested power is $P_{EH,tot} = \hat{N}_{EH} P_{EH,1}(P_R/\hat{N}_{EH})$, where $\hat{N}_{EH}$ is the

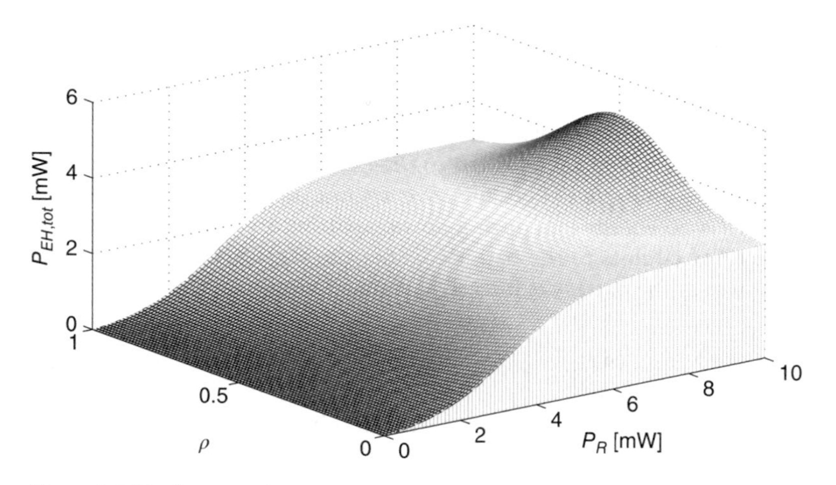

Figure 3.8 The harvested power depending on the received power P_R and ρ.

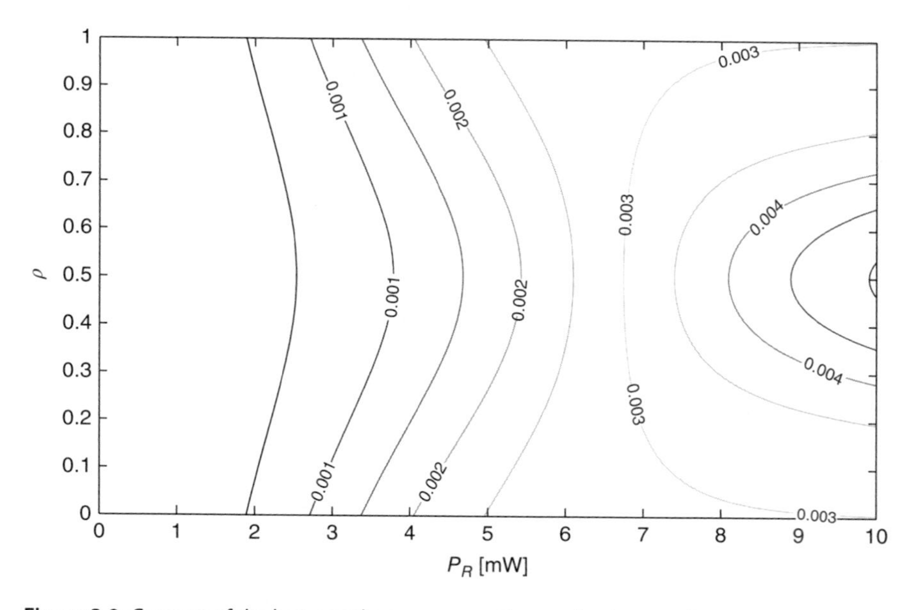

Figure 3.9 Contour of the harvested power versus the received power P_R.

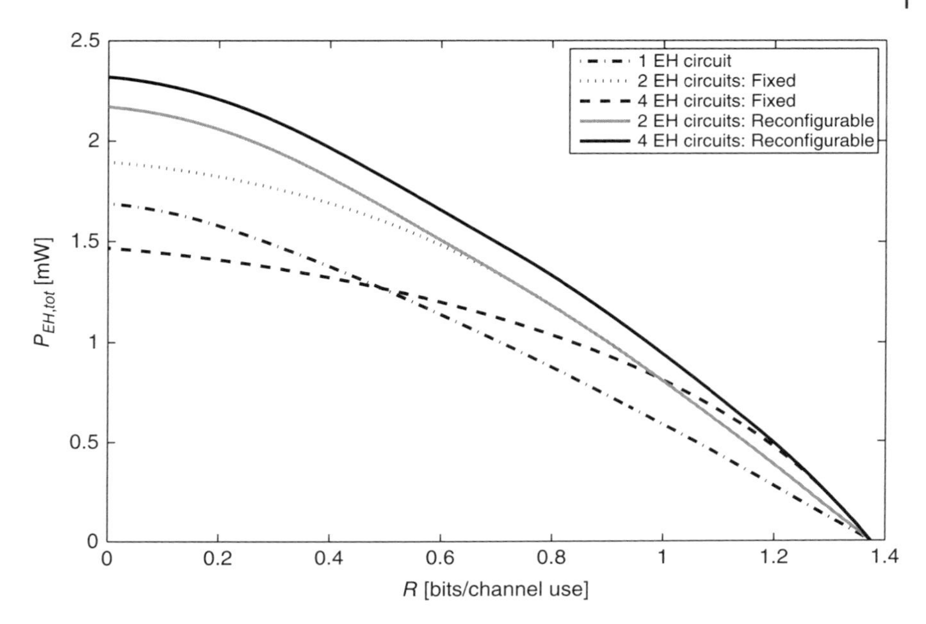

Figure 3.10 The rate–energy tradeoff.

number of activated EH circuits depending on the received power. $\hat{N}_{EH}$ can be determined by $\hat{N}_{EH} = \arg\max_{N_{EH}}\{P_{EH,1}(P_R), ..., N_{EH}P_{EH,1}(P_R/N_{EH})\}$, yielding the optimal power-splitting ratio $\rho_i^* = 1/\hat{N}_{EH}, i \in \{1, ..., \hat{N}_{EH}\}$.

Evaluation. To validate the performance improvements by the reconfigurable energy harvester, we evaluate the rate–energy tradeoff as

$$C_{R-E} = \bigcup_{\alpha}\{(R, P_{EH}) : R \leq \mathbf{E}\{R(\alpha)\}, P_{EH} \leq \mathbf{E}\{P_{EH,tot}(\alpha, \boldsymbol{\rho}^*)\}\} \tag{3.34}$$

where $R(\alpha) = (1-\alpha)\log_2(1 + P_R/\sigma^2)$. The parameters for each EH circuit are set to $P_{sat} = 3mW$, $c_0 = 0.0033$, and $c_1 = 1000$. In the simulation, we have used $B = 10^5$ fading blocks, each of which was generated from the Rician fading channel. We assume that the power gain of the LOS channel is -30 dB with Rician factor 1. The noise variance is set to yield an average SNR of 3 dB, regardless of the received power, to ensure a fair comparison.

Figure 3.10 shows the rate–energy tradeoff when the transmit power is 5 W. We first compare the receiver with fixed multiple EH circuits ($N_{EH} = 2, 4$) with that with a single EH circuit ($N_{EH} = 1$). We see that using two EH circuits yields more harvested energy compared to using a single or four EH circuits. If the reconfigurable receiver ($N_{EH} = 2, 4$) is used, using more circuits always yields better performance. Thus, the reconfigurable receiver always has better performance than the receiver with fixed multiple EH circuits. Their harvested power gap decreases as the rate increases. Consequently, the former has much better performance when the rate is low or the energy constraint is tight, suitable for low-power smart devices that are self-powering.

3.5 Discussion and Conclusion

First, we implemented multi-tone SWIPT using PAPR based signaling and a combined receiver that realizes the RF-to-DC conversion at once, both for ID and EH, with no channel estimation. In particular, the PAPR based ID over a selective sub-band offers a *reliable* BER performance regardless of the specific channel gain, as long as the sub-band has an FF channel. Using multi-tone waveforms was shown to be the prerequisite for efficient WPT, and in conjunction with PAPR based ID it leads to a promising solution for realizing low-power smart devices, eventually with batteryless operation, facilitating the massive and dense deployment of MTC for IoT.

Second, we have designed a dual mode SWIPT transceiver alternating between single and multi-tone waveforms. We also developed the new concept of adaptive MS based on the nonlinear EH model. The dual mode operation mitigates the energy causality constraint and increases the operational range. It was demonstrated that dual mode SWIPT provides significant gains in achievable rate distribution over the conventional one. In particular, the adaptive MS algorithm was shown to be a viable solution for enlarging the operational range, leading to self-powering for energy neutral operation.

Third, we have proposed a reconfigurable energy harvester with multiple EH circuits implemented in parallel before the battery. We showed how the practical energy harvester can be reconfigured based on the received power. To verify the performance improvement, we have investigated the rate–energy tradeoff. The reconfigurable receiver was compared with two other receivers: one with a single EH circuit and the other with fixed multiple EH circuits. The performance evaluations showed that the reconfigurable receiver achieves the best rate–energy tradeoff, with which SWIPT can self-power smart devices effectively while ensuring energy neutral operation.

Acknowledgements. This work was supported by a National Research Foundation of Korea (NRF) Grant funded by the Korean Government under Grant 2014R1A5A1011478.

Bibliography

1 X. Lu, P. Wang, D. Niyato, D.I. Kim, and Z. Han (2015) Wireless networks with RF energy harvesting: A contemporary survey. *IEEE Commun. Surveys & Tutorials*, **17** (2): 757–789.

2 X. Lu, P. Wang, D. Niyato, D.I. Kim, and Z. Han (2016) Wireless charging technologies: Fundamentals, standards, and network applications. *IEEE Commun. Surveys & Tutorials*, **18** (2): 1413–1452.
3 V. Liu, A. Parks, V. Talla, S. Gollakota, D. Wetherall, and J.R. Smith (2013) Ambient backscatter: Wireless communication out of thin air. *ACM SIGCOMM 2013*, Hong Kong, China, pp. 39–50.
4 V. Talla, M. Hessar, B. Kellogg, A. Najafi, J.R. Smith, and S. Gollakota (2017) LoRa backscatter: Enabling the vision of ubiquitous connectivity. *ACM UBICOMM 2017*, **1** (3): 105:1–105:24.
5 R. Eletreby, D. Zhang, S. Kumar, and O. Yagan (2017) Empowering low-power wide area networks in urban settings. *ACM SIGCOMM 2017*, Los Angeles, CA, pp. 309–321.
6 S. Bi, C.K. Ho, and R. Zhang (2015) Wireless powered communication: Opportunities and challenges. *IEEE Commun. Mag.*, **53** (4): 117–125.
7 K. Huang and X. Zhou (2015) Cutting the last wires for mobile communications by microwave power transfer. *IEEE Commun. Mag.*, **53** (6): 86–93.
8 A.S. Boaventura and N.B. Carvalho (2011) Maximizing DC power in energy harvesting circuits using multisine excitation. *2011 IEEE MTT-S International Microwave Symposium Digest (MTT)*, Baltimore, MD, pp. 1–4.
9 B. Clerckx and E. Bayguzina (2016) Waveform design for wireless power transfer. *IEEE Trans. Signal Process.*, **64** (23): 6313–6328.
10 D.I. Kim, J.H. Moon, and J.J. Park (2016) New SWIPT using PAPR: How it works. *IEEE Wireless Commun. Lett.*, **5** (6): 672–675.
11 R. Zhang and C.K. Ho (2013) MIMO broadcasting for simultaneous wireless information and power transfer. *IEEE Trans. Wireless Commun.*, **12** (5): 1989–2001.
12 X. Zhou, R. Zhang, and C.K. Ho (2013) Wireless information and power transfer: Architecture design and rate–energy tradeoff. *IEEE Trans. Commun.*, **11**: 4754–4767.
13 L. Liu, R. Zhang, and K.C. Chua (2013) Wireless information transfer with opportunistic energy harvesting. *IEEE Trans. Wireless Commun.*, **12** (1): 288–300.
14 E. Boshkovska, D.W.K. Ng, N. Zlatanov, and R. Schober (2015) Practical nonlinear energy harvesting model and resource allocation for SWIPT systems. *IEEE Commun. Lett.*, **19** (12): 2082–2085.

4

Industrial SWIPT

Backscatter Radio and RFIDs

*Panos N. Alevizos** *and Aggelos Bletsas*

School of ECE, Technical University of Crete, Greece

4.1 Introduction

Radio frequency identification (RFID) is a concrete, industrial example of simultaneous wireless information and power transfer (SWIPT) and dates back to the early 1940s [1]. It is based on backscatter radio, i.e. communications by means of reflection, due to its extraordinary ultra-low power nature, recently exploited in digital sensor networking [2–5], analog sensor networking [6–8], and IoT [9–11]. Thus, it is interesting to examine such paradigm, under the prism of recent findings in the SWIPT community.

One of the most important problems in state-of-the-art RFID systems is the limited range, due to RF energy harvesting issues at the tag. Far-field RF energy harvesting circuitry includes one or more diodes for rectification of the incoming RF signal and conversion to DC; such rectification circuits demonstrate highly nonlinear characteristics, as well as *limited* RF harvesting sensitivity, i.e. output power is zero when the power of the input RF signal is below the sensitivity threshold [12]. Thus, offering an accurate, yet-tractable model of the tag's harvesting circuit and corresponding output harvested power as a function of input RF power is of vital importance in the SWIPT community. Linear RF harvesting models have been largely incorporated, while nonlinear models have been recently proposed (e.g., [13, 14]). Still, there is no authoritative model that incorporates both nonlinearity as well as limited sensitivity of the RF harvester. Sensitivity is slowly improved by a factor of 2 approximately every 5 years [15], exhibiting values orders of magnitude worse than the sensitivity of typical communication receivers. Thus, *simultaneous* information and power transfer may not always be feasible [16].

*Corresponding author: Aggelos Bletsas; aggelos@telecom.tuc.gr

Wireless Information and Power Transfer: Theory and Practice, First Edition.
Edited by Derrick Wing Kwan Ng, Trung Q. Duong, Caijun Zhong, and Robert Schober.

This work compares various far-field RF harvesting models in the context of industrial SWIPT, focusing on nonlinearity and *limited* sensitivity of the RF harvesting circuit. A specific, yet-tractable modeling methodology is proposed, it is shown that the inherent non-ideal characteristics of the RF harvesting circuitry should be explicitly taken into account, and signals that convey information are not always appropriate for power transfer under realistic modeling of the RF harvester.

Section 4.2 presents the adopted wireless signal model for RFID. Section 4.3 overviews the basic building blocks of passive (i.e., batteryless) RFID tags. Section 4.4 examines the event of bit error rate (BER) at the RFID reader. Section 4.5 studies and analyzes the compound event of successful tag reception, including the case where RF power transfer fails. Numerical results under different RF energy harvesting models are presented in Section 4.6. Finally, work is concluded in Section 4.7.

4.2 Wireless Signal Model

A *monostatic* backscatter radio system consists of a reader and a tag. In that case, the reader acts as both the illuminating carrier emitter and receiver of the information reflected from the tag. In general, there are three prominent operating architectures in backscatter communications:

1) Monostatic architecture and reader with single antenna for both transmission and reception, where path loss and small-scale fading are the same for both reader-to-tag and tag-to-reader links.
2) Monostatic architecture and reader consisting of different transmit and receive antennas, where path loss remains the same, while small-scale fading coefficients are, in general, different.
3) Bistatic architecture, where carrier emitter (CE) and receiver are not co-located, the path loss is different for the CE-to-tag and tag-to-reader links, while the corresponding small-scale fading coefficients are, in general, independent.

For easier exposition, the first architecture is considered, depicted in Figure 4.1. Due to the relatively small distance d involved between reader and tag, the following path-loss model is considered:

$$\mathsf{L} \equiv \mathsf{L}(d) = \left(\frac{\lambda}{4\pi d_0}\right)^2 \left(\frac{d_0}{d}\right)^\nu, \tag{4.1}$$

where d_0 is a reference distance (assumed unit thereinafter), λ is the carrier wavelength, and ν is the path-loss exponent.

Tag communication bandwidth and channel delay spread are assumed to be relatively small and thus frequency non-selective (flat) fading is assumed, i.e.

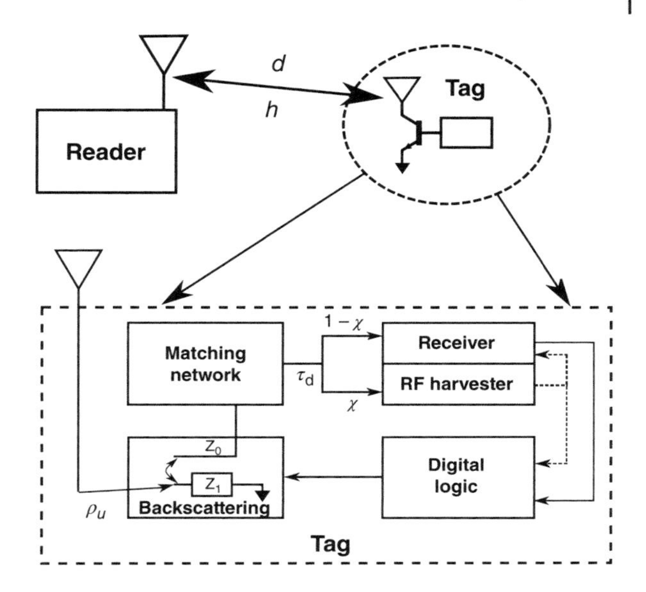

Figure 4.1 Monostatic backscatter architecture consisting of a reader and a passive (i.e., batteryless) RFID tag. The reader acts as transmitter of the illuminating signal, as well as receiver of the tag reflected (backscattered) information.

the wireless small-scale fading channel for the tag-to-reader and reader-to-tag link is given by $h = a\mathrm{e}^{-\mathrm{j}\phi}$, where $a \in \mathbb{R}_+$ is the fading amplitude and $\phi \in [0, 2\pi)$ is the fading phase. Due to potentially strong line-of-sight (LOS) signals in scatter radio environments, Nakagami small-scale fading is assumed with $\mathbb{E}\,[a^2] = 1$, where $\mathbb{E}[\cdot]$ denotes the expectation operation, and probability density function (PDF) [17], p. 79]:

$$\mathsf{f}_a(x) = 2\,\mathtt{M}^{\mathtt{M}}\,\frac{x^{2\mathtt{M}-1}}{\Gamma(\mathtt{M})}\,\mathrm{e}^{-\mathtt{M}x^2}, \quad x \geq 0, \tag{4.2}$$

where $\mathtt{M} \geq \frac{1}{2}$ is the Nakagami parameter and $\Gamma(x) = \int_0^{\infty} t^{x-1}\mathrm{e}^{-t}\mathrm{d}t$ is the Gamma function. For the special cases of $\mathtt{M} = 1$ and $\mathtt{M} = \infty$, Rayleigh and no fading (i.e., $a = 1$) is obtained, respectively. For $\mathtt{M} = \frac{(\kappa+1)^2}{2\kappa+1}$, the distribution in (4.2) approximates Rician fading with Rician parameter κ, where commonly $\kappa \in [0, 20]$ [17].

The reader emits a carrier with frequency F_c and passband representation,

$$\mathsf{c}_\mathrm{R}(t) = \sqrt{2\,\mathtt{P}_\mathrm{R}}\,\Re\{\mathrm{e}^{2\pi F_\mathrm{c} t}\}, \tag{4.3}$$

where $\Re\{\cdot\}$ denotes the real part operator and $\mathtt{P}_\mathrm{R}$ is the reader's transmit power. The signal propagates along the reader-to-tag link and thus the passband received signal at the tag antenna is given by:

$$\mathsf{c}_\mathrm{T}(t) = \sqrt{2\,\mathsf{L}\,\mathtt{P}_\mathrm{R}}\,\Re\{h\,\mathrm{e}^{2\pi F_\mathrm{c} t}\}. \tag{4.4}$$

4.3 RFID Tag Operation

The operational modules of a passive RFID tag are graphically illustrated in Figure 4.1 and explained in the following subsections. The tag's circuitry does not include any type of signal conditioning units (e.g., amplifiers, filters or mixers); instead, the tag communicates through the use of the reflection coefficient between the tag antenna and the connected load, through a controllable switch. Two loads are typically used and hence binary modulation with two different reflection coefficients. Information is conveyed through reflection and backscattering of the illuminating reader's signal. The absence of signal conditioning at the tag offers ultra-low power consumption and is the main advantage of scatter radio.

The received (input) power at the tag is given by:

$$P_{\text{in}} = \mathsf{L}\ \mathsf{P}_{\text{R}}\ |h|^2 = \mathsf{L}\ \mathsf{P}_{\text{R}}\ a^2. \tag{4.5}$$

As a is a unit power Nakagami RV ($\mathbb{E}\ [a^2] = 1$)), P_{in} follows Gamma distribution with shaping parameters $\left(\mathsf{M}, \frac{\mathsf{L}\ \mathsf{P}_{\text{R}}}{\mathsf{M}}\right)$, i.e.[1] [18]

$$\mathsf{f}_{P_{\text{in}}}(x) = \left(\frac{\mathsf{M}}{\mathsf{L}\ \mathsf{P}_{\text{R}}}\right)^{\mathsf{M}} \frac{x^{\mathsf{M}-1}}{\Gamma(\mathsf{M})}\ \mathsf{e}^{-\frac{\mathsf{M}}{\mathsf{L}\ \mathsf{P}_{\text{R}}}x}, \quad x \geq 0. \tag{4.6}$$

4.3.1 RF Harvesting and Powering for RFID Tag

The tag is able to harvest energy in order to power its electronics and start operating only if the input power is above the tag's harvesting sensitivity, P_{sen}, i.e. $P_{\text{in}} > \mathsf{P}_{\text{sen}}$. In that case, the RF input power, incident on the tag's rectifier circuit, is converted to DC and powers the tag's electronics. The RF harvesting sensitivity is a vital performance parameter in backscatter communications with passive (i.e., batteryless) tags, given the fact that the majority of current far-field RF harvesting circuits in the literature have limited sensitivity.

As ground-truth, the harvested power as a function of the input power is given by the following nonlinear equation:

$$\mathsf{p}(x) \triangleq \begin{cases} 0, & x \in [0, \mathsf{P}_{\text{sen}}], \\ \eta(x) \cdot x, & x \in [\mathsf{P}_{\text{sen}}, \mathsf{P}_{\text{sat}}], \\ \eta(\mathsf{P}_{\text{sat}}) \cdot \mathsf{P}_{\text{sat}} & x \in [\mathsf{P}_{\text{sat}}, \infty). \end{cases} \tag{4.7}$$

where $\eta\ (\cdot)$ is the harvesting efficiency as a function of input power, defined over the interval $\mathcal{P} \triangleq [\mathsf{P}_{\text{sen}}, \mathsf{P}_{\text{sat}}]$, and P_{sat} denotes the saturation power threshold of the harvester, after which the harvested power is constant. Function $\eta(x)$ is

1 A random variable x, which adheres to Gamma distribution with parameters (κ, θ), has a PDF $\mathsf{f}_x(x) = \frac{1}{\theta^{\kappa}\Gamma(k)} x^{\kappa-1} \mathsf{e}^{-x/\theta}$.

non-negative and continuous over $\mathcal{P}$ and obtains the value zero for $x = \mathrm{P}_{\mathrm{sen}}$. The harvested power function $\mathsf{p} : \mathbb{R}_+ \to \mathbb{R}_+$ is assumed:

1) non-decreasing, i.e. $x < y \Rightarrow \mathsf{p}(x) \leq \mathsf{p}(y)$, and
2) continuous, i.e. $x \to x_0 \Rightarrow \mathsf{p}(x) \to \mathsf{p}(x_0)$.

Note that the assumptions above, even though mild, are in full accordance with nonlinear the harvested power curves reported in the RF energy harvesting circuits prior art, e.g. [12, 19–21].

A fraction τ_{d} of time the antenna load is at Z_0 (absorbing state at Figure 4.1), while the rest $1 - \tau_{\mathrm{d}}$ corresponds to the fraction of time at load Z_1 (reflection state). Assume that χ is the fraction of receiving input power (when tag's antenna load is in the absorbing state) dedicated to the RF energy harvesting operation; thus, the total $\zeta_{\mathrm{har}} = \chi\ \tau_{\mathrm{d}}$ percentage of input power is dedicated to RF energy harvesting, with $\zeta_{\mathrm{har}} \in (0, 1)$. The remaining $(1 - \chi)\tau_{\mathrm{d}}$ input signal power is exploited by the reader-to-tag downlink communication circuitry.

Thus, if the total harvested power, $\mathsf{p}(\zeta_{\mathrm{har}}\ P_{\mathrm{in}})$, is larger than the tag's power consumption P_{c}, the tag starts the backscattering operation, otherwise the tag is not able to power its electronics. This is critical, given the fact that batteryless RFIDs typically incorporate no energy storage element, e.g., (super)capacitor, due to size and cost limitations.

4.3.2 RFID Tag Backscatter (Uplink) Radio

Suppose that the tag's antenna is terminated between two load values Z_0 and Z_1 (Figure 4.1). When the antenna is terminated at Z_0, the tag antenna is matched to the (antenna) input load and the tag absorbs the power from the incident signal. When the antenna is terminated at load Z_1, the tag reflects the incoming signal, i.e. scatters back information, provided that it has a sufficient amount of energy. When the antenna is terminated at Z_i, the tag reflection coefficient, $\Gamma_i \in \{0, 1\}$, is given by the modified reflection coefficient [22]:

$$\Gamma_i = \frac{Z_i - \mathrm{Z}_{\mathrm{a}}^*}{Z_i + \mathrm{Z}_{\mathrm{a}}}, \quad i \in \{0, 1\}, \tag{4.8}$$

where Z_{a} is the tag antenna impedance. The baseband equivalent of the tag-reflected signal is given by [22]:

$$\mathrm{A}_s - \Gamma_i, \tag{4.9}$$

which in turn depends on the (load-independent) tag antenna structural mode A_s and the transmitted bit $i \in \{0, 1\}$.

A fraction of $\rho_{\mathrm{u}} \leq 1 - \tau_{\mathrm{d}}$ incident input power is used for uplink (reflection) scatter radio operation. This number depends on the tag scattering efficiency and incorporates non-idealities of the tag's hardware (e.g., mismatch).

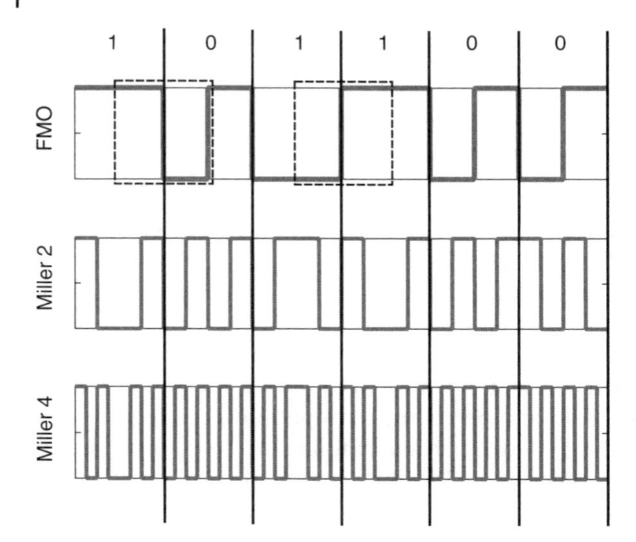

Figure 4.2 Examples of the two-line codes used in industrial RFID.

Thus, the backscattered baseband signal, for a duration of N tag bits, reflected by the tag antenna, is given by [23]:

$$\mathsf{b}(t) = \sqrt{\mathsf{L}\ \rho_{\mathrm{u}}\ \mathrm{P}_{\mathrm{R}}}\ h\left(\mathbb{A}_s - \Gamma_0 + (\Gamma_0 - \Gamma_1)\sum_{n=1}^{N} \mathsf{s}_{b_n}(t-(n-1)T)\right), \quad (4.10)$$

where $b_n \in \{0, 1\}$ is the nth reflected bit of duration T, while function $\mathsf{s}_{b_n}(\cdot)$ is the reflected signal basis function, when bit b_n is transmitted.

In order to (i) balance the time for which the tag is absorbing energy, independently of the tag's data bits, and (ii) avoid *ghost* tag reception, i.e. the reader misinterpreting thermal noise as tag emitting information, a *line* code is used in commercial GEN2 RFIDs [24], selecting between FM0 and Miller line codes. Specifically, the FM0 line code adheres to the following rules: (i) the line level always changes at the bit boundaries and (ii) the line changes at the middle of the bit, only when bit 0 is transmitted; a 6-bit example is shown in the top row of Figure 4.2. For Miller-M, with $M = 2, 4, 8$, there are $M/2$ line level transitions per half-bit and additionally the following rules are imposed: (i) the line level changes at the middle of the bit only when bit 0 is transmitted, (ii) the line level at the beginning of the bit must be different from the line level at the beginning of the previous bit, and (iii) there is an exception to the previous rule when transition from 0 to 1 bit in the data bits occurs, i.e. the line level at the beginning of bit 1 is the same as the line level at the beginning of the previous bit 0; examples for $M = 2$ and $M = 4$ are shown in middle and last rows of Figure 4.2, respectively.

Notice in the top row of Figure 4.2 that for the signal half-bit before and the signal half-bit after the bit boundary, there are only two possible, s-type waveforms, depicted with discontinuous rectangles; these two T-duration waveforms are orthogonal. Thus, for the FM0 line coding, observation of a $2T$ signal duration for each bit (of duration T) suffices for (differential) detection and $\mathsf{s}_{b_n}(\cdot)$ is a $T/2$-shifted waveform given by [25]:

$$\mathsf{s}_0(t) \triangleq \begin{cases} 1, & 0 \leq t < \frac{T}{2}, \\ 0, & \text{otherwise}, \end{cases} \tag{4.11}$$

and

$$\mathsf{s}_1(t) \triangleq \begin{cases} 1, & \frac{T}{2} \leq t < T, \\ 0, & \text{otherwise}. \end{cases} \tag{4.12}$$

The backscattered signal is propagated through the tag-to-reader link; the received signal becomes:

$$\tilde{\mathsf{y}}(t) = \sqrt{\mathsf{L}}\, h\, \mathsf{b}(t) + \mathsf{n}(t), \tag{4.13}$$

where $\mathsf{n}(t)$ is a circularly-symmetric complex Gaussian noise process with flat power spectral density (taking the value $\mathtt{N}_0$) in the frequency band $[-\mathtt{W}, \mathtt{W}]$. Some of the waveforms $\{\mathsf{s}_{b_n}(\cdot)\}_{n=1}^{N}$ are known to the reader to enable estimation of the leakage term at the received symbols and the compound uplink wireless channel, and for correlation preamble-based synchronization. After removing the leakage term, the resulting signal for a duration of N symbols at the reader follows:

$$\mathsf{y}(t) = \mathsf{L}\sqrt{\rho_{\mathrm{u}}\, \mathtt{P}_{\mathrm{R}}}\, h^2 (\Gamma_0 - \Gamma_1) \sum_{n=1}^{N} \mathsf{s}_{b_n}(t - (n-1)T) + \mathsf{n}(t), \tag{4.14}$$

where the complex term h^2 exists due to (i) the round-trip nature of backscatter communication and (ii) the monostatic setup of Figure 4.1. Assuming perfect synchronization, the optimal demodulator projects the received signal onto the basis function subspace using two correlators. The baseband signal at the output of the correlators for a duration of N symbols becomes:

$$\mathbf{y}_n = \mathsf{L}\sqrt{\rho_{\mathrm{u}}\, \mathtt{P}_{\mathrm{R}}}\, h^2 (\Gamma_0 - \Gamma_1)\mathbf{s}_n + \mathbf{w}_n = g\,\mathbf{s}_n + \mathbf{w}_n, n = 1, 2, \ldots, N, \tag{4.15}$$

where

$$g \triangleq \mathsf{L}\sqrt{\rho_{\mathrm{u}}\, \mathtt{P}_{\mathrm{R}}}\, h^2 (\Gamma_0 - \Gamma_1), \tag{4.16}$$

and $\mathbf{s}_n$ is the vector representation for the nth transmitted signal, while vector $\mathbf{w}_n$ is a circularly-symmetric complex Gaussian vector with diagonal covariance matrix, i.e. $\mathbf{w}_n \sim \mathcal{CN}(\mathbf{0}_2, \sigma^2 \mathbf{I}_2)$, with σ^2 denoting the variance of each noise component and quantities $\mathbf{0}_2$ and $\mathbf{I}_2$ are the two-dimensional all-zero vector and the 2×2 identity matrix, respectively. For RFID systems, which employ $T/2$-shifted FM0 line coding, the signal satisfies $\mathbf{s}_n \in \{[1 \quad 0]^\top, [0 \quad 1]^\top\}$ [26, 27].

4.4 Reader BER for Operational RFID

For the baseband signal in Eq. (4.15), given known channel g, and coherent maximum likelihood (ML) differential detection, the conditional bit error probability (BER) for RFID systems employing FM0 line coding follows from [27, 28]:

$$\mathbb{P}(\text{error}\,|g) = 2\mathsf{Q}\left(\frac{|g|}{\sigma}\right)\left(1 - \mathsf{Q}\left(\frac{|g|}{\sigma}\right)\right), \tag{4.17}$$

where $\mathsf{Q}(x) = \frac{1}{\sqrt{2\pi}}\int_x^{\infty} \mathsf{e}^{-\frac{t^2}{2}}\mathsf{d}t$ is the Gaussian Q-function. Interestingly, a similar expression applies to Miller line coding, when the receiver performs coherent (ML) bit-by-bit detection (and not Viterbi). Harnessing the following recent approximation for the Q-function given in [29],

$$\mathsf{Q}(x) \approx \frac{1}{2}\,\mathsf{e}^{-0.374x^2-0.777x},\ x \geq 0, \tag{4.18}$$

the conditional BER in (4.17) is approximated as follows:

$$\mathbb{P}(\text{error}\,|g) \approx \mathsf{e}^{-0.374\left(\frac{|g|}{\sigma}\right)^2-0.777\left(\frac{|g|}{\sigma}\right)} - \frac{1}{2}\mathsf{e}^{-0.748\left(\frac{|g|}{\sigma}\right)^2-1.554\left(\frac{|g|}{\sigma}\right)}. \tag{4.19}$$

Using the fact $|g| = |\mathsf{L}\sqrt{\rho_{\mathrm{u}}\,\mathrm{P_R}}\,h^2(\Gamma_0 - \Gamma_1)| \triangleq \frac{|\Gamma_0-\Gamma_1|\,\sqrt{\rho_{\mathrm{u}}}\,P_{\mathrm{in}}}{\sqrt{\mathrm{P_R}}}$, according to Eq. (4.5), the PDF $\mathsf{f}_{|g|}(x)$ of $|g|$ is a Gamma distribution with shaping parameters $\left(\mathsf{M}, \frac{\mathsf{L}\,|\Gamma_0-\Gamma_1|\,\mathrm{P_R}\,\sqrt{\rho_{\mathrm{u}}}}{\sqrt{\mathrm{P_R}}\,\mathsf{M}}\right) = \left(\mathsf{M}, \frac{\mathsf{L}\,|\Gamma_0-\Gamma_1|\,\sqrt{\mathrm{P_R}\,\rho_{\mathrm{u}}}}{\mathsf{M}}\right)$. Thus,

$$\begin{aligned}
\mathbb{P}(\text{error}) &= \mathbb{E}_{|g|}[\mathbb{P}(\text{error}|g)] \\
&\approx \int_0^{\infty}\left(\mathsf{e}^{-0.374\left(\frac{x}{\sigma}\right)^2-0.777\left(\frac{x}{\sigma}\right)} - \frac{1}{2}\mathsf{e}^{-0.748\left(\frac{x}{\sigma}\right)^2-1.554\left(\frac{x}{\sigma}\right)}\right)\mathsf{f}_{|g|}(x)\mathsf{d}x \\
&\overset{(a)}{=} \left(\frac{\mathsf{M}}{\mathsf{L}\,|\Gamma_0-\Gamma_1|\,\sqrt{\rho_{\mathrm{u}}\,\mathrm{P_R}}}\right)^{\mathsf{M}} 2^{-\mathsf{M}} \\
&\times\left[\left(\frac{0.374}{\sigma^2}\right)^{-\frac{\mathsf{M}}{2}}\mathsf{U}\left(\frac{\mathsf{M}}{2},\frac{1}{2},\frac{\left(\frac{0.777}{\sigma}+\frac{\mathsf{M}}{\mathsf{L}\,|\Gamma_0-\Gamma_1|\,\sqrt{\rho_{\mathrm{u}}\,\mathrm{P_R}}}\right)^2\sigma^2}{1.496}\right)\right. \\
&\left.-\frac{1}{2}\left(\frac{0.748}{\sigma^2}\right)^{-\frac{\mathsf{M}}{2}}\mathsf{U}\left(\frac{\mathsf{M}}{2},\frac{1}{2},\frac{\left(\frac{1.554}{\sigma}+\frac{\mathsf{M}}{\mathsf{L}\,|\Gamma_0-\Gamma_1|\,\sqrt{\rho_{\mathrm{u}}\,\mathrm{P_R}}}\right)^2\sigma^2}{2.992}\right)\right],
\end{aligned} \tag{4.20}$$

where $\mathsf{U}(\cdot,\cdot,\cdot)$ is the confluent hypergeometric function [30], Eq. (13.4.4)]. Step (a) above utilized first [31], Eqs. (3.462.1), (9.240)] and then [30], Eq. (13.14.3)] to simplify the final formula.

4.5 RFID Reader SWIPT Reception

The reader receives successfully the RFID tag's information when: (i) the input RF power at the tag antenna is above the tag's RF harvesting sensitivity, (ii) the RF harvested power is above the tag's power consumption, given that the RFID tag does not include energy storage elements (e.g., capacitors), and (iii) the BER at the reader is below a threshold β. The probability of these events is analyzed below.

4.5.1 Harvesting Sensitivity Outage

Given the definition of input power in Eq. (4.5), the tag's harvesting sensitivity outage metric is defined as follows:

$$\mathbb{P}(\mathcal{A}) \triangleq \mathbb{P}(\mathcal{P}_{\text{in}} \leq \mathtt{P}_{\text{sen}}) = \mathsf{F}_{\mathcal{P}_{\text{in}}}(\mathtt{P}_{\text{sen}}), \tag{4.21}$$

i.e. the probability that the input power P_{in} at the RFID tag antenna (which depends on the wireless channel/fading) is below the tag RF harvester's sensitivity $\mathtt{P}_{\text{sen}}$, where $\mathsf{F}_{\mathcal{P}_{\text{in}}}(\cdot)$ is the cumulative distribution function (CDF) of P_{in}. The above outage event defines the fraction of time the tag's rectenna cannot harvest RF energy due to inadequate incident input RF power.

Figure 4.3 examines the above outage probability as a function of the tag RF harvester's sensitivity $\mathtt{P}_{\text{sen}}$. Under Nakagami fading such outage is given by

$$\mathsf{F}_{\mathcal{P}_{\text{in}}}(\mathtt{P}_{\text{sen}}) = 1 - \int_{\mathtt{P}_{\text{sen}}}^{\infty} \mathsf{f}_{\mathcal{P}_{\text{in}}}(y)\text{d}y = 1 - \frac{\Gamma\left(\mathtt{M}, \frac{\mathtt{M}}{\mathtt{L}\ \mathtt{P}_{\text{R}}}\mathtt{P}_{\text{sen}}\right)}{\Gamma(\mathtt{M})}, \tag{4.22}$$

where $\Gamma(\alpha, z) = \int_z^{\infty} t^{\alpha-1}\text{e}^{-t}\text{d}t$. The path-loss model in Eq. (4.1) is employed with $\nu = 2.1$ and $\lambda = 0.3456$, and Nakagami parameter $\mathtt{M} = 5$. It can be clearly seen that less-sensitive RF harvesters, i.e. harvesters that require higher levels of input power and have higher values of $\mathtt{P}_{\text{sen}}$, suffer from higher outage probabilities.

Two harvesters from prior art are further discussed. The first RF harvester is the commercial PowerCast module [21], with offered sensitivity of $\mathtt{P}_{\text{sen}} = 10^{-1.2}$ mW (-12 dBm); the probability of sensitivity outage due to limited input power is almost 1 for transmission power $P_{\text{R}} = 20$ dBm and tag-reader distance d more than 4 m, while for $P_{\text{R}} = 35$ dBm and $d = 4$ m the outage event becomes 10%. The second RF harvester is the sensitive rectenna in [20] with measured sensitivity at $\mathtt{P}_{\text{sen}} = 10^{-4.25}$ mW, thus the corresponding sensitivity outage event becomes almost 0 for all studied parameters P_{R} and d. The RF harvesting sensitivity is commonly neglected in SWIPT research, even though it tremendously impacts the *power transfer* part and thus overall performance. It is also emphasized that RF receiver sensitivity for communication purposes can attain values

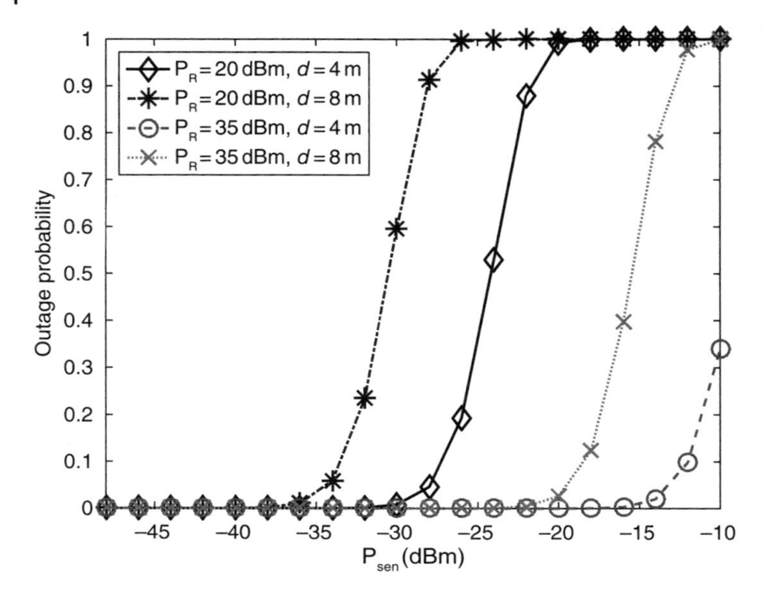

Figure 4.3 Probability of sensitivity outage event as a function of the tag's harvesting sensitivity.

of −80 dBm or less, while state-of-the-art rectennas offer harvesting sensitivities in the order of around −40 to −35 dBm. Clearly, signals useful for communication may not be useful for power transfer.

4.5.2 Power Consumption Outage

When the input power is above the tag's harvesting sensitivity, the next type of outage comes into play if the harvested power, $\mathsf{p}(\zeta_{\text{har}}\, P_{\text{in}})$ is *not enough*, i.e. below the tag's power consumption $\mathtt{P}_{\text{c}}$. This is critical for devices that cannot store harvested energy. Thus, power consumption outage is described as follows:

$$\mathbb{P}(\mathsf{p}(\zeta_{\text{har}}\, P_{\text{in}}) \leq \mathtt{P}_{\text{c}}), \tag{4.23}$$

depending on: (i) the input power (due to the wireless channel), (ii) the type of harvester, and (iii) the tag's power consumption $\mathtt{P}_{\text{c}}$; such probability describes the fraction of time the RF scavenged power is not adequate for tag powering. If $\mathsf{p}(\cdot)$ is strictly increasing and continuous around $\mathtt{P}_{\text{c}}$ [32], the event in Eq. (4.23) can be simplified as follows:

$$\mathbb{P}(\mathcal{B}) \triangleq \mathbb{P}\left(P_{\text{in}} \leq \frac{\mathsf{p}^{-\infty}(\mathtt{P}_{\text{c}})}{\zeta_{\text{har}}}\right) = \mathsf{F}_{P_{\text{in}}}\left(\frac{\mathsf{p}^{-\infty}(\mathtt{P}_{\text{c}})}{\zeta_{\text{har}}}\right), \tag{4.24}$$

where $\mathsf{p}^{-1}(\mathtt{P}_{\text{c}})$ is the inverse function of $\mathsf{p}(\cdot)$ at point $\mathtt{P}_{\text{c}}$.

4.5.3 Information Outage

RFID tag information outage at the reader is defined when BER in (4.17) is below a predefined precision β. Setting $\mathsf{R}(x) \triangleq 2\,\mathsf{Q}(x)\,(1-\mathsf{Q}(x)), x \in (0,\infty)$, this event can be mathematically expressed as follows [16]:

$$\mathbb{P}(\mathcal{C}) \triangleq \mathbb{P}\left(\mathsf{R}\left(\frac{|\Gamma_0-\Gamma_1|\,\sqrt{\rho_\mathrm{u}}\,P_\mathrm{in}}{\sqrt{\mathrm{P_R}}\,\sigma}\right) \leq \beta\right) = \mathbb{P}\left(P_\mathrm{in} \leq \frac{\sqrt{\mathrm{P_R}}\,\sigma\,\mathsf{R}^{-1}(\beta)}{|\Gamma_0-\Gamma_1|\,\sqrt{\rho_\mathrm{u}}}\right)$$
$$= \mathsf{F}_{P_\mathrm{in}}\left(\frac{\sqrt{\mathrm{P_R}}\,\sigma\,\mathsf{R}^{-1}(\beta)}{|\Gamma_0-\Gamma_1|\,\sqrt{\rho_\mathrm{u}}}\right), \tag{4.25}$$

where $\mathsf{R}^{-1}(\cdot)$ is the inverse function of $\mathsf{R}(\cdot)$ given by $\mathsf{R}^{-1}(x) = \mathsf{Q}^{-1}\left(\frac{1-\sqrt{1-2\,x}}{2}\right)$, defined for $x \in (0, 0.5)$, with $\mathsf{Q}^{-1}(\cdot)$ denoting the inverse of the Q-function, tabulated in several software packages.

4.5.4 Successful SWIPT Reception

Tag information is unsuccessfully received when *either* of the events $\mathcal{A}$, $\mathcal{B}$, $\mathcal{C}$ discussed in Sections 4.5.1, 4.5.2, and 4.5.3, respectively, occurs. Assuming that function $\mathsf{p}(\cdot)$ is strictly increasing and continuous around $\mathrm{P_c}$ and denoting for an event $\mathcal{D}$ its complement as $\mathcal{D}^\mathrm{C}$, the probability of unsuccessful SWIPT reception of tag information at the reader, denoted as event $\mathcal{F}$, is given as follows:

$$\mathbb{P}(\mathcal{F}) = 1 - \mathbb{P}(\mathcal{F}^\mathrm{C}) = \infty - \mathbb{P}(\mathcal{A}^\mathrm{C} \cap \mathcal{B}^\mathrm{C} \cap \mathcal{C}^\mathrm{C})$$
$$\equiv \mathbb{P}(\mathcal{A}) + \mathbb{P}(\mathcal{B}|\mathcal{A}^\mathrm{C}) \cdot \mathbb{P}(\mathcal{A}^\mathrm{C}) + \mathbb{P}(\mathcal{C}|\mathcal{A}^\mathrm{C} \cap \mathcal{B}^\mathrm{C}) \cdot \mathbb{P}(\mathcal{B}^\mathrm{C} \cap \mathcal{A}^\mathrm{C})$$
$$= 1 - \mathbb{P}\left(\{P_\mathrm{in} > \mathrm{P_{sen}}\} \cap \left\{P_\mathrm{in} > \frac{\mathsf{p}^{-1}(\mathrm{P_c})}{\zeta_\mathrm{har}}\right\} \cap \left\{P_\mathrm{in} > \frac{\sqrt{\mathrm{P_R}}\,\sigma\,\mathsf{R}^{-1}(\beta)}{|\Gamma_0-\Gamma_1|\,\sqrt{\rho_\mathrm{u}}}\right\}\right)$$
$$= 1 - \mathbb{P}\left(P_\mathrm{in} > \max\left\{\mathrm{P_{sen}}, \frac{\mathsf{p}^{-1}(\mathrm{P_c})}{\zeta_\mathrm{har}}, \frac{\sqrt{\mathrm{P_R}}\,\sigma\,\mathsf{R}^{-1}(\beta)}{|\Gamma_0-\Gamma_1|\,\sqrt{\rho_\mathrm{u}}}\right\}\right)$$
$$= \mathsf{F}_{P_\mathrm{in}}\left(\max\left\{\mathrm{P_{sen}}, \frac{\mathsf{p}^{-1}(\mathrm{P_c})}{\zeta_\mathrm{har}}, \frac{\sqrt{\mathrm{P_R}}\,\sigma\,\mathsf{R}^{-1}(\beta)}{|\Gamma_0-\Gamma_1|\,\sqrt{\rho_\mathrm{u}}}\right\}\right). \tag{4.26}$$

Consequently, successful SWIPT reception of RFID tag information at the reader, under Nakagami fading, is given in closed form as follows:

$$\mathbb{P}(\text{SWIPT success}) \equiv \mathbb{P}(\mathcal{F}^\mathrm{C}) = \frac{\Gamma\left(\mathrm{M}, \frac{\mathrm{M}}{\mathrm{L}\,\mathrm{P_R}}\theta_\mathcal{F}\right)}{\Gamma(\mathrm{M})}, \tag{4.27}$$

where $\theta_\mathcal{F} \triangleq \max\left\{\mathrm{P_{sen}}, \frac{\mathsf{p}^{-1}(\mathrm{P_c})}{\zeta_\mathrm{har}}, \frac{\sqrt{\mathrm{P_R}}\,\sigma\,\mathsf{R}^{-1}(\beta)}{|\Gamma_0-\Gamma_1|\,\sqrt{\rho_\mathrm{u}}}\right\}$.

4.6 Numerical Results

The RF harvesting model is critical for accurate SWIPT performance analysis. Thus, a state-of-the-art rectenna from the microwave engineering community [20] is utilized to determine the tag's harvested power response in Eq. (4.7) through parameter fitting.

Furthermore, several linear and nonlinear RF harvesting power models, from recent SWIPT prior art, are also compared. Specifically, the following models for $\mathsf{p}(\cdot)$ are considered:

- A usual linear model (L) $\mathsf{p}_1(x) \triangleq \eta_{\mathrm{L}} x, x \geq 0$, where η_{L} is the single parameter of the model, representing the harvesting efficiency. This is the most common model in recent SWIPT literature, and is linear and does not account for RF harvesting sensitivity.
- A linear model, accounting for harvesting sensitivity, called the constant-linear model (CL) [16]:

$$\mathsf{p}_2(x) \triangleq \begin{cases} 0, & x \in [0, \mathtt{P}_{\mathrm{sen}}], \\ \eta_{\mathrm{CL}} \cdot (x - \mathtt{P}_{\mathrm{sen}}), & x \in [\mathtt{P}_{\mathrm{sen}}, \infty), \end{cases} \tag{4.28}$$

where η_{CL} is the harvesting efficiency parameter of the model.
- A nonlinear normalized sigmoid harvested power model:

$$\mathsf{p}_3(x) \triangleq \frac{\frac{\mathtt{c}_0}{1+\exp(-\mathtt{a}_0(x-\mathtt{b}_0))} - \frac{\mathtt{c}_0}{1+\exp(\mathtt{a}_0\ \mathtt{b}_0)}}{1 - \frac{1}{1+\exp(\mathtt{a}_0\ \mathtt{b}_0)}}. \tag{4.29}$$

This model was proposed in [33] and subsequently adopted in [34–37]. It does not account for the tag's harvesting sensitivity and assumes $\mathtt{P}_{\mathrm{sen}} = 0$. It is parametrized by three real numbers $\mathtt{a}_0$, $\mathtt{b}_0$, and $\mathtt{c}_0$, determining the final shape of function $\mathsf{p}_3(\cdot)$.
- A nonlinear harvested power model, based on a second-order polynomial (in mW scale) [38]:

$$\mathsf{p}_4(x) \triangleq \mathtt{a}_1\ x^2 + \mathtt{b}_1\ x + \mathtt{c}_1. \tag{4.30}$$

This nonlinear model is described by three real numbers $\mathtt{a}_1, \mathtt{b}_1, \mathtt{c}_1$ and does not account for the tag's harvesting sensitivity, i.e. $\mathtt{P}_{\mathrm{sen}} = 0$.
- A nonlinear harvested power model, based on a normalized sigmoid and accounting for harvesting sensitivity $\mathtt{P}_{\mathrm{sen}}$ [39]:

$$\mathsf{p}_5(x) \triangleq \max\left\{ \frac{\mathtt{c}_2}{\exp(-\mathtt{a}_2 \mathtt{P}_{\mathrm{sen}} + \mathtt{b}_2)} \left(\frac{1 + \exp(-\mathtt{a}_2 \mathtt{P}_{\mathrm{sen}} + \mathtt{b}_2)}{1 + \exp(-\mathtt{a}_2 x + \mathtt{b}_2)} - 1 \right), 0 \right\}. \tag{4.31}$$

It is described by $\mathtt{P}_{\mathrm{sen}}$ and three real numbers $\mathtt{a}_2$, $\mathtt{b}_2$, $\mathtt{c}_2$.

- A nonlinear model that models harvesting efficiency function as a high-order polynomial in the dBm scale:

$$\eta(x) = w_0 + \sum_{i=1}^{W} w_i(10\log_{10}(x))^i, x \in \mathcal{P}, \tag{4.32}$$

  This function is described by the $W+1$ (real) coefficients of the polynomial, where W is the degree of the polynomial. Once $\eta(\cdot)$ is specified over $\mathcal{P}$, the harvested power function for that model, $\mathsf{p}_6(\cdot)$, can be obtained through (4.7).
- The last model stems directly from the measured data for a given rectenna (or other type of) RF harvesting circuit [16]. Specifically, given a set of $J+1$ data pairs of harvested power and corresponding input power, denoted as $\{q_j\}_{j=0}^{J}$ and $\{v_j\}_{j=0}^{J}$, respectively, slopes $l_j \triangleq \frac{v_j - v_{j-1}}{q_j - q_{j-1}}, j \in [J]$ are defined, where $[J] \triangleq \{1, 2, \ldots, J\}$. Points q_0 and q_J satisfy $q_0 = \mathsf{P}_{\text{sen}}$ and $q_J = \mathsf{P}_{\text{sat}}$. The final harvested power function interpolates with a piecewise linear function the $J+1$ points as follows:

$$\mathsf{p}_7(x) \triangleq \begin{cases} 0 & x \in [0, q_0], \\ l_j(x - q_{j-1}) + v_{j-1}, & x \in (q_{j-1}, q_j], \forall j \in [J], \\ v_J, & x \in [q_J, \infty). \end{cases} \tag{4.33}$$

  Note that function $\mathsf{p}_7(\cdot)$ is described by $2(J+1)$ real positive numbers. Given the fact that the points are available to the system designer from the harvesting circuit specification rather easily, determining function $\mathsf{p}_7(\cdot)$ is straightforward and requires no tuning.

The parameter fitting procedure for the nonlinear RF harvested power models $\mathsf{p}_3(\cdot)$, $\mathsf{p}_4(\cdot)$, and $\mathsf{p}_5(\cdot)$ was done through Matlab's fitting toolbox, while for the model $\mathsf{p}_6(\cdot)$, standard constrained convex optimization fitting methods are employed [16].

Figure 4.4 demonstrates the fitting accuracy of the above nonlinear models $\mathsf{p}_3(\cdot)$, $\mathsf{p}_4(\cdot)$, $\mathsf{p}_5(\cdot)$, and $\mathsf{p}_6(\cdot)$. The required parameters for each model are obtained through fitting from $J+1 = 118$ given data pairs (harvested power, corresponding input RF power) of the rectenna in [20]. It can be seen that $\mathsf{p}_6(\cdot)$ fits to the data, in contrast to the rest; in the following, $\mathsf{p}_6(\cdot) \equiv \mathsf{p}(\cdot)$ is considered.

The probability of successful SWIPT reception at the reader of an RF-powered tag is studied next. The path-loss model of Eq. (4.1) is considered with $\nu = 2.1$ and $\lambda = 0.3456$ (UHF carrier frequency). Tag antenna reflection coefficients Γ_0 and Γ_1 satisfying $|\Gamma_0 - \Gamma_1| = 1$, $\tau_{\text{d}} = 0.5$, $\chi = 0.5$, and $\rho_{\text{u}} = 0.01$ are utilized for RF harvesting and backscattering at the tag, while the BER threshold is set as $\beta = 10^{-5}$. The variance of noise at the reader was set to 10^{-11}.

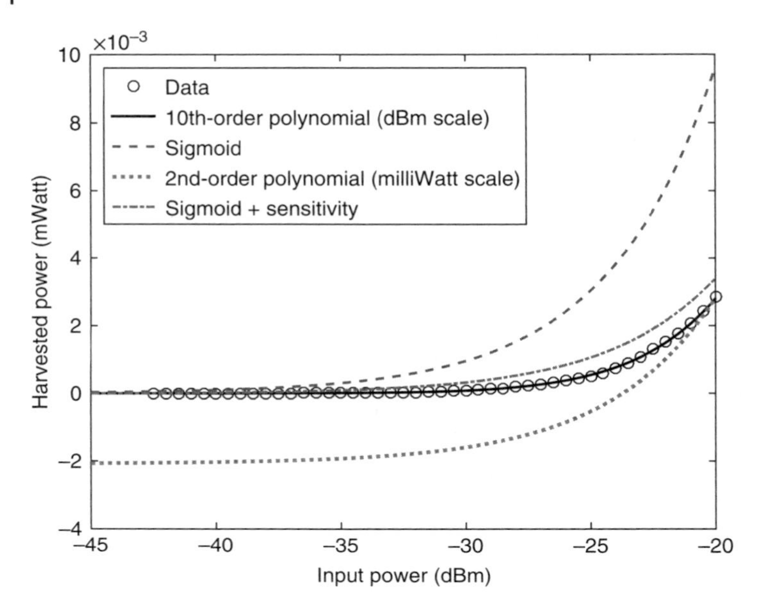

Figure 4.4 Harvested power (in mW) versus input power (in dBm) for all nonlinear prior art RF harvesting models. Input power ranges in [−45, −20] dBm.

The methodology of Section 4.5.2 to determine the event of successful SWIPT reception, through Eq. (4.26), requires functions $\mathsf{p}_j(\cdot), j = 1, 2, \ldots, 7$ to be strictly increasing at $\mathtt{P}_\mathrm{c}$ and continuous in a neighborhood of $\mathtt{P}_\mathrm{c}$ (in order to have a well-defined inverse around that point). For the studied rectenna model in [20], the above requirement is satisfied for $\mathsf{p}_l(\cdot), l = 1, 2, \ldots, 7$ and for the studied values of $\mathtt{P}_\mathrm{c}$.

Figure 4.5 (top) depicts the probability of successful SWIPT reception at the reader, as a function of the tag's power consumption, in a strong LOS scenario (Nakagami parameter $\mathtt{M} = 10$), $d = 5$ m, and $P_\mathrm{R} = 1.5$ W. Figure 4.5 (bottom) examines the same relationship in a non-line-of-sight (NLOS) scenario (Nakagami parameter $\mathtt{M} = 2$), $d = 8$ m, and $P_\mathrm{R} = 2.5$ W.

Both figures clearly show that the performance of the piecewise linear model $\mathsf{p}_7(\cdot)$ coincides with the exact (ground-truth), data-driven harvested power model. The performance of the $\mathsf{p}_1(\cdot)$ (L), as well as the $\mathsf{p}_2(\cdot)$ (CL) model both deviate from reality, even though the best values for the efficiency parameters were utilized in the above plots (i.e., the efficiency values that offered a performance as close as possible to the real, ground-truth model $\mathsf{p}(\cdot)$). Both nonlinear sigmoid models $\mathsf{p}_3(\cdot)$ and $\mathsf{p}_5(\cdot)$ tend to overestimate the event, while the one with incorporated harvesting sensitivity ($\mathsf{p}_5(\cdot)$) offers closer-to-reality results in the LOS scenario and deviates further in the NLOS scenario. Finally, the second-order polynomial $\mathsf{p}_4(\cdot)$ underestimates performance, with

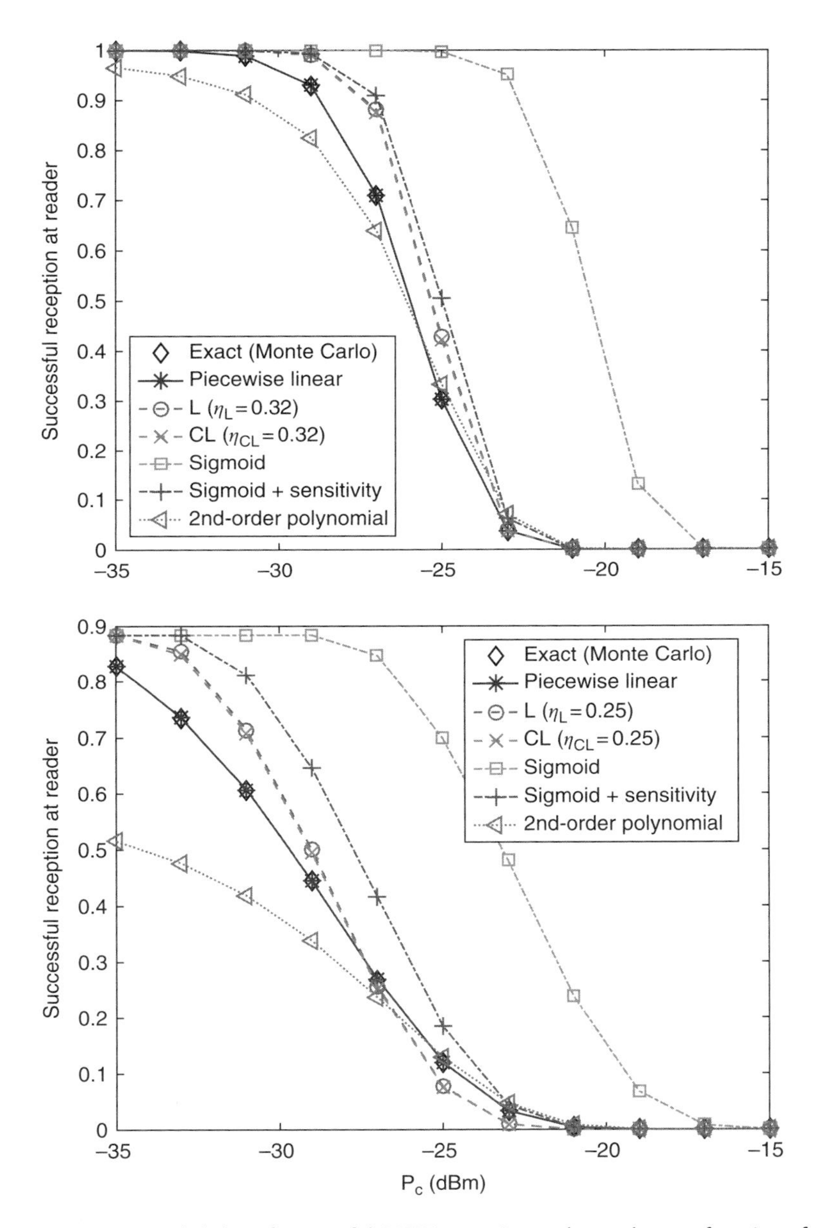

Figure 4.5 Probability of successful SWIPT reception at the reader, as a function of the tag power consumption. Top: strong LOS; bottom: NLOS scenario.

a performance gap that depends on the scenario (LOS vs NLOS) and the tag power consumption value.

In short, SWIPT research requires accurate harvesting modeling.

4.7 Conclusion

It is evident that SWIPT research is interdisciplinary and should be always carefully coupled with fundamentals in the electronics and microwave community; otherwise, relevant protocol and algorithmic design may be impractical. We touched upon the sensitivity and nonlinearity of the harvester. Other important building blocks, such as the boost converter/maximum power point tracking (MPPT) module's impact, important in any energy harvesting method, should be also examined.

Acknowledgements. This research is implemented through the Operational Program Human Resources Development, Education and Lifelong Learning and is co-financed by the European Union (European Social Fund) and Greek national funds.

Bibliography

1 H. Stockman (1948) Communication by means of reflected power. *Proceedings of the IRE*, 1196–1204.
2 G. Vannucci, A. Bletsas, and D. Leigh (2008) A software-defined radio system for backscatter sensor networks. *IEEE Trans. Wireless Commun.*, **7** (6): 2170–2179.
3 J. Kimionis, A. Bletsas, and J.N. Sahalos (2012) Design and implementation of RFID systems with software defined radio. In *Proceedings of the IEEE EuCAP*, Prague.
4 J. Kimionis, A. Bletsas, and J.N. Sahalos (2012) Bistatic backscatter radio for tag read-range extension. *In Proceedings of the IEEE RFID-TA*, Nice.
5 P.N. Alevizos, K. Tountas, and A. Bletsas (2018) Multistatic scatter radio sensor networks for extended coverage. *IEEE Trans. Wireless Commun.*, **17** (17): 4522–4535.
6 S.N. Daskalakis, S.D. Assimonis, E. Kampianakis, and A. Bletsas (2014) Soil moisture wireless sensing with analog scatter radio, low power, ultra-low cost and extended communication ranges. In *Proceedings of the IEEE Sensors Conference (Sensors)*, Valencia.
7 E. Kampianakis, J. Kimionis, K. Tountas, C. Konstantopoulos, E. Koutroulis, and A. Bletsas (2014) Wireless environmental sensor networking with analog scatter radio & timer principles. *IEEE Sensors J.*, **14** (10): 3365–3376.

8 S.N. Daskalakis, S.D. Assimonis, E. Kampianakis, and A. Bletsas (2016) Soil moisture scatter radio networking with low power. *IEEE Trans. Microwave Theory Techniques,* **64** (7): 2338–2346.

9 P.N. Alevizos and A. Bletsas (2015) Noncoherent composite hypothesis testing receivers for extended range bistatic scatter radio WSNs. In *Proceedings of the IEEE ICC,* London.

10 P.N. Alevizos, A. Bletsas, and G.N. Karystinos (2017) Noncoherent short packet detection and decoding for scatter radio sensor networking. *IEEE Trans. Commun.,* **65** (5): 1–13.

11 P.N. Alevizos (2017) Intelligent scatter radio, RF harvesting analysis, and resource allocation for ultra-low-power Internet-of-Things. Ph.D. dissertation, Technical University of Crete, Chania.

12 C.R. Valenta and G.D. Durgin (2014) Harvesting wireless power: Survey of energy-harvester conversion efficiency in far-field, wireless power transfer systems. *IEEE Microwave Mag.,* **15** (4): 108–120.

13 B. Clerckx and E. Bayguzina, “Waveform design for wireless power transfer,” *IEEE Trans. Signal Processing,* vol. 64, no. 23, pp. 6313–6328, Dec. 2016.

14 B. Clerckx (2018) Wireless information and power transfer: Nonlinearity, waveform design and rate–energy tradeoff. *IEEE Trans. Signal Process.,* **66** (4): 847–862.

15 G.D. Durgin (2016) RF thermoelectric generation for passive RFID. In *Proceedings of the IEEE RFID,* Orlando, FL, pp. 1–8.

16 P.N. Alevizos and A. Bletsas (2018) Sensitive and nonlinear far field RF energy harvesting in wireless communications. *IEEE Trans. Wireless Commun.,* **17** (6): 3670–3685.

17 A. Goldsmith (2005) *Wireless Communications.* Cambridge University Press, New York.

18 A. Pappoulis and S.U. Pillai (2002) *Probability, Random Variables and Stochastic Processes,* 4th edn. McGraw-Hill, New York.

19 Z. Popović, E.A. Falkenstein, D. Costinett, and R. Zane (2013) Low-power far-field wireless powering for wireless sensors. *Proceedings of the IEEE,* **101** (6): 1397–1409.

20 S.D. Assimonis, S.-N. Daskalakis, and A. Bletsas (2016) Sensitive and efficient RF harvesting supply for batteryless backscatter sensor networks. *IEEE Trans. Microwave Theory Techniques,* **64** (4): 1327–1338.

21 PowerCast Module, http://www.mouser.com/ds/2/329/P2110B-Datasheet-Rev-3-1091766.pdf.

22 A. Bletsas, A.G. Dimitriou, and J. Sahalos (2010) Improving backscatter radio tag efficiency. *IEEE Trans. Microwave Theory Techniques,* **58** (6): 1502–1509.

23 J. Kimionis, A. Bletsas, and J.N. Sahalos (2014) Increased range bistatic scatter radio. *IEEE Trans. Commun.,* **62** (3): 1091–1104.

24 EPC Global (2015) EPC Radio-Frequency Identity Protocols, Class-1 Generation-2 UHF RFID Protocol for Communications at 860 MHZ-960 MHZ, version 2.0.1.

25 P.N. Alevizos, Y. Fountzoulas, G.N. Karystinos, and A. Bletsas (2016) Log-linear-complexity GLRT-optimal noncoherent sequence detection for orthogonal and RFID-oriented modulations. *IEEE Trans. Commun.*, **64** (4): 1600–1612.

26 N. Fasarakis-Hilliard, P.N. Alevizos, and A. Bletsas (2015) Coherent detection and channel coding for bistatic scatter radio sensor networking. *IEEE Trans. Commun.*, **63**: 1798–1810.

27 N. Kargas, F. Mavromatis, and A. Bletsas (2015) Fully-coherent reader with commodity SDR for Gen2 FM0 and computational RFID. *IEEE Wireless Commun. Lett.*, **4** (6): 617–620.

28 M. Simon and D. Divsalar (2006) Some interesting observations for certain line codes with application to RFID. *IEEE Trans. Commun.*, **54** (4): 583–586.

29 A. Mastin and P. Jaillet (2013) Log-quadratic bounds for the Gaussian Q-function. *arXiv preprint arXiv:1304.2488*.

30 F.W.J. Olver, D.W. Lozier, R.F. Boisvert, and C.W. Clark (2010) *NIST Handbook of Mathematical Functions*. Cambridge University Press, New York.

31 I.S. Gradshteyn and I.M. Ryzhik (2007) *Table of Integrals, Series, and Products*, 7th edn. Elsevier/Academic Press, Amsterdam.

32 G.B. Folland (1999) *Real Analysis: Modern techniques and their applications*, 2nd edm. John Wiley & Sons, Inc., New York.

33 E. Boshkovska, D.W.K. Ng, N. Zlatanov, and R. Schober (2015) Practical non-linear energy harvesting model and resource allocation for SWIPT systems. *IEEE Commun. Lett.*, **19** (12): 2082–2085.

34 E. Boshkovska, N. Zlatanov, L. Dai, D.W.K. Ng, and R. Schober (2017) Secure SWIPT networks based on a non-linear energy harvesting model. In *Proceedings of the IEEE WCNC*, San Francisco, CA, pp. 1–6.

35 E. Boshkovska, X. Chen, L. Dai, D.W.K. Ng, and R. Schober (2017) Max-min fair beamforming for SWIPT systems with non-linear EH model. *2017 IEEE 86th Vehicular Technology Conference (VTC-Fall)*, Toronto, Ontario, pp. 1–6.

36 E. Boshkovska, D.W.K. Ng, L. Dai, and R. Schober (2018) Power-efficient and secure WPCNs with hardware impairments and non-linear EH circuit. *IEEE Trans. Commun.*, **66** (6): 2642–2657.

37 E. Boshkovska, D.W.K. Ng, N. Zlatanov, A. Koelpin, and R. Schober (2017) Robust resource allocation for MIMO wireless powered communication networks based on a non-linear EH model. *IEEE Trans. Commun.*, **65** (5): 1984–1999.

38 X. Xu, A. Özçelikkale, T. McKelvey, and M. Viberg (2017) Simultaneous information and power transfer under a non-linear RF energy harvesting model. In *Proceeedings of the IEEE ICC*, Paris.

39 S. Wang, M. Xia, K. Huang, and Y.C. Wu (2017) Wirelessly powered two-way communication with nonlinear energy harvesting model: Rate regions under fixed and mobile relay. *IEEE Trans. Wireless Commun.*, **16** (12): 8190–8204.

5

Multi-antenna Energy Beamforming for SWIPT

Jie Xu[1*] *and Rui Zhang*[2]

[1] *School of Information Engineering, Guangdong University of Technology, China*
[2] *Department of Electrical and Computer Engineering, National University of Singapore, Singapore*

5.1 Introduction

Energy-constrained wireless networks, such as sensor networks, are typically powered by batteries that have limited operation time. Although replacing or recharging the batteries can prolong the lifetime of networks to a certain extent, it usually incurs high costs and is inconvenient, hazardous (say, in toxic environments), or even impossible (e.g., for sensors embedded in building structures or inside human bodies). A more convenient, safer, as well as "greener" alternative is thus to harvest energy from the environment, which virtually provides perpetual energy supplies to wireless devices. In addition to other commonly used energy sources such as solar and wind, ambient radio-frequency (RF) signals can be a viable new source for energy scavenging. It is worth noting that RF-based energy harvesting is typically suitable for low-power applications (e.g., sensor networks), but also can be applied for scenarios with more substantial power consumptions if dedicated wireless power transmission is implemented.

On the other hand, since RF signals carrying energy can at the same time be used as a vehicle for transporting information, *simultaneous wireless information and power transfer* (SWIPT) becomes an interesting new area of research that has attracted increasing attention. In [1], Varshney first proposed a *capacity-energy* function to characterize the fundamental tradeoffs in simultaneous information and energy transfer. For the single-antenna or single-input single-output (SISO) additive white Gaussian noise (AWGN) channel with amplitude-constrained inputs, it was shown in [1] that there exist nontrivial tradeoffs in maximizing the transferred information rate

*Corresponding author: Jie Xu; jiexu@gdut.edu.cn

Wireless Information and Power Transfer: Theory and Practice, First Edition.
Edited by Derrick Wing Kwan Ng, Trung Q. Duong, Caijun Zhong, and Robert Schober.

versus (vs.) power by optimizing the input distribution. However, if the average transmit-power constraint is considered instead, the above two goals can be shown to be aligned for the SISO AWGN channel with Gaussian input signals, and thus there is no nontrivial tradeoff. In [2], Grover and Sahai extended [1] to frequency-selective single-antenna AWGN channels with the average power constraint, by showing that a nontrivial tradeoff exists in frequency-domain power allocation for maximal information vs. energy transfer.

As a matter of fact, *wireless power transfer* (WPT), which generally refers to the transmissions of electrical energy from a power source to one or more electrical loads without any interconnecting wires, has been investigated and implemented with a long history. Generally speaking, WPT is carried out using either the near-field electromagnetic (EM) induction (e.g., inductive coupling, capacitive coupling) for short-distance (say, less than a meter) applications such as passive radio-frequency identification (RFID) [3], or the far-field EM radiation in the form of microwaves or lasers for long-range (say, a few kilometers) applications such as the transmissions of energy from orbiting solar power satellites to nearby spacecraft [4]. However, prior research on EM radiation based WPT, in particular over the RF band, has been pursued independently from that on wireless information transfer (WIT) or radio communication. This is not surprising since these two lines of work in general have very different research goals: WIT is used to maximize the *information transmission capacity* of wireless channels subject to channel impairments such as the fading and receiver noise, while WPT is used to maximize the *energy transmission efficiency* (defined as the ratio of the energy harvested and stored at the receiver to that consumed by the transmitter) over a wireless medium. Nevertheless, it is worth noting that the design objectives for WPT and WIT systems can be aligned, since given a transmitter energy budget maximizing the signal power received (for WPT) is also beneficial in maximizing the channel capacity (for WIT) against the receiver noise.

Hence, a unified study on WIT and WPT has been pursued for emerging wireless applications with such a dual usage. An example of such wireless dual networks is envisaged in Figure 5.1, where a fixed access point (AP) coordinates the two-way communications to/from a set of distributed user terminals (UTs). However, unlike the conventional wireless network in which both the AP and UTs draw energy from constant power supplies (e.g., by connecting to the grid or a battery), in our model only the AP is assumed to have a constant power source, while all UTs need to replenish energy from the received signals sent by the AP via the far-field RF-based WPT. Consequently, the AP needs to coordinate the wireless information and energy transfer to UTs in the downlink, in addition to the information transfer from UTs in the uplink. How to characterize the fundamental information–energy transmission tradeoff in such dual networks is an important and fundamental problem.

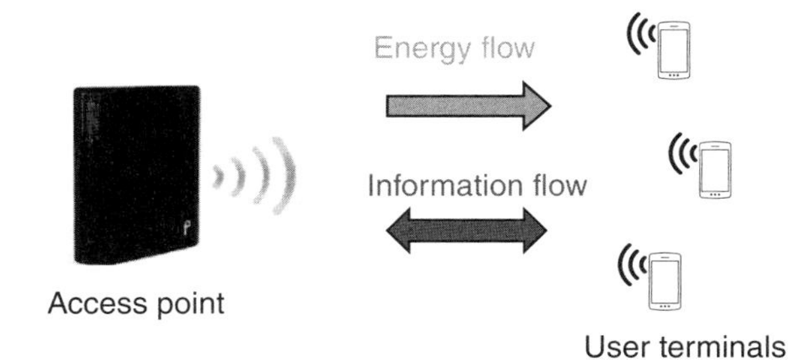

Figure 5.1 A wireless network with dual information and energy transfer.

In this chapter, we focus our study on the downlink case with simultaneous WIT and WPT (a.k.a. SWIPT) from the AP to UTs. In the generic system model depicted in Figure 5.1, each UT can in general harvest energy and decode information at the same time. However, from an implementation viewpoint, one particular design whereby each UT operates as either an information receiver or an energy receiver at any given time may be desirable, which is referred to as *time switching*. This scheme is practically appealing since state-of-the-art wireless information and energy receivers are typically designed to operate separately with very different power sensitivities (e.g., −50 dBm for information receivers vs. −10 dBm for energy receivers). As a result, if time switching is employed at each UT jointly with the near-far based transmission scheduling at the AP, i.e. UTs that are close to the AP and thus receive high power from the AP are scheduled for WET, whereas those that are more distant from the AP and thus receive lower power are scheduled for WIT, then SWIPT systems can be efficiently implemented with existing information and energy receivers and an additional time-switching device at each receiver.

In particular, in this chapter we consider simplified scenarios with only one or two active UTs in the network. For the case of two UTs, we assume time switching, i.e. the two UTs take turns to receive energy or (independent) information from the AP over different time blocks. As a result, when one UT receives information from the AP, the other UT can opportunistically harvest energy from the same signal broadcast by the AP, and vice versa. Hence, at each block, one UT operates as an information decoding (ID) receiver, and the other UT as an energy harvesting (EH) receiver. Without loss of generality, we focus on a multi-antenna or multiple-input multiple-output (MIMO) system as shown in Figure 5.2, in which the AP transmitter is equipped with multiple antennas, and each UT receiver is equipped with one or more antennas, for enabling both the high-performance wireless energy and information transmissions (as it is well known that for WIT only, MIMO systems can achieve substantial array/capacity gains over SISO systems by spatial beamforming/multiplexing [5]). Under this setup, we design the optimal transmission strategy to achieve different tradeoffs between maximal information rate vs. energy transfer, which are characterized by the boundary of a so-called *rate–energy* (R-E) region. We derive a semi-closed-form expression for the optimal transmit covariance matrix (for the joint precoding and power allocation) to achieve different rate–energy pairs on the Pareto boundary of the R-E region. After that, we further discuss some extensions of the SWIPT design to other system setups under various practical considerations.

Notation: For a square matrix $\boldsymbol{S}$, $\mathtt{tr}\,(\boldsymbol{S})$, $|\boldsymbol{S}|$, $\boldsymbol{S}^{-1}$, and $\boldsymbol{S}^{\frac{1}{2}}$ denote its trace, determinant, inverse, and square-root, respectively, while $\boldsymbol{S} \succeq \boldsymbol{0}$ and $\boldsymbol{S} \succ \boldsymbol{0}$ mean that $\boldsymbol{S}$ is positive semi-definite and positive definite, respectively. For an arbitrary-size matrix $\boldsymbol{M}$, $\boldsymbol{M}^H$, and $\boldsymbol{M}^T$ denote the conjugate transpose and transpose of $\boldsymbol{M}$, respectively. $\mathtt{diag}(x_1, \dots, x_M)$ denotes an $M \times M$ diagonal matrix with $x_1, \dots, x_M$ being the diagonal elements. $\boldsymbol{I}$ and $\boldsymbol{0}$ denote an identity matrix and an all-zero vector/matrix, respectively, with appropriate dimensions. $\mathbb{E}[\cdot]$ denotes the statistical expectation. The distribution of a circularly symmetric complex Gaussian (CSCG) random vector with mean $\boldsymbol{x}$ and covariance matrix Σ is denoted by $\mathcal{CN}(\boldsymbol{x}, \Sigma)$, and $\sim$ stands for "distributed as". $\mathbb{C}^{x \times y}$ denotes the space of $x \times y$ matrices with complex entries. $\| \boldsymbol{z} \|$ is the Euclidean norm of a complex vector $\boldsymbol{z}$, and $|z|$ is the absolute value of a complex scalar z. $\max(x, y)$ and $\min(x, y)$ denote the maximum and minimum between two real numbers, x and y, respectively, and $(x)^+ = \max(x, 0)$. All the $\log(\cdot)$ functions have base-2 by default.

5.2 System Model

As shown in Figure 5.2, we consider a wireless broadcast system consisting of one transmitter, one EH receiver, and one ID receiver. It is assumed that the transmitter is equipped with $M \geq 1$ transmitting antennas, and the EH receiver and the ID receiver are equipped with $N_{\rm EH} \geq 1$ and $N_{\rm ID} \geq 1$ receiving antennas, respectively. In addition, it is assumed that the transmitter and both receivers

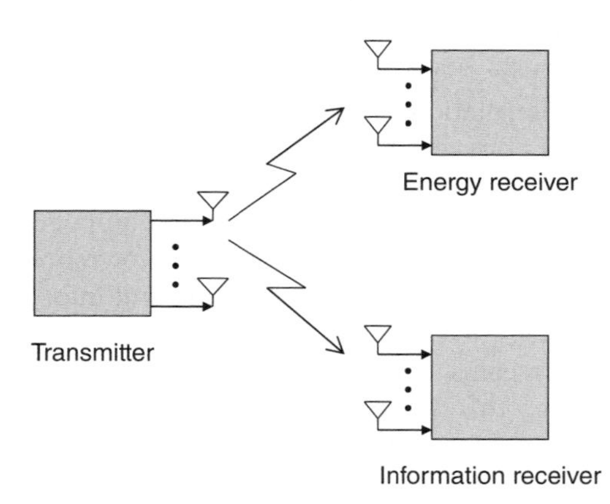

Figure 5.2 A MIMO broadcast system for simultaneous wireless information and power transfer.

operate over the same frequency band. Assuming a narrow-band transmission over quasi-static fading channels, the baseband equivalent channels from the transmitter to the EH receiver and ID receiver can be modeled by matrices $\boldsymbol{G} \in \mathbb{C}^{N_{\text{EH}} \times M}$ and $\boldsymbol{H} \in \mathbb{C}^{N_{\text{ID}} \times M}$, respectively. It is assumed that at each fading state, $\boldsymbol{G}$ and $\boldsymbol{H}$ are both known at the transmitter, and separately known at the corresponding receiver.

It is worth noting that the EH receiver does not need to convert the received signal from the RF band to the baseband in order to harvest the carried energy. Nevertheless, thanks to the law of energy conservation, it can be assumed that the total harvested RF-band power (energy normalized by the baseband symbol period), denoted by Q, from all receiving antennas at the EH receiver is proportional to that of the received baseband signal, i.e.

$$Q = \zeta \mathbb{E}[\|\boldsymbol{G}\boldsymbol{x}(n)\|^2], \tag{5.1}$$

where ζ is a constant that accounts for the loss in the energy transducer for converting the harvested energy to electrical energy to be stored; for the convenience of analysis, it is assumed that $\zeta = 1$ in this chapter unless stated otherwise. We use $\boldsymbol{x}(n) \in \mathbb{C}^{M \times 1}$ to denote the baseband signal broadcast by the transmitter at the nth symbol interval, which is assumed to be random over n, without loss of generality. The expectation in (5.1) is thus used to compute the average power harvested by the EH receiver at each fading state. Note that for simplicity we assumed in (5.1) that the harvested energy due to the background noise at the EH receiver is negligible and thus can be ignored.

On the other hand, the baseband transmission from the transmitter to the ID receiver can be modeled by

$$\boldsymbol{y}(n) = \boldsymbol{H}\boldsymbol{x}(n) + \boldsymbol{z}(n), \tag{5.2}$$

where $\boldsymbol{y}(n) \in \mathbb{C}^{N_{\text{ID}} \times 1}$ denotes the received signal at the nth symbol interval, and $\boldsymbol{z}(n) \in \mathbb{C}^{N_{\text{ID}} \times 1}$ denotes the receiver noise vector. It is assumed that $\boldsymbol{z}(n)$ values are independent over n and $\boldsymbol{z}(n) \sim \mathcal{CN}(\boldsymbol{0}, \boldsymbol{I})$. Under the assumption that $\boldsymbol{x}(n)$ is random over n, we use $\boldsymbol{S} = \mathbb{E}[\boldsymbol{x}(n)\boldsymbol{x}^H(n)]$ to denote the covariance matrix of $\boldsymbol{x}(n)$. In addition, we assume that there is an average power constraint at the transmitter across all transmitting antennas denoted by $\mathbb{E}[\|\boldsymbol{x}(n)\|^2] = \texttt{tr}(\boldsymbol{S}) \leq P$. In the following, we examine the optimal transmit covariance $\boldsymbol{S}$ to maximize the transported energy efficiency and information rate to the EH and ID receivers, respectively.

Before investigating the rate–energy tradeoff in the case where both the EH and ID receivers are present in Section 5.3, we first consider two special cases when only the EH receiver or ID receiver is present.

Consider first the MIMO link from the transmitter to the EH receiver when the ID receiver is not present. In this case, the design objective for $\boldsymbol{S}$ is to maximize the power Q received at the EH receiver. Since from (5.1) it follows that $Q = \mathtt{tr}(\boldsymbol{G}\boldsymbol{S}\boldsymbol{G}^H)$ with $\zeta = 1$, the aforementioned design problem can be formulated as

$$\begin{aligned}\text{(P1)}\quad & \text{maximize}_{S} \quad Q := \mathtt{tr}(\boldsymbol{G}\boldsymbol{S}\boldsymbol{G}^H) \\ & \mathtt{subject\ to} \quad \mathtt{tr}(\boldsymbol{S}) \leq P, \boldsymbol{S} \succeq \boldsymbol{0}.\end{aligned}$$

Let $T_1 = \min(M, N_{\mathrm{EH}})$ and the (reduced) singular value decomposition (SVD) of $\boldsymbol{G}$ be denoted by $\boldsymbol{G} = \boldsymbol{U}_G \boldsymbol{\Gamma}_G^{1/2} \boldsymbol{V}_G^H$, where $\boldsymbol{U}_G \in \mathbb{C}^{N_{\mathrm{EH}} \times T_1}$ and $\boldsymbol{V}_G \in \mathbb{C}^{M \times T_1}$, each of which consists of orthogonal columns with unit norm, and $\boldsymbol{\Gamma}_G = \mathtt{diag}(g_1, \dots, g_{T_1})$ with $g_1 \geq g_2 \geq \dots \geq g_{T_1} \geq 0$. Furthermore, let $\boldsymbol{v}_1$ denote the first column of $\boldsymbol{V}_G$. Then, we have the following proposition. Note that all the proofs of propositions, theorems, and corollaries in this chapter are omitted for brevity, and can be found in [11].

Proposition 5.1 The optimal solution to (P1) is $\boldsymbol{S}_{\mathrm{EH}} = P\boldsymbol{v}_1\boldsymbol{v}_1^H$.

Given $\boldsymbol{S} = \boldsymbol{S}_{\mathrm{EH}}$, it follows that the maximum harvested power at the EH receiver is given by $Q_{\max} = g_1 P$. It is worth noting that since $\boldsymbol{S}_{\mathrm{EH}}$ is a rank-one matrix, the maximum harvested power is achieved by *beamforming* at the transmitter, which aligns with the strongest eigenmode of the matrix $\boldsymbol{G}^H\boldsymbol{G}$, i.e. the transmitted signal can be written as $\boldsymbol{x}(n) = \sqrt{P}\boldsymbol{v}_1 s(n)$, where $s(n)$ is an arbitrary random signal over n with zero mean and unit variance, and $\boldsymbol{v}_1$ is the transmit beamforming vector. For convenience, we name the above transmit beamforming scheme to maximize the efficiency of WPT as "energy beamforming".

Next, consider the MIMO link from the transmitter to the ID receiver without the presence of any EH receiver. Assuming the optimal Gaussian codebook at the transmitter, i.e. $\boldsymbol{x}(n) \sim \mathcal{CN}(\boldsymbol{0}, \boldsymbol{S})$, the transmit covariance $\boldsymbol{S}$ to maximize the transmission rate over this MIMO channel can be obtained by solving the following problem [6]:

$$\begin{aligned}\text{(P2)}\quad & \text{maximize}_{S} \quad R := \log|\boldsymbol{I} + \boldsymbol{H}\boldsymbol{S}\boldsymbol{H}^H| \\ & \mathtt{subject\ to} \quad \mathtt{tr}(\boldsymbol{S}) \leq P, \boldsymbol{S} \succeq \boldsymbol{0}.\end{aligned}$$

The optimal solution to the above problem is known to have the following form [6]: $\boldsymbol{S}_{\mathrm{ID}} = \boldsymbol{V}_H \boldsymbol{\Lambda} \boldsymbol{V}_H^H$, where $\boldsymbol{V}_H \in \mathbb{C}^{M \times T_2}$ is obtained from the (reduced) SVD of $\boldsymbol{H}$ expressed by $\boldsymbol{H} = \boldsymbol{U}_H \boldsymbol{\Gamma}_H^{1/2} \boldsymbol{V}_H^H$, with $T_2 = \min(M, N_{\mathrm{ID}})$, $\boldsymbol{U}_H \in \mathbb{C}^{N_{\mathrm{ID}} \times T_2}$, $\boldsymbol{\Gamma}_H = \mathtt{diag}(h_1, \dots, h_{T_2})$, $h_1 \geq h_2 \geq \dots \geq h_{T_2} \geq 0$, and $\boldsymbol{\Lambda} = \mathtt{diag}(p_1, \dots, p_{T_2})$ with the diagonal elements obtained from the standard "water-filling (WF)" power allocation solution [6]:

$$p_i = \left(\nu - \frac{1}{h_i}\right)^+, \qquad i = 1, \dots, T_2, \tag{5.3}$$

with ν being the so-called (constant) water-level that makes $\sum_{i=1}^{T_2} p_i = P$. The corresponding maximum transmission rate is then given by $R_{\max} = \sum_{i=1}^{T_2} \log(1 + h_i p_i)$. The maximum rate is achieved in general by *spatial multiplexing* [5] over up to T_2 spatially decoupled AWGN channels, together with the Gaussian codebook, i.e. the transmitted signal can be expressed as $\boldsymbol{x}(n) = \boldsymbol{V}_H \boldsymbol{\Lambda}^{1/2} \boldsymbol{s}(n)$, where $\boldsymbol{s}(n)$ is a CSCG random vector with $\boldsymbol{s}(n) \sim \mathcal{CN}(\boldsymbol{0}, \boldsymbol{I})$, and $\boldsymbol{V}_H$ and $\boldsymbol{\Lambda}^{1/2}$ denote the precoding matrix and the (diagonal) power allocation matrix, respectively.

Remark 5.1 It is worth noting that in problem (P1) it is assumed that the transmitter sends to the EH receiver continuously. Now suppose that the transmitter only transmits a fraction of the total time denoted by α with $0 < \alpha \leq 1$. Furthermore, assume that the transmit power level can be adjusted flexibly provided that the consumed average power is bounded by P, i.e. $\alpha \cdot \mathtt{tr}(\boldsymbol{S}) + (1 - \alpha) \cdot 0 \leq P$ or $\mathtt{tr}(\boldsymbol{S}) \leq P/\alpha$. In this case, it can be easily shown that the transmit covariance $\boldsymbol{S} = (P/\alpha)\boldsymbol{v}_1\boldsymbol{v}_1^H$ also achieves the maximum harvested power $Q_{\max} = g_1 P$ for any $0 < \alpha \leq 1$, which suggests that the maximum power delivered is independent of transmission time. However, unlike the case of maximum power transfer, the maximum information rate reliably transmitted to the ID receiver requires that the transmitter send signals continuously, i.e. $\alpha = 1$, as assumed in problem (P2). This can be easily verified by observing that for any $0 < \alpha \leq 1$ and $\boldsymbol{S} \succeq \boldsymbol{0}$, $\alpha \log |\boldsymbol{I} + \boldsymbol{H}(\boldsymbol{S}/\alpha)\boldsymbol{H}^H| \leq \log |\boldsymbol{I} + \boldsymbol{HSH}^H|$ where the equality holds only when $\alpha = 1$, since R is a nonlinear concave function of $\boldsymbol{S}$. Thus, to maximize both power and rate transfer at the same time, the transmitter should broadcast to the EH and ID receivers all the time. Furthermore, note that the assumed Gaussian distribution for transmitted signals is necessary for achieving the maximum rate transfer, but not necessary for the maximum power transfer. In fact, for any arbitrary complex number c that satisfies $|c| = 1$, even a deterministic transmitted signal $\boldsymbol{x}(n) = \sqrt{P}\boldsymbol{v}_1 c, \forall n$, achieves the maximum transferred power $Q_{\max}$ in problem (P1). However, to maximize simultaneous power and information transfer with the same transmitted signal, the Gaussian input distribution is sufficient as well as necessary.

5.3 Rate–Energy Region Characterization

5.3.1 Problem Formulation

In this section, we consider the case where both the EH and ID receivers are present. From the above results in Section 5.2, it is seen that the optimal transmission strategies for maximal power transfer and information transfer are in general different, and are energy beamforming and information spatial multiplexing, respectively. It thus motivates our investigation of the following

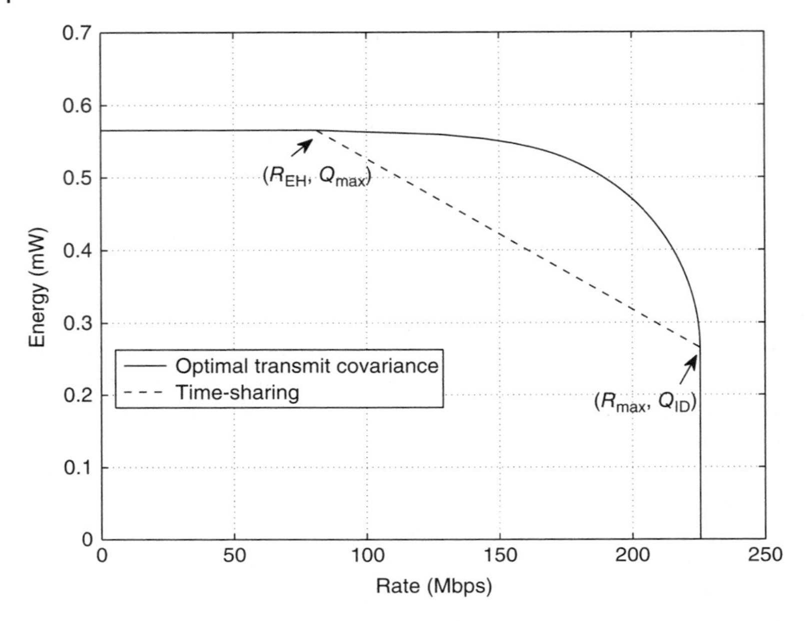

Figure 5.3 Rate–energy tradeoff for a MIMO broadcast system with separated EH and ID receivers, and $M = N_{EH} = N_{ID} = 4$.

question: What is the optimal broadcasting strategy for simultaneous wireless power and information transfer (SWIPT)? To answer this question, we propose to use the R-E region (defined below) to characterize all the achievable rate (in bits/s/Hz or bps/Hz for information transfer) and energy (in J/s or W for power transfer) pairs under a given transmit power constraint. Without loss of generality, assuming that the transmitter sends Gaussian signals continuously (cf. Remark 5.1), the R-E region is defined as

$$\begin{aligned} C_{\text{R-E}}(P) \triangleq \{(R, Q) \, : \, R \le \log |\boldsymbol{I} + \boldsymbol{H}\boldsymbol{S}\boldsymbol{H}^H|, \\ Q \le \texttt{tr}(\boldsymbol{G}\boldsymbol{S}\boldsymbol{G}^H), \texttt{tr}(\boldsymbol{S}) \le P, \boldsymbol{S} \succeq \boldsymbol{0}\}. \end{aligned} \tag{5.4}$$

In Figure 5.3, an example of the above defined R-E region (see Section 5.3.2 for the algorithm to compute the boundary of this region) is shown for a practical MIMO broadcast system with separated EH and ID receivers (i.e., $\boldsymbol{G} \neq \boldsymbol{H}$). It is assumed that $M = N_{\text{EH}} = N_{\text{ID}} = 4$. The transmitter power is assumed to be $P =$ 1 W or 30 dBm. The distances from the transmitter to the EH and ID receivers are assumed to be 1 m and 10 m, respectively; thus, we can exploit the near-far based energy and information transmission scheduling, which may correspond to, for example, a dedicated energy transfer system (to "near" users) with opportunistic information transmission (to "far" users), or vice versa. Assuming a carrier frequency of $f_c = 900$ MHz and the power path-loss exponent to be 4, the distance-dependent signal attenuation from the AP to EH/ID receiver can be estimated as 40 dB and 80 dB, respectively. Accordingly, the average signal power at the EH/ID receiver is thus 30 dBm−40 dB= −10 dBm and 30 dBm−80 dB= −50 dBm, respectively. It is further assumed that in addition to signal path loss, Rayleigh fading is present,[1] as such each element of channel matrices $\boldsymbol{G}$ and $\boldsymbol{H}$ is independently drawn from the CSCG distribution with zero mean and variance −10 dBm (for the EH receiver) and −50 dBm (for the ID receiver), respectively (to be consistent with the signal path loss previously assumed). Furthermore, the bandwidth of the transmitted signal is assumed to be 10 MHz, while the receiver noise is assumed to be white Gaussian with power spectral density −140 dBm/Hz (which is dominated by the receiver processing noise rather than the background thermal noise) or average power −70 dBm over the bandwidth of 10 MHz. As a result, considering all of transmit power, signal attenuation, fading and receiver noise, the per-antenna average SNR at the ID receiver is equal to $30 - 80 - (-70) = 20$ dB, which corresponds to $P = 100$ in the equivalent signal model for the ID receiver given in (5.2) with unit-norm noise. In addition, we assume that for the EH receiver, the energy conversion efficiency is ζ=50%. Considering this together with transmit power and signal attenuation, the average per-antenna signal power at the EH receiver is thus 0.5×(30 dBm−40 dB) = 50 μW.

From Figure 5.3, it is observed that with energy beamforming, the maximum harvested energy rate for the EH receiver is around $Q_{\max} = 0.57$ mW, while with spatial multiplexing, the maximum information rate for the ID receiver is around $R_{\max} = 225$ Mbps. It is easy to identify two boundary points of this R-E region denoted by $(R_{\rm EH}, Q_{\max})$ and $(R_{\max}, Q_{\rm ID})$. For the former boundary point, the transmit covariance is $\boldsymbol{S}_{\rm EH}$, which corresponds to transmit energy beamforming and achieves the maximum transferred power $Q_{\max}$ to the EH receiver, while the resultant information rate for the ID receiver is given by $R_{\rm EH} = \log(1 + \| \boldsymbol{H}\boldsymbol{v}_1 \|^2 P)$. On the other hand, for the latter boundary point, the transmit covariance is $\boldsymbol{S}_{\rm ID}$, which corresponds to transmit spatial multiplexing and achieves the maximum information rate transferred to the ID receiver $R_{\max}$, while the resultant power transferred to the EH receiver is given by $Q_{\rm ID} = \texttt{tr}(\boldsymbol{G}\boldsymbol{S}_{\rm ID}\boldsymbol{G}^H)$.

Since the optimal tradeoff between the maximum energy and information transfer rates is characterized by the boundary of the R-E region, it is important to characterize all the boundary rate–power pairs of $C_{\text{R-E}}(P)$ for any $P > 0$. From Figure 5.3, it is easy to observe that if $R \leq R_{\rm EH}$, the maximum harvested power $Q_{\max}$ is achievable with the same transmit covariance that achieves the rate–power pair $(R_{\rm EH}, Q_{\max})$; similarly, the maximum information rate $R_{\max}$ is

1 Rayleigh fading is considered here for illustration. As the distance from the AP to the EH receiver is short, in practice there may exist a strong line-of-sight (LOS) component between them, which is beneficial for further increasing the harvested energy at the EH receiver.

achievable provided that $Q \leq Q_{\text{ID}}$. Thus, the remaining boundary of $C_{\text{R-E}}(P)$ yet to be characterized is over the intervals $R_{\text{EH}} < R < R_{\max}$ and $Q_{\text{ID}} < Q < Q_{\max}$. We thus consider the following optimization problem:

$$\begin{aligned}
\text{(P3)} \quad \text{maximize}_{S} \quad & \log|\boldsymbol{I} + \boldsymbol{HSH}^H| \\
\texttt{subject to} \quad & \texttt{tr}(\boldsymbol{GSG}^H) \geq \overline{Q}, \texttt{tr}(\boldsymbol{S}) \leq P, \boldsymbol{S} \succeq \boldsymbol{0}.
\end{aligned}$$

Note that if $\overline{Q}$ takes values from $Q_{\text{ID}} < \overline{Q} < Q_{\max}$, the corresponding optimal rate solutions of the above problems are the boundary rate points of the R-E region over $R_{\text{EH}} < R < R_{\max}$. Notice that the transmit covariance solutions to the above problems in general yield larger rate–power pairs than those by simply "time-sharing" the optimal transmit covariance matrices $\boldsymbol{S}_{\text{EH}}$ and $\boldsymbol{S}_{\text{ID}}$ for EH and ID receivers separately (see the dashed line in Figure 5.3)[2].

Problem (P3) is a convex optimization problem, since its objective function is concave over $\boldsymbol{S}$ and its constraints specify a convex set of $\boldsymbol{S}$. Note that (P3) resembles a similar problem formulated in [7, 8] (see also [9] and references therein) under the cognitive radio (CR) setup, where the rate of a secondary MIMO link is maximized subject to a set of so-called *interference power constraints* to protect the co-channel primary receivers. However, there is a key difference between (P3) and the problem in [9]: the harvested power constraint in (P3) has the reversed inequality of that of the interference power constraint in [8], since in our case it is desirable for the EH receiver to harvest more power from the transmitter, as opposed to that in [8] where the interference power at the primary receiver should be minimized. As such, it is not immediately clear whether the solution in [8] can be directly applied for solving (P3) with the reversed power inequality.

5.3.2 Optimal Solution

In this subsection, we first solve problem (P3) with arbitrary $\boldsymbol{G}$ and $\boldsymbol{H}$ and derive a semi-closed-form expression for the optimal transmit covariance. Then we examine the optimal solution for the special case of MISO channels from the transmitter to ID and/or EH receivers.

Since problem (P3) is convex and satisfies the Slater's condition [10], it has a zero duality gap and thus can be solved using the Lagrange duality method.[3] Thus, we introduce two non-negative dual variables, λ and μ, associated with

2 By time-sharing, we mean that the AP transmits simultaneously to both EH and ID receivers with the energy-maximizing transmit covariance $\boldsymbol{S}_{\text{EH}}$ (i.e., energy beamforming) for β portion of each block time, and the information-rate-maximizing transmit covariance $\boldsymbol{S}_{\text{ID}}$ (i.e., spatial multiplexing) for the remaining $1 - \beta$ portion of each block time, with $0 \leq \beta \leq 1$.

3 It is worth noting that problem (P3) is convex and thus can be solved efficiently by the interior point method [11]. In this chapter, we apply the Lagrange duality method for this problem mainly to reveal the optimal precoder structure.

the harvested power constraint and transmit power constraint in (P3), respectively. The optimal solution to problem (P3) is then given by the following theorem in terms of λ^* and μ^*, which are the optimal dual solutions of problem (P3). Note that for problem (P3), given any pair of $\overline{Q}$ ($Q_{\text{ID}} < \overline{Q} < Q_{\max}$) and $P > 0$, there exists one unique pair of $\lambda^* > 0$ and $\mu^* > 0$.

Theorem 5.1 The optimal solution to problem (P3) has the following form:

$$S^* = A^{-1/2}\tilde{V}\tilde{\Lambda}\tilde{V}^H A^{-1/2}, \tag{5.5}$$

where $A = \mu^* I - \lambda^* G^H G$, $\tilde{V} \in \mathbb{C}^{M\times T_2}$ is obtained from the (reduced) SVD of the matrix $HA^{-1/2}$ given by $HA^{-1/2} = \tilde{U}\tilde{\Gamma}^{1/2}\tilde{V}^H$, with $\tilde{\Gamma} = \texttt{diag}(\tilde{h}_1, \dots, \tilde{h}_{T_2})$, $\tilde{h}_1 \geq \tilde{h}_2 \geq \dots \geq \tilde{h}_{T_2} \geq 0$, and $\tilde{\Lambda} = \texttt{diag}(\tilde{p}_1, \dots, \tilde{p}_{T_2})$, with $\tilde{p}_i = (1 - 1/\tilde{h}_i)^+, i = 1, \dots, T_2$.

Note that this theorem requires that $A = \mu^* I - \lambda^* G^H G \succ \mathbf{0}$, implying that $\mu^* > \lambda^* g_1$ (recall that g_1 is the largest eigenvalue of matrix $G^H G$), which is not present for a similar result in [9] under the CR setup with the reversed interference power constraint. Please refer to [11] for one algorithm that can be used to solve (P3). From Theorem 5.1, the maximum transmission rate for problem (P3) can be shown to be $R^* = \log|I + HS^*H^H| = \sum_{i=1}^{T_2} \log(1 + \tilde{h}_i\tilde{p}_i)$.

Next, we examine the optimal solution to problem (P3) for the special case where the ID receiver has one single antenna, i.e. $N_{\text{ID}} = 1$, and thus the MIMO channel H reduces to a row vector h^H with $h \in \mathbb{C}^{M\times 1}$. Suppose that the EH receiver is still equipped with $N_{\text{EH}} \geq 1$ antennas, and thus the MIMO channel G remains unchanged. From Theorem 5.1, we obtain the following corollary.

Corollary 5.1 In the case of MISO channel from the transmitter to ID receiver, i.e. $H \equiv h^H$, the optimal solution to problem (P3) reduces to the following form:

$$S^* = A^{-1}h\left(\frac{1}{\| A^{-1/2}h\|^2} - \frac{1}{\| A^{-1/2}h\|^4}\right)^+ h^H A^{-1} \tag{5.6}$$

where $A = \mu^* I - \lambda^* G^H G$, with λ^* and μ^* denoting the optimal dual solutions of problem (P3). Correspondingly, the optimal value of (P3) is $R^* = (2\log(\| A^{-1/2}h \|))^+$.

From (5.6), it is observed that the optimal transmit covariance is a *rank-one* matrix, from which it follows that *beamforming* is the optimal transmission strategy in this case, where the transmit beamforming vector should be aligned with the vector $A^{-1}h$. Moreover, consider the case where both channels from the transmitter to ID/EH receivers are MISO, i.e. $H \equiv h^H$, and $G \equiv g^H$ with $g \in \mathbb{C}^{M\times 1}$. From Corollary 5.1, it follows immediately that the optimal covariance solution to problem (P3) is still beamforming. In the following theorem, we show a closed-form solution of the optimal beamforming vector at the transmitter for this special case, which differs from the semi-closed-form solution (5.6) that was expressed in terms of dual variables.

Theorem 5.2 In the case of MISO channels from transmitter to both ID and EH receivers, i.e. $\boldsymbol{H} \equiv \boldsymbol{h}^H$ and $\boldsymbol{G} \equiv \boldsymbol{g}^H$, the optimal solution to problem (P3) can be expressed as $\boldsymbol{S}^* = P\boldsymbol{v}\boldsymbol{v}^H$, where the beamforming vector $\boldsymbol{v}$ has a unit-norm and is given by

$$\boldsymbol{v} = \begin{cases} \hat{\boldsymbol{h}} & 0 \le \overline{Q} \le |\boldsymbol{g}^H \hat{\boldsymbol{h}}|^2 P \\ \sqrt{\frac{\overline{Q}}{P\|\boldsymbol{g}\|^2}} e^{j\angle\alpha_{gh}} \hat{\boldsymbol{g}} & \\ +\sqrt{1 - \frac{\overline{Q}}{P\|\boldsymbol{g}\|^2}} \hat{\boldsymbol{h}}_{g^\perp} & |\boldsymbol{g}^H \hat{\boldsymbol{h}}|^2 P < \overline{Q} \le P \parallel \boldsymbol{g} \|^2 \end{cases} \tag{5.7}$$

where $\hat{\boldsymbol{h}} = \boldsymbol{h}/\parallel \boldsymbol{h} \parallel$, $\hat{\boldsymbol{g}} = \boldsymbol{g}/\parallel \boldsymbol{g} \parallel$, $\hat{\boldsymbol{h}}_{g^\perp} = \boldsymbol{h}_{g^\perp}/\parallel \boldsymbol{h}_{g^\perp} \parallel$ with $\boldsymbol{h}_{g^\perp} = \boldsymbol{h} - (\hat{\boldsymbol{g}}^H \boldsymbol{h})\hat{\boldsymbol{g}}$, and $\alpha_{gh} = \hat{\boldsymbol{g}}^H \boldsymbol{h}$ with $\angle\alpha_{gh} \in [0, 2\pi)$ denoting the phase of complex number α_{gh}. Correspondingly, the optimal value of (P3) is

$$R^* = \begin{cases} \log(1+\parallel \boldsymbol{h}\|^2 P) & 0 \le \overline{Q} \le |\boldsymbol{g}^H \hat{\boldsymbol{h}}|^2 P \\ \log\left(1+\left(\sqrt{\frac{\overline{Q}}{\|\boldsymbol{g}\|^2}}|\alpha_{gh}| + \sqrt{P-\frac{\overline{Q}}{\|\boldsymbol{g}\|^2}}\sqrt{\parallel \boldsymbol{h}\|^2 - |\alpha_{gh}|^2}\right)^2\right) & |\boldsymbol{g}^H \hat{\boldsymbol{h}}|^2 P < \overline{Q} \le P \parallel \boldsymbol{g}\|^2. \end{cases} \tag{5.8}$$

It is worth noting that in (5.7), if $\overline{Q} \le |\boldsymbol{g}^H \hat{\boldsymbol{h}}|^2 P$, the optimal transmit beamforming vector is based on the principle of maximal-ratio-transmission (MRT) with respect to the MISO channel $\boldsymbol{h}^H$ from the transmitter to the ID receiver, and in this case the harvested power constraint in problem (P3) is indeed not active; however, when $\overline{Q} > |\boldsymbol{g}^H \hat{\boldsymbol{h}}|^2 P$, the optimal beamforming vector is a linear combination of the two vectors $\hat{\boldsymbol{g}}$ and $\hat{\boldsymbol{h}}_{g^\perp}$, and the combining coefficients are designed such that the harvested power constraint is satisfied with equality.

In Figure 5.4, we show the achievable R-E regions for the case of MISO channels from the transmitter to both EH and ID receivers. We set $P = 10$. For the purpose of exposition, it is assumed that $\|\boldsymbol{h}\| = \|\boldsymbol{g}\| = 1$ and $|\alpha_{gh}|^2 = \rho$, with $0 \le \rho \le 1$ denoting the correlation between the two unit-norm vectors $\boldsymbol{h}$ and $\boldsymbol{g}$. This channel setup may correspond to the practical scenario where the EH and ID receivers are equipped at a single device (but still physically separated), and as a result their respective MISO channels from the transmitter have the same power gain but are spatially correlated due to the insufficient spacing between two separate receiving antennas. From Theorem 5.2, the R-E regions for the three cases of $\rho = 0.1, 0.5$, and 0.9 are obtained, as shown in Figure 5.4. Interestingly, it is observed that increasing ρ enlarges the achievable R-E region, which indicates that the antenna correlation between the EH and ID receivers can be a

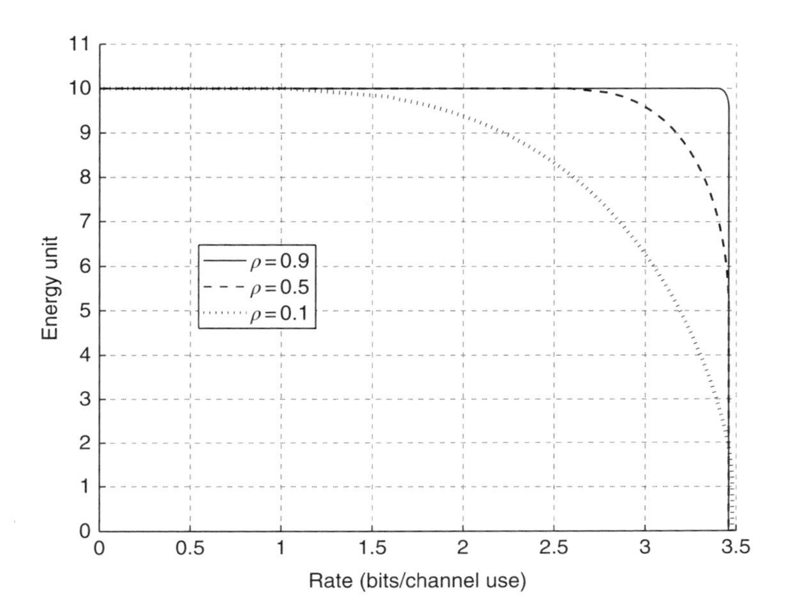

Figure 5.4 Rate–energy tradeoff for a MISO broadcast system with correlated MISO channels to the (separated) EH and ID receivers.

beneficial factor for simultaneous information and power transfer. Note that in this figure, we express energy and rate in terms of energy unit and bits/channel use, respectively, since their practical values can be obtained by appropriate scaling based on the realistic system parameters, as for Figure 5.3.

5.4 Extensions

In the preceding sections, we have focused on the multi-antenna design for R-E region characterization in a two-user SWIPT system with separated ID and EH receivers, by assuming perfect channel state information (CSI) at the transmitter. Such designs are extendable to different system setups under various practical considerations. In the following, we discuss some extensions that have been investigated in the literature to motivate future work.

Besides the separated ID and EH receivers, the authors in [11] also considered the case with co-located ID and EH receivers, such that the active UT can harvest energy as well as decode information from the same signal sent by the AP. Under practical time-switching and power-splitting receiver structures, the AP should optimize the transmit covariance matrices jointly with the time-switching or power-splitting factors at the receiver, for balancing the rate–energy tradeoff. Such designs have been further extended to other SWIPT setups, such as the broadcast channels with more than two users [12, 13], the interference channels [14–17], the relay channels [18], secrecy communications [19–21], and the case with nonlinear EH models [22–24].

In order to practically implement the transmit optimization, it is crucial for the multi-antenna AP to have accurate CSI to the EH and ID receivers. However, this is practically a very challenging task, especially for the EH receivers, since conventional methods for channel estimation in wireless communications [25] may not be directly applicable due to the energy and hardware limitations of EH receivers. To tackle this issue, there have been various channel acquisition methods that have been proposed for multi-antenna WPT systems (see, e.g., [26–30]). In [26–28], the authors proposed channel acquisition methods based on the energy measurement feedback, in which the EH receiver measures its harvested energy amounts over different training intervals using an energy meter, and sends such information back to the AP per interval. Based on the feedback information collected over intervals, the AP adjusts its transmit beamforming in subsequent training intervals and obtains refined estimates of the MIMO channel. In [29, 30], the authors developed channel learning methods that exploit the channel reciprocity, where the AP obtains the CSI by performing a reverse-link channel estimation based on the training signals sent by the EH receivers.

In addition to the SWIPT system, another related and interesting research topic is the so-called wireless powered communication network (WPCN), where RF signals are used to power UTs for their information transmission [31–35]. Specifically, under the setup of one multi-antenna AP and multiple single-antenna UTs, a harvest-then-transmit protocol is investigated in [31], where the AP first uses energy beamforming in the downlink to charge all UTs, then the UTs use their harvested energy to send an independent message to the AP simultaneously in the uplink. Such a WPCN design is extended in [32] to the scenario when the AP is equipped with a large number of antennas to improve the efficiency of WPCN. Furthermore, in [36], the idea of WPCN has been further extended to a wireless powered edge computing system, in which a multi-antenna AP uses energy beamforming to power a set of UTs for their local computing and computation offloading.

5.5 Conclusion

This chapter investigated the performance limits of emerging wireless-powered communication networks by means of dedicated wireless power transfer. Under a simplified three-node setup, our study revealed fundamental trade-offs in designing wireless MIMO systems for maximizing the efficiency of simultaneous information and energy transmission. Furthermore, we discussed some extensions of such a SWIPT design to different setups by considering various practical considerations.

Bibliography

1 L.R. Varshney (2008) Transporting information and energy simultaneously. In *Proceedings of the IEEE International Symposium Information Theory (ISIT)*, pp. 1612–1616.

2 P. Grover and A. Sahai (2010) Shannon meets Tesla: wireless information and power transfer. In *Proceedings of the IEEE International Symposium on Inference Theory (ISIT)*, pp. 2363–2367.

3 R. Want (2004) Enabling ubiquitous sensing with RFID. *IEEE Computer*, **37**: 84–86.

4 G. Landis, M. Stavnes, S. Oleson, and J. Bozek (1992) Space transfer with ground-based laser/electric propulsion. *NASA Technical Memorandum*, TM-106060.

5 I.E. Telatar (1999) Capacity of multi-antenna Gaussian channels. *Eur. Trans. Telecommun.*, **10** (6): 585–595.

6 T. Cover and J. Thomas (1991) *Elements of information theory*. Wiley, New York.

7 M. Gastpar (2007) On capacity under receive and spatial spectrum-sharing constraints. *IEEE Trans. Inf. Theory*, **53** (2): 471–487.

8 R. Zhang and Y.C. Liang (2008) Exploiting multi-antennas for opportunistic spectrum sharing in cognitive radio networks. *IEEE J. Selected Topics Signal Process.*, **2** (1): 88–102.

9 R. Zhang, Y.C. Liang, and S. Cui (2010) Dynamic resource allocation in cognitive radio networks. *IEEE Signal Process. Mag.*, **27** (3): 102–114.

10 S. Boyd and L. Vandenberghe (2004) *Convex optimization*. Cambridge University Press.

11 R. Zhang and C.K. Ho (2013) MIMO broadcasting for simultaneous wireless information and power transfer. *IEEE Trans. Wireless Commun.*, **12** (5): 1989–2001.

12 J. Xu, L. Liu, and R. Zhang (2014) Multiuser MISO beamforming for simultaneous wireless information and power transfer. *IEEE Trans. Signal Process.*, **62** (1): 4798–4810.

13 Q. Shi, L. Liu, W. Xu, and R. Zhang (2014) Joint transmit beamforming and receive power splitting for MISO SWIPT systems. *IEEE Trans. Wireless Commun.*, **13** (6): 3269–3280.

14 S. Lee, L. Liu, and R. Zhang (2015) Collaborative wireless energy and information transfer in interference channel. *IEEE Trans. Wireless Commun.*, **14** (1): 545–557.

15 C. Shen, W.-Q. Li, and T.-H. Chang (2014) Wireless information and power transfer in multi-antenna interference channel. *IEEE Trans. Signal Process.*, **62** (23): 6249–6246.

16 S. Timotheou, I. Krikidis, G. Zheng, and B. Ottersten (2014) Beamforming for MISO interference channels with QoS and RF energy transfer. *IEEE Trans. Wireless Commun.*, **13** (5): 2646–2658.

17 Q. Shi, W. Xu, T.H. Chang, Y. Wang, and E. Song (2014) Joint beamforming and power splitting for MISO interference channel with SWIPT: An SOCP relaxation and decentralized algorithm. *IEEE Trans. Signal Process.*, **62** (23): 6194–6208.

18 G. Amarasuriya, E.G. Larsson, and H.V. Poor (2016) Wireless information and power transfer in multiway massive MIMO relay networks. *IEEE Trans. Wireless Commun.*, **15** (6): 3837–3855.

19 L. Liu, R. Zhang, and K.C. Chua (2014) Secrecy wireless information and power transfer with MISO beamforming. *IEEE Trans. Signal Process.*, **62** (7): 1850–1863.

20 D.W.K. Ng, E.S. Lo, and R. Schober (2014) Robust beamforming for secure communication in systems with wireless information and power transfer. *IEEE Trans. Wireless Commun.*, **13**: 4599–4615.

21 Y. Liu, J. Xu, and R. Zhang (2018) Exploiting interference for secrecy wireless information and power transfer. *IEEE Wireless Commun.*, **5** (1): 133–139.

22 E. Boshkovska, D.W.K. Ng, N. Zlatanov, and R. Schober (2015) Practical non-linear energy harvesting model and resource allocation for SWIPT systems. *IEEE Commun. Lett.*, **19** (12): 2082–2085.

23 B. Clerckx and E. Bayguzina (2016) Waveform design for wireless power transfer. *IEEE Trans. Signal Process.*, **64** (23): 6313–6328.

24 Y. Zeng, B. Clerckx, and R. Zhang (2017) Communications and signals design for wireless power transmission. *IEEE Trans. Commun.*, **65** (5): 2264–2290.

25 D.J. Love, R.W. Heath Jr, V.K.N. Lau, D. Gesbert, B.D. Rao, and M. Andrews (2008) An overview of limited feedback in wireless communication systems. *IEEE J. Selected Areas Commun.*, **26** (8): 1341–1365.

26 J. Xu and R. Zhang (2014) Energy beamforming with one-bit feedback. *IEEE Trans. Signal Process.*, **62** (20): 5370–5381.

27 J. Xu and R. Zhang (2016) A general design framework for MIMO wireless energy transfer with limited feedback. *IEEE Trans. Signal Process.*, **64** (10): 2475–2488.

28 S. Lee and R. Zhang (2017) Distributed wireless power transfer with energy feedback. *IEEE Trans. Signal Process.*, **65** (7): 1685–1699.

29 Y. Zeng and R. Zhang (2015) Optimized training design for wireless energy transfer. *IEEE Trans. Commun.*, **63** (2): 536–550.

30 Y. Zeng and R. Zhang (2015) Optimized training for net energy maximization in multi-antenna wireless energy transfer over frequency-selective channel. *IEEE Trans. Commun.*, **63** (6): 2360–2373.

31 L. Liu, R. Zhang, and K.C. Chua (2014) Multi-antenna wireless powered communication with energy beamforming. *IEEE Trans. Commun.*, **62** (12): 4349–4361.

32 G. Yang, C.K. Ho, R. Zhang, and Y.L. Guan (2015) Throughput optimization for massive MIMO systems powered by wireless energy transfer. *IEEE J. Selected Areas Commun.*, **33** (8): 1640–1650.

33 S. Bi, C.K. Ho, and R. Zhang (2015) Wireless powered communication: Opportunities and challenges. *IEEE Commun. Mag.*, **53** (4): 117–125.

34 H. Ju and R. Zhang (2014) Throughput maximization in wireless powered communication networks. *IEEE Trans. Wireless Commun.*, **13** (1): 418–428.

35 S. Bi, Y. Zeng, and R. Zhang (2016) Wireless powered communication networks: An overview. *IEEE Wireless Commun.*, **23** (4): 10–18.

36 F. Wang, J. Xu, X. Wang, and S. Cui (2017) Joint offloading and computing optimization in wireless powered mobile-edge computing systems. *IEEE Trans. Wireless Commun.*, submitted. [Online] Available: https://arxiv.org/abs/1702.00606.

6

On the Application of SWIPT in NOMA Networks

Yuanwei Liu and Maged Elkashlan*

Electronic Engineering and Computer Science, Queen Mary University of London, United Kingdom

6.1 Introduction

Due to the increasing number of devices to be connected in fifth-generation (5G) and Internet of Things (IoT) networks, spectrum efficiency becomes a very challenging issue even though extensive research work has been carried out to solve the spectral scarcity problem [1, 2]. Non-orthogonal multiple access (NOMA) is an effective solution to improve spectral efficiency and has recently received significant attention for its promising application in 5G networks [3–7, 12, 13, 30]. The key idea of NOMA is to realize multiple access (MA) in the power domain, which is fundamentally different from conventional orthogonal MA technologies (e.g., time/frequency/code division MA). The motivation behind this approach lies in the fact that NOMA can use spectrum more efficiently by opportunistically exploring users' channel conditions [8]. In [9], the authors investigated the performance of a downlink NOMA scheme with randomly deployed users. An uplink NOMA transmission scheme was proposed in [10], and its performance was evaluated systematically. In [8], the impact of user pairing was characterized by analyzing the sum rates in two NOMA systems, namely, fixed power allocation NOMA and cognitive radio inspired NOMA. In [11], a new cooperative NOMA scheme was proposed and analyzed in terms of outage probability and diversity gain. Additionally, the performance of NOMA in large-scale cognitive radio networks was investigated in [14], where the positions of the primary and secondary users were modeled using stochastic geometry. In [15], the fairness issues of the MIMO-NOMA scenario were addressed by applying appropriate user allocation algorithms among the clusters and dynamic power allocation algorithms within each cluster.

*Corresponding author: Yuanwei Liu; yuanwei.liu@qmul.ac.uk

Wireless Information and Power Transfer: Theory and Practice, First Edition.
Edited by Derrick Wing Kwan Ng, Trung Q. Duong, Caijun Zhong, and Robert Schober.

In addition to improving spectral efficiency, which is the motivation of NOMA, another key objective of future 5G networks is to maximize energy efficiency. Simultaneous wireless information and power transfer (SWIPT), which was initially proposed in [16], has rekindled the interest of researchers to explore more energy efficient networks [17, 18]. In [16], it was assumed that both information and energy could be extracted from the same radio frequency signals at the same time, which does not hold in practice. Motivated by this issue, two practical receiver architectures, namely time switching (TS) receiver and power splitting (PS) receiver, were proposed in a multi-input and multi-output (MIMO) system in [19]. Since point-to-point communication systems with SWIPT are well established in the existing literature, recent research on SWIPT has focused on two common cooperative relaying systems: amplify-and-forward (AF) and decode-and-forward (DF). On the one hand, for AF relaying, a TS-based relaying protocol and a PS-based relaying protocol were proposed in [20]. On the other hand, for DF relaying, a new antenna switching SWIPT protocol was proposed in [21] to lower the implementation complexity. Additionally, a novel wireless energy harvesting DF relaying protocol was proposed in [22] for underlay cognitive networks to enable secondary users to harvest energy from the primary users. In [23], the application of SWIPT to DF cooperative networks with randomly deployed relays was investigated using stochastic geometry in a cooperative scenario with multiple source nodes and a single destination. A scenario in which multiple source–destination pairs are randomly deployed and communicate with each other via a single energy harvesting relay was considered in [24].

6.1.1 Motivation

The aforementioned two communication concepts, NOMA and SWIPT, can be naturally linked together as a new spectrum and energy efficient wireless energy harvesting multiple access protocol, which is the focus of this chapter. In this chapter, the near NOMA users that are close to the source are used as relays to help the far NOMA users with weaker channel conditions. To improve the reliability of these far NOMA users without draining the near users' batteries, we consider the application of SWIPT to NOMA, where SWIPT is performed at the near NOMA users. Therefore a natural question arises: which near NOMA user is to help which far NOMA user? An effective solution is to design sophisticated user pairing, and such user pairing is also important to the implementation of NOMA in practice, as explained in the following. Since NOMA is co-channel interference limited, it is realistic to implement NOMA with conventional MA technologies to achieve a new MA network. For example, we can first schedule users in pairs to perform NOMA, and then use conventional time/frequency/code division MA to serve the different user pairs.

In this chapter, users are spatially randomly deployed in two groups via the homogeneous Poisson point process (PPP). Here, the near users are grouped together and deployed in an area close to the base station (BS). The far users are in the other group and are deployed close to the edge of the cell controlled by the BS. Three opportunistic user selection schemes are proposed based on the locations of users to perform NOMA as follows: (i) random near user and random far user (RNRF) selection, where both the near and far users are randomly selected from the two groups, (ii) nearest near user and nearest far user (NNNF) selection, where a near user and a far user closest to the BS are selected from the two groups, and (iii) nearest near user and farthest far user (NNFF) selection, where a near user which is closest to the BS is selected and a far user which is farthest from the BS is selected.

6.2 Network Model

We consider a network with a single source S (i.e., BS) and two groups of randomly deployed users $\{A_i\}$ and $\{B_i\}$. We assume that the users in group $\{B_i\}$ are deployed within disc D_B with radius R_{D_B}. The far users $\{A_i\}$ are deployed within ring D_A with radius R_{D_C} and R_{D_A} (assuming $R_{D_C} \gg R_{D_B}$), as shown in Figure 6.1. Note that the BS is located at the origin of both the disc D_B and the ring D_A. The locations of the near and far users are modeled as homogeneous PPPs Φ_κ ($\kappa \in \{A,B\}$) with densities λ_{Φ_κ}. Here the near users are uniformly distributed within the disc and the far users are uniformly distributed within the ring. The number of users in R_{D_κ}, denoted by N_κ, follows a Poisson distribution $\Pr(N_\kappa = k) = (\mu_\kappa^k / k!) e^{-\mu_\kappa}$, where μ_κ is the mean measure, i.e. $\mu_A = \pi(R_{D_A}^2 - R_{D_C}^2)\lambda_{\Phi_A}$ and $\mu_B = \pi R_{D_B}^2 \lambda_{\Phi_B}$. All channels are assumed to be quasi-static Rayleigh fading, where the channel coefficients are constant for each transmission block but vary independently between different blocks. In the proposed network, we consider that the users in $\{B_i\}$ are energy harvesting relays that harvest energy from the BS and forward the information to $\{A_i\}$ using the harvested energy as their transmit powers. The DF strategy is applied at $\{B_i\}$ and the cooperative NOMA system consists of two phases, detailed in the following. In this work, without loss of generality, it is assumed that the two phases have the same transmission periods, the same as in [20, 23, 24]. It is worth pointing out that dynamic time allocation for the two phases may further improve the performance of the proposed cooperative NOMA scheme, but consideration of this issue is beyond the scope of the chapter.

6.2.1 Phase 1: Direct Transmission

Prior to transmission, the two users denoted by A_i and B_i, are selected to perform NOMA, where the selection criterion will be discussed in the next section.

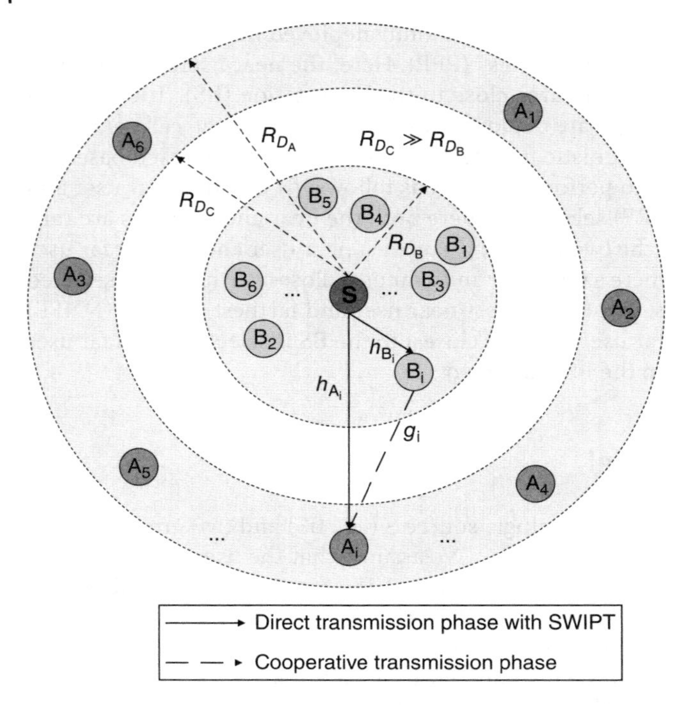

Figure 6.1 An illustration of a downlink SWIPT NOMA system with a base station S. The spatial distributions of the near users and the far users follow homogeneous PPPs.

During the first phase, the BS sends two messages $p_{i1}x_{i1} + p_{i2}x_{i2}$ to two selected users $\mathrm{A_i}$ and $\mathrm{B_i}$ based on NOMA [9], where p_{i1} and p_{i2} are the power allocation coefficients and x_{i1} and x_{i2} are the messages of $\mathrm{A_i}$ and $\mathrm{B_i}$, respectively. The observation at $\mathrm{A_i}$ is given by

$$y_{\mathrm{A_i},1} = \sqrt{P_\mathrm{S}} \sum_{k\in\{1,2\}} p_{ik}x_{ik}\frac{h_{\mathrm{A_i}}}{\sqrt{1+d_{\mathrm{A_i}}^\alpha}} + n_{\mathrm{A_i},1}, \tag{6.1}$$

where P_S is the transmit power at the BS, $h_{\mathrm{A_i}}$ models the small-scale Rayleigh fading from the BS to $\mathrm{A_i}$ with $h_{\mathrm{A_i}} \sim \mathcal{CN}(0,1)$, $n_{\mathrm{A_i},1}$ is additive Gaussian white noise (AWGN) at $\mathrm{A_i}$ with variance $\sigma_{\mathrm{A_i}}^2$, $d_{\mathrm{A_i}}$ is the distance between BS and $\mathrm{A_i}$, and α is the path-loss exponent.

Without loss of generality, we assume that $|p_{i1}|^2 > |p_{i2}|^2$ with $|p_{i1}|^2 + |p_{i2}|^2 = 1$. The received signal to interference and noise ratio (SINR) at $\mathrm{A_i}$ to detect x_{i1} is given by

$$\gamma_{\mathrm{S,A_i}}^{x_{i1}} = \frac{\rho|h_{\mathrm{A_i}}|^2|p_{i1}|^2}{\rho|p_{i2}|^2|h_{\mathrm{A_i}}|^2 + 1 + d_{\mathrm{A_i}}^\alpha}, \tag{6.2}$$

where $\rho = \frac{P_S}{\sigma^2}$ is the transmit signal to noise ratio (SNR) (assuming $\sigma_{A_i}^2 = \sigma_{B_i}^2 = \sigma^2$).

We consider that the near users have rechargeable storage energy ability [20] and the power splitting receiver [19] is adopted to perform SWIPT. From the implementation point of view, this rechargeable storage unit can be a supercapacitor or a short-term high-efficiency battery [21, 25]. The power splitting approach is applied as explained in the following: the observation at B_i is divided into two parts. One part is used for information decoding by directing the observation flow to the detection circuit and the remaining part is used for energy harvesting to powers B_i for helping A_i. Thus,

$$y_{B_i,1} = \sqrt{P_S} \sum_{k \in \{1,2\}} p_{ik} x_{ik} \frac{\sqrt{1-\beta_i} h_{B_i}}{\sqrt{1+d_{B_i}^{\alpha}}} + n_{B_i,1}, \tag{6.3}$$

where β_i is the power splitting coefficient which is detailed in (6.7), h_{B_i} models the small-scale Rayleigh fading from the BS to B_i with $h_{B_i} \sim \mathcal{CN}(0,1)$, n_{B_i} is AWGN at $n_{B_i,1}$ with variance $\sigma_{B_i}^2$, and d_{B_i} is the distance between the BS and B_i. We use the bounded path-loss model to ensure that the path loss is always larger than one even for small distances [23].

Applying NOMA, successive interference cancellation (SIC) [26] is carried out at B_i. In particular, B_i first decodes the message of A_i, then subtracts this component from the received signal to detect its own information. Therefore, the received SINR at B_i to detect x_{i1} of A_i is given by

$$\gamma_{S,B_i}^{x_{i1}} = \frac{\rho |h_{B_i}|^2 |p_{i1}|^2 (1-\beta_i)}{\rho |h_{B_i}|^2 |p_{i2}|^2 (1-\beta_i) + 1 + d_{B_i}^{\alpha}}. \tag{6.4}$$

The received SNR at B_i to detect x_{i2} of B_i is given by

$$\gamma_{S,B_i}^{x_{i2}} = \frac{\rho |h_{B_i}|^2 |p_{i2}|^2 (1-\beta_i)}{1 + d_{B_i}^{\alpha}}. \tag{6.5}$$

The power splitting coefficient β_i is used to determine the amount of harvested energy. Based on (6.4), the data rate supported by the channel from the BS to B_i for decoding x_{i1} is given by

$$R_{x_{i1}} = \frac{1}{2} \log \left(1 + \frac{\rho |h_{B_i}|^2 |p_{i1}|^2 (1-\beta_i)}{\rho |h_{B_i}|^2 |p_{i2}|^2 (1-\beta_i) + 1 + d_{B_i}^{\alpha}} \right), \tag{6.6}$$

where $\frac{1}{2}$ is due to the fact that we use two phases to complete the communications. We assume that the energy required to receive/process information is negligible compared to the energy required for information transmission [20, 25]. In this work, we apply the dynamic power splitting protocol, which means that the power splitting coefficient β_i is a variable and opportunistically tuned to support the relay transmission. Our aim is to first guarantee the detection of the message of the far NOMA user, A_i, at the near NOMA user B_i, then B_i can harvest the remaining energy. In this case, based on (6.6), in order to ensure that B_i can successfully decode the information of A_i, we have a rate, i.e. $R_1 = R_{x_{i1}}$. Therefore, the power splitting coefficient is set as follows:

$$\beta_i = \max\left\{0, 1 - \frac{\tau_1(1 + d_{B_i}^{\alpha})}{\rho(|p_{i1}|^2 - \tau_1|p_{i2}|^2)|h_{B_i}|^2}\right\}, \tag{6.7}$$

where $\tau_1 = 2^{2R_1} - 1$. Here $\beta_i = 0$ means that all the energy is used for information decoding and no energy remains for energy harvesting.

Based on (6.3), the energy harvested at B_i is given by

$$E_{B_i} = \frac{T\eta P_S \beta_i |h_{B_i}|^2}{2(1 + d_{B_i}^{\alpha})}, \tag{6.8}$$

where T is the time period for the entire transmission, including the direct transmission phase and the cooperative transmission phase, and η is the energy harvesting coefficient. We assume that the two phases have the same transmission period, and therefore the transmit power at B_i can be expressed as follows:

$$P_t = \frac{\eta P_S \beta_i |h_{B_i}|^2}{1 + d_{B_i}^{\alpha}}. \tag{6.9}$$

6.2.2 Phase 2: Cooperative Transmission

During this phase, B_i forwards x_{i1} to A_i by using the harvested energy during the direct transmission phase. In this case, A_i observes

$$y_{A_i,2} = \frac{\sqrt{P_t} x_{i1} g_i}{\sqrt{1 + d_{C_i}^{\alpha}}} + n_{A_i,2}, \tag{6.10}$$

where g_i models the small-scale Rayleigh fading from B_i to A_i with $g_i \sim \mathcal{CN}(0, 1)$, $n_{Ai,2}$ is AWGN at A_i with variance $\sigma_{A_i}^2$, $d_{C_i} = \sqrt{d_{A_i}^2 + d_{B_i}^2 - 2d_{A_i} d_{B_i} \cos(\theta_i)}$ is the distance between B_i and A_i, and θ_i denotes the angle $\angle A_i S B_i$.

Based on (6.9) and (6.10), the received SNR for A_i to detect x_{i1} forwarded from B_i is given by

$$\gamma_{A_i,B_i}^{x_{i1}} = \frac{P_t |g_i|^2}{(1 + d_{C_i}^{\alpha})\sigma^2} = \frac{\eta\rho\beta_i |h_{B_i}|^2 |g_i|^2}{(1 + d_{C_i}^{\alpha})(1 + d_{B_i}^{\alpha})}. \tag{6.11}$$

At the end of this phase, A_i combines the signals from the BS and B_i using maximal-ratio combining (MRC). Combining the SNR of the direct transmission phase (6.2) and the SINR of the cooperative transmission phase (6.11), we obtain the received SINR at A_i as follows:

$$\gamma_{A_i,MRC}^{x_{i1}} = \frac{\rho |h_{A_i}|^2 |p_{i1}|^2}{\rho |h_{A_i}|^2 |p_{i2}|^2 + 1 + d_{A_i}^{\alpha}} + \frac{\eta \rho \beta_i |h_{B_i}|^2 |g_i|^2}{(1 + d_{B_i}^{\alpha})(1 + d_{C_i}^{\alpha})}. \tag{6.12}$$

6.3 Non-Orthogonal Multiple Access with User Selection

In this section, the performance of three user selection schemes are characterized.

6.3.1 RNRF Selection Scheme

In this scheme, the BS randomly selects a near user B_i and a far user A_i. This selection scheme provides a fair opportunity for each user to access the source with the NOMA protocol. The advantage of this user selection scheme is that it does not require the knowledge of instantaneous channel state information (CSI). To make meaningful conclusions in the rest of the chapter, we only focus on $\beta_i > 0$ and the number of near users and far users satisfies $N_B \geq 1, N_A \geq 1$.

6.3.1.1 Outage Probability of the Near Users of RNRF

In the NOMA protocol, an outage of B_i can occur for two reasons. The first is that B_i cannot detect x_{i1}. The second is that B_i can detect x_{i1} but cannot detect x_{i2}. To guarantee that the NOMA protocol can be implemented, the condition $|p_{i1}|^2 - |p_{i2}|^2 \tau_1 > 0$ should be satisfied [9]. Based on this, the outage probability of B_i can be expressed as follows:

$$P_{B_i} = \Pr\left(\frac{\rho |h_{B_i}|^2 |p_{i1}|^2}{\rho |h_{B_i}|^2 |p_{i2}|^2 + 1 + d_{B_i}^{\alpha}} < \tau_1 \right) + \Pr\left(\frac{\rho |h_{B_i}|^2 |p_{i1}|^2}{\rho |h_{B_i}|^2 |p_{i2}|^2 + 1 + d_{B_i}^{\alpha}} > \tau_1, \gamma_{S,B_i}^{x_{i2}} < \tau_2 \right), \tag{6.13}$$

where $\tau_2 = 2^{2R_2} - 1$, with R_2 being the target rate at which B_i can detect x_{i2}.

The following theorem provides the outage probability of the near users in RNRF for an arbitrary choice of α.

Theorem 6.1 Conditioned on the PPPs, the outage probability of the near users B_i can be approximated as follows:

$$P_{B_i} \approx \frac{1}{2} \sum_{n=1}^{N} \omega_N \sqrt{1 - {\phi_n}^2} (1 - e^{-c_n \epsilon_{A_i}})(\phi_n + 1), \tag{6.14}$$

if $\varepsilon_{\mathrm{A}_i} \geq \varepsilon_{\mathrm{B}_i}$, otherwise $P_{\mathrm{B_i}} = 1$, where $\varepsilon_{\mathrm{A}_i} = \frac{\tau_1}{\rho(|p_{i1}|^2 - |p_{i2}|^2\tau_1)}$ and $\varepsilon_{\mathrm{B}_i} = \frac{\tau_2}{\rho|p_{i2}|^2}$, N is a parameter to ensure a complexity-accuracy tradeoff, $c_n = 1 + \left(\frac{R_{D_\mathrm{B}}}{2}(\phi_n + 1)\right)^\alpha$, $\omega_N = \frac{\pi}{N}$, and $\phi_n = \cos\left(\frac{2n-1}{2N}\pi\right)$.

Proof: Define $X_i = \frac{|h_{\mathrm{A_i}}|^2}{1+d_{\mathrm{A_i}}^\alpha}$, $Y_i = \frac{|h_{\mathrm{B_i}}|^2}{1+d_{\mathrm{B_i}}^\alpha}$, and $Z_i = \frac{|g_i|^2}{1+d_{\mathrm{C_i}}^\alpha}$. Substituting (6.4) and (6.5) into (6.13), the outage probability of the near users is given by

$$P_{\mathrm{B_i}} = \Pr(Y_i < \varepsilon_{\mathrm{A}_i}) + \Pr(Y_i > \varepsilon_{\mathrm{A}_i}, \varepsilon_{\mathrm{A}_i} < \varepsilon_{\mathrm{B}_i}). \tag{6.15}$$

If $\varepsilon_{\mathrm{A}_i} < \varepsilon_{\mathrm{B}_i}$, the outage probability at the near users is always one.

For the case $\varepsilon_{\mathrm{A}_i} \geq \varepsilon_{\mathrm{B}_i}$, note that the users are deployed in D_B and D_A according to homogeneous PPPs. Therefore, the NOMA users are modeled as independently and identically distributed (i.i.d.) points in D_B and D_A, denoted by W_{κ_i} ($\kappa \in \{\mathrm{A}, \mathrm{B}\}$), which contain the location information about $\mathrm{A_i}$ and $\mathrm{B_i}$, respectively. The probability density functions (PDFs) of $W_{\mathrm{A_i}}$ and $W_{\mathrm{B_i}}$ are given by

$$f_{W_{\mathrm{B_i}}}(\omega_{\mathrm{B_i}}) = \frac{\lambda_{\Phi_\mathrm{B}}}{\mu_{R_{D_\mathrm{B}}}} = \frac{1}{\pi R_{D_\mathrm{B}}^2} \tag{6.16}$$

and

$$f_{W_{\mathrm{A_i}}}(\omega_{\mathrm{A_i}}) = \frac{\lambda_{\Phi_\mathrm{A}}}{\mu_{R_{D_\mathrm{A}}}} = \frac{1}{\pi(R_{D_\mathrm{A}}^2 - R_{D_\mathrm{C}}^2)}, \tag{6.17}$$

respectively.

Therefore, for the case $\varepsilon_{\mathrm{A}_i} \geq \varepsilon_{\mathrm{B}_i}$, the cumulative distribution function (CDF) of Y_i is given by

$$\begin{aligned} F_{Y_i}(\varepsilon) &= \int_{D_\mathrm{B}} (1 - e^{-(1+d_{\mathrm{B_i}}^\alpha)\varepsilon}) f_{W_{\mathrm{B_i}}}(\omega_{\mathrm{B_i}}) d\omega_{\mathrm{B_i}} \\ &= \frac{2}{R_{D_\mathrm{B}}^2} \int_0^{R_{D_\mathrm{B}}} (1 - e^{-(1+r^\alpha)\varepsilon}) r dr. \end{aligned} \tag{6.18}$$

For many communication scenarios $\alpha > 2$, and it is challenging to obtain exact closed-from expressions for the above. In this case, we can use Gaussian–Chebyshev quadrature [27] to find the approximation of (6.18) as follows:

$$F_{Y_i}(\varepsilon) \approx \frac{1}{2} \sum_{n=1}^{N} \omega_N \sqrt{1 - \phi_n{}^2} (1 - e^{-c_n\varepsilon})(\phi_n + 1). \tag{6.19}$$

Applying $\varepsilon_{A_i} \rightarrow \varepsilon$ into (6.19), (6.14) is obtained, and the proof of the theorem is completed.

6.3.1.2 Outage Probability of the Far Users of RNRF

With the proposed cooperative SWIPT NOMA protocol, outage experienced by A_i can occur in two situations. The first is when B_i can detect x_{i1} but the overall received SNR at A_i cannot support the targeted rate. The second is when neither A_i nor B_i can detect x_{i1}. Based on this, the outage probability can be expressed as follows:

$$P_{A_i} = \Pr(\gamma^{x_{i1}}_{A_i,MRC} < \tau_1, \gamma^{x_{i1}}_{S,B_i}|_{\beta_i=0} > \tau_1) + \Pr(\gamma^{x_{i1}}_{S,A_i} < \tau_1, \gamma^{x_{i1}}_{S,B_i}|_{\beta_i=0} < \tau_1). \tag{6.20}$$

The following theorem provides the outage probability of the far users in RNRF for an arbitrary choice of α.

Theorem 6.2 Conditioned on the PPPs, and assuming $R_{D_C} \gg R_{D_B}$, the outage probability of A_i can be approximated as follows:

$$\begin{aligned} P_{A_i} \approx\ & \zeta_1 \sum_{n=1}^{N} (\phi_n + 1)\sqrt{1-\phi_n{}^2} c_n \sum_{k=1}^{K} \sqrt{1-\psi_k^2} s_k (1+s_k^\alpha)^2 \\ & \times \sum_{m=1}^{M} \sqrt{1-\varphi_m^2} e^{-(1+s_k^\alpha)t_m} \chi_{t_m} \left(\ln \frac{\chi_{t_m}(1+s_k^\alpha)}{\eta\rho} c_n + 2c_0 \right) \\ & + a_1 \sum_{n=1}^{N} \sqrt{1-\phi_n{}^2} c_n (\phi_n+1) \sum_{k=1}^{K} \sqrt{1-\psi_k{}^2}(1+s_k^\alpha) s_k, \end{aligned} \tag{6.21}$$

where M and K are parameters to ensure a complexity-accuracy trade-off, $\zeta_1 = -\frac{\varepsilon_{A_i} R_{D_{B_i}} \omega_N \omega_K \omega_M}{8(R_{D_{A_i}} + R_{D_{C_i}})\eta\rho}$, $\chi_{t_m} = \tau_1 - \frac{\rho t_m |p_{i1}|^2}{\rho t_m |p_{i2}|^2 + 1}$, $t_m = \frac{\varepsilon_{A_i}}{2}(\varphi_m + 1)$, $\omega_M = \frac{\pi}{M}$, $\varphi_m = \cos\left(\frac{2m-1}{2M}\pi\right)$, $s_k = \frac{R_{D_A} - R_{D_C}}{2}(\psi_k + 1) + R_{D_C}$, $\omega_K = \frac{\pi}{K}$, $\psi_k = \cos\left(\frac{2k-1}{2K}\pi\right)$, $c_0 = -\frac{\varphi(1)}{2} - \frac{\varphi(2)}{2}$, and $a_1 = \frac{\omega_K \omega_N \varepsilon^2_{A_1}}{2(R_{D_A} + R_{D_C})}$.

Proof: Please refer to Appendix A of [29].

6.3.1.3 Diversity Analysis of RNRF

To obtain further insights into the derived outage probability, we provide a diversity analysis of both the near and far users of RNRF.

Near users: For the near users, based on the analytical results, we carry out high SNR approximations as follows. When $\varepsilon \rightarrow 0$, a high SNR approximation of (6.19) with $1 - e^{-x} \approx x$ is given by

$$F_{Y_i}(\varepsilon) \approx \frac{1}{2} \sum_{n=1}^{N} \omega_N \sqrt{1 - {\phi_n}^2} c_n \varepsilon_{A_i}(\phi_n + 1). \tag{6.22}$$

The diversity gain is defined as follows:

$$d = -\lim_{\rho \to \infty} \frac{\log P(\rho)}{\log \rho}. \tag{6.23}$$

Substituting (6.22) into (6.23), we find that the diversity gain for the near users is one, which means that using NOMA with energy harvesting will not decrease the diversity gain.

Far users: For the far users, substituting (6.21) into (6.23), we obtain

$$\begin{aligned} d &= -\lim_{\rho \to \infty} \frac{\log\left(-\frac{1}{\rho^2}\log\frac{1}{\rho}\right)}{\log \rho} \\ &= -\lim_{\rho \to \infty} \frac{\log\log\rho - \log\rho^2}{\log \rho} = 2. \end{aligned} \tag{6.24}$$

As we can see from (6.24), the diversity gain of RNRF is two, which is the same as that of the conventional cooperative network [28]. This result indicates that using NOMA with an energy harvesting relay will not affect the diversity gain. In addition, we see that at high SNRs, the dominant factor for the outage probability is $\frac{1}{\rho^2} \ln \rho$. Therefore we conclude that the outage probability of using NOMA with SWIPT decays at a rate of $\frac{\ln SNR}{SNR^2}$. However, for a conventional cooperative system without energy harvesting, a faster decreasing rate of $\frac{1}{SNR^2}$ can be achieved.

6.3.1.4 System Throughput in Delay-Sensitive Transmission Mode of RNRF

In this chapter, we will focus on the delay-sensitive throughput. In this mode, the transmitter sends information at a fixed rate and the throughput is determined by evaluating the outage probability.

Based on the analytical results for the outage probability of the near and far users, the system throughput of RNRF in the delay-sensitive transmission mode is given by

$$R_{\tau_{\mathrm{RNRF}}} = (1 - P_{A_i})R_1 + (1 - P_{B_i})R_2, \tag{6.25}$$

where P_{A_i} and P_{B_i} are obtained from (6.21) and (6.14), respectively.

6.3.2 NNNF Selection Scheme

In this subsection, we characterize the performance of NNNF, which exploits the users' CSI opportunistically. We first select a user within the disc D_B which has the shortest distance to the BS as the near NOMA user (denoted by B_{i^*}). This is because the near users also act as energy harvesting relays to help the far users. The NNNF scheme can enable the selected near user to harvest more energy. Then we select a user within the ring D_{A}, which has the shortest distance to the BS as the far NOMA user (denoted by $\mathrm{A}_{\mathrm{i}*}$). The advantage of the NNNF scheme is that it can minimize the outage probability of both the near and far users.

6.3.2.1 Outage Probability of the Near Users of NNNF

Using the same definition of the outage probability as the near users of NOMA, we can characterize the outage probability of the near users of NNNF.

The following theorem provides the outage probability of the near users of NNNF for an arbitrary choice of α.

Theorem 6.3 Conditioned on the PPPs, the outage probability of $\mathrm{B}_{\mathrm{i}*}$ can be approximated as follows:

$$P_{\mathrm{B}_{\mathrm{i}*}} \approx b_1 \sum_{n=1}^{N} \sqrt{1-{\phi_n}^2}(1-e^{-(1+c_{n*}^{\alpha})\varepsilon_{\mathrm{A}_i}})c_{n*}e^{-\pi\lambda_{\Phi_{\mathrm{B}}}c_{n*}^2}, \tag{6.26}$$

if $\varepsilon_{\mathrm{A}_i} \geq \varepsilon_{\mathrm{B}_i}$, otherwise $P_{\mathrm{B}_{\mathrm{i}*}} = 1$, where $c_{n*} = \frac{R_{D_{\mathrm{B}}}}{2}(\phi_n + 1)$, $b_1 = \frac{\xi_{\mathrm{B}}\omega_N R_{D_{\mathrm{B}}}}{2}$, and $\xi_{\mathrm{B}} = \frac{2\pi\lambda_{\Phi_{\mathrm{B}}}}{1-e^{-\pi\lambda_{\Phi_{\mathrm{B}}}R_{D_{\mathrm{B}}}^2}}$.

Proof: Similar to (6.15), the outage probability of $\mathrm{B}_{\mathrm{i}*}$ can be expressed as follows:

$$P_{\mathrm{B}_{\mathrm{i}*}} = \Pr(Y_{i^*} < \varepsilon_{\mathrm{A}_i}|N_{\mathrm{B}} \geq 1) = F_{Y_{i^*}}(\varepsilon_{\mathrm{A}_i}), \tag{6.27}$$

where $Y_{i^*} = \frac{|h_{\mathrm{B}_i}|^2}{1+d_{\mathrm{B}_{\mathrm{i}*}}^{\alpha}}$ and $d_{\mathrm{B}_{\mathrm{i}*}}$ is the distance from the nearest $\mathrm{B}_{\mathrm{i}*}$ to the BS.

The CDF of Y_{i^*} can be written as follows:

$$F_{Y_{i^*}}(\varepsilon) = \int_0^{R_{D_{\mathrm{B}}}} (1-e^{-(1+r_{\mathrm{B}}^{\alpha})\varepsilon}) f_{d_{\mathrm{B}_{\mathrm{i}*}}}(r_{\mathrm{B}})dr_{\mathrm{B}}, \tag{6.28}$$

where $f_{d_{\mathrm{B}_{\mathrm{i}*}}}$ is the PDF of the shortest distance from $\mathrm{B}_{\mathrm{i}*}$ to the BS.

The probability $\Pr\{d_{\mathrm{B}_{\mathrm{i}}*} > r|N_{\mathrm{B}} \geq 1\}$ conditioned on $N_{\mathrm{B}} \geq 1$ is the event that there is no point located in the disc. Therefore we can express this probability as follows:

$$\begin{aligned}\Pr\{d_{\mathrm{B}_{\mathrm{i}}*} > r|N_{\mathrm{B}} \geq 1\} &= \frac{\Pr\{d_{\mathrm{B}_{\mathrm{i}}*} > r\} - \Pr\{d_{\mathrm{B}_{\mathrm{i}}*} > r, N_{\mathrm{B}} = 0\}}{\Pr\{N_{\mathrm{B}} \geq 1\}} \\ &= \frac{e^{-\pi\lambda_{\Phi_{\mathrm{B}}}r^2} - e^{-\pi\lambda_{\Phi_{\mathrm{B}}}R_{D_{\mathrm{B}}}^2}}{1-e^{-\pi\lambda_{\Phi_{\mathrm{B}}}R_{D_{\mathrm{B}}}^2}}.\end{aligned} \tag{6.29}$$

Then the corresponding PDF of $\mathrm{B}_{\mathrm{i}*}$ is given by

$$f_{d_{\mathrm{B}_{\mathrm{i}*}}}(r_{\mathrm{B}}) = \xi_{\mathrm{B}} r_{\mathrm{B}} e^{-\pi\lambda_{\Phi_{\mathrm{B}}}r_{\mathrm{B}}^2}. \tag{6.30}$$

Substituting (6.30) into (6.28), we obtain

$$F_{Y_{i*}}(\varepsilon) = \xi_{\mathrm{B}} \int_0^{R_{D_{\mathrm{B}}}} (1 - e^{-(1+r_{\mathrm{B}}^{\alpha})\varepsilon}) r_{\mathrm{B}} e^{-\pi \lambda_{\Phi_{\mathrm{B}}} r_{\mathrm{B}}^2} dr_{\mathrm{B}}. \tag{6.31}$$

Applying the Gaussian–Chebyshev quadrature approximation to (6.19), we obtain

$$F_{Y_{i*}}(\varepsilon) \approx \frac{\xi_{\mathrm{B}} \omega_N R_{D_{\mathrm{B}}}}{2} \sum_{n=1}^{N} \sqrt{1 - {\phi_n}^2} (1 - e^{-(1+c_{n*}^{\alpha})\varepsilon}) c_{n*} e^{-\pi \lambda_{\Phi_{\mathrm{B}}} c_{n*}^2}. \tag{6.32}$$

Applying $\varepsilon_{\mathrm{A}_i} \rightarrow \varepsilon$, we obtain the approximate outage probability of B_{i*} in (6.26).

6.3.2.2 Outage Probability of the Far Users of NNNF

Using the same definition of the outage probability for the far users of NOMA, and similar to (6.20), we can characterize the outage probability of the far users in NNNF. The following theorem provides the outage probability of the far users in NNNF for an arbitrary choice of α.

Theorem 6.4 Conditioned on the PPPs and assuming $R_{D_{\mathrm{C}}} \gg R_{D_{\mathrm{B}}}$, the outage probability of A_{i*} can be approximated as follows:

$$\begin{aligned} P_{\mathrm{A}_{i*}} \approx\ & \varsigma^* \sum_{n=1}^{N} \sqrt{1 - {\phi_n}^2} (1 + c_{n*}^{\alpha}) c_{n*} e^{-\pi \lambda_{\Phi_{\mathrm{B}}} c_{n*}^2} \\ & \times \sum_{k=1}^{K} \sqrt{1 - \psi_k^2} (1 + s_k^{\alpha})^2 s_k e^{-\pi \lambda_{\Phi_{\mathrm{A}}} (s_k^2 - R_{D_{\mathrm{C}}}^2)} \sum_{m=1}^{M} \sqrt{1 - \varphi_m^2} \\ & \times e^{-(1+s_k^{\alpha}) t_m} \chi_{t_m} \left(\ln \frac{\chi_{t_m} (1 + s_k^{\alpha})(1 + c_{n*}^{\alpha})}{\eta \rho} + 2c_0 \right) \\ & + b_2 b_3 \sum_{k=1}^{K} \sqrt{1 - {\psi_k}^2} (1 + s_k^{\alpha}) s_k e^{-\pi \lambda_{\Phi_{\mathrm{A}}} s_k^2} \\ & \times \sum_{n=1}^{N} (\sqrt{1 - {\phi_n}^2} (1 + c_{n*}^{\alpha}) c_{n*} e^{-\pi \lambda_{\Phi_{\mathrm{B}}} c_{n*}^2}), \end{aligned} \tag{6.33}$$

where $\varsigma^* = -\frac{\xi_{\mathrm{B}} \xi_{\mathrm{A}} \omega_N \omega_K \omega_M \varepsilon_{\mathrm{A}_i} R_{D_{\mathrm{B}}} (R_{D_{\mathrm{A}}} - R_{D_{\mathrm{C}}})}{8 \eta \rho}$, $b_2 = \frac{\xi_{\mathrm{A}} e^{\pi \lambda_{\Phi_{\mathrm{A}}} R_{D_{\mathrm{C}}}^2} \omega_K \varepsilon_{\mathrm{A}_i}}{R_{D_{\mathrm{A}}} + R_{D_{\mathrm{C}}}}$, and $b_3 = \frac{\xi_{\mathrm{B}} \omega_N R_{D_{\mathrm{B}}} \varepsilon_{\mathrm{A}_i}}{2}$.

Proof: Please refer to Appendix B of [29].

6.3.2.3 Diversity Analysis of NNNF

Similarly, we provide diversity analysis of both the near and far users of NNNF.

Near users: For the near users, based on the analytical results, we carry out the high SNR approximation as follows. When $\varepsilon \rightarrow 0$, a high SNR approximation of (6.26) with $1 - e^{-x} \approx x$ is given by

$$P_{\mathrm{B}_{i*}} \approx b_1 \epsilon_{\mathrm{A}_i} \sum_{n=1}^{N} (\sqrt{1-\phi_n{}^2}(1+c_{n*}^{\alpha})c_{n*}e^{-\pi\lambda_{\Phi_{\mathrm{B}}}c_{n*}^2}). \tag{6.34}$$

Substituting (6.34) into (6.23), we find that the diversity gain for the near users of NNNF is one, which indicates that using NNNF will not affect the diversity gain.

Far users: For the far users, substituting (6.33) into (6.23), we find that the diversity gain is still two. This indicates that NNNF will not affect the diversity gain.

6.3.2.4 System Throughput in Delay-Sensitive Transmission Mode of NNNF

Based on the analytical results for the outage probability of the near and far users, the system throughput of NNNF in the delay-sensitive transmission mode is given by

$$R_{\tau_{\mathrm{NNNF}}} = (1-P_{\mathrm{A}_{i*}})R_1 + (1-P_{\mathrm{B}_{i*}})R_2, \tag{6.35}$$

where $P_{\mathrm{A}_{i*}}$ and $P_{\mathrm{B}_{i*}}$ are obtained from (6.33) and (6.26), respectively.

6.3.3 NNFF Selection Scheme

In this scheme, we first select a user within disc D_{B}, which has the shortest distance to the BS as a near NOMA user. Then we select a user within ring D_{A}, which has the farthest distance to the BS as a far NOMA user (denoted by $\mathrm{A}_{i'}$). The use of this selection scheme is inspired by an interesting observation described in [9] that NOMA can offer a larger performance gain over conventional MA when user channel conditions are more distinct.

6.3.3.1 Outage Probability of the Far Users of NNFF

Using the same definition of the outage probability of the far users, and similar to (6.20), we can characterize the outage probability of the far users of NNFF. The following theorem provides the outage probability of the far user of NNFF for an arbitrary choice of α.

Theorem 6.5 Conditioned on the PPPs and assuming $R_{D_{\mathrm{C}}} \gg R_{D_{\mathrm{B}}}$, the outage probability of $\mathrm{A}_{i'}$ can be approximated as follows:

$$P_{\mathrm{A}_{i'}} \approx \varsigma^* \sum_{n=1}^{N} \sqrt{1-\phi_n{}^2}(1+c_{n*}^{\alpha})c_{n*}e^{-\pi\lambda_{\Phi_{\mathrm{B}}}c_{n*}^2}$$
$$\times \sum_{k=1}^{K} \sqrt{1-\psi_k^2}(1+s_k^{\alpha})^2 s_k e^{-\pi\lambda_{\Phi_{\mathrm{A}}}(R_{D_{\mathrm{A}}}^2 - s_k^2)} \sum_{m=1}^{M} \sqrt{1-\varphi_m^2}$$
$$\times e^{-(1+s_k^{\alpha})t_m} \chi_{t_m} \left(\ln \frac{\chi_{t_m}(1+s_k^{\alpha})(1+c_{n*}^{\alpha})}{\eta\rho} + 2c_0 \right)$$

$$
+ b_3 b_4 \sum_{k=1}^{K} \sqrt{1 - {\psi_k}^2}(1 + s_k^{\alpha}) s_k e^{\pi \lambda_{\Phi_A} s_k^2}
$$

$$
\times \sum_{n=1}^{N} (\sqrt{1 - {\phi_n}^2}(1 + c_{n*}^{\alpha}) c_{n*} e^{-\pi \lambda_{\Phi_B} c_{n*}^2}), \tag{6.36}
$$

where $b_4 = \frac{\xi_A e^{-\pi \lambda_{\Phi_A} R_{D_A}^2} \omega_K \varepsilon_{A_i}}{R_{D_A} + R_{D_C}}$.

Proof: Please refer to Appendix D of [29].

6.3.3.2 Diversity Analysis

Similarly, we provide diversity analysis of both the near and far users in NNFF.

Near users: Since the same criterion for selecting a near user is used, the diversity gain is one, which is the same as for NNNF.

Far users: Substituting (6.36) into (6.23), we find that the diversity gain is still two. Therefore, we conclude that using opportunistic user selection schemes (NNNF and NNFF) based on distances will not affect the diversity gain.

6.3.3.3 System Throughput in Delay-Sensitive Transmission Mode of NNFF

Based on the analytical results for the outage probability of the near and far users, the system throughput of NNFF in the delay-sensitive transmission mode is given by

$$
R_{\tau_{\text{NNFF}}} = (1 - P_{A_{i'}}) R_1 + (1 - P_{B_{i*}}) R_2, \tag{6.37}
$$

where $P_{A_{i'}}$ and $P_{B_{i*}}$ are obtained from (6.36) and (6.26), respectively.

6.4 Numerical Results

In the considered network, we assume that the energy conversion efficiency of SWIPT is $\eta = 0.7$ and the power allocation coefficients of NOMA are $|p_{i1}|^2 = 0.8$ and $|p_{i1}|^2 = 0.2$. In the following figures, we use red, blue, and black colored lines to represent the RNRF, NNNF, and NNFF user selection schemes, respectively.

6.4.1 Outage Probability of the Near Users

In this subsection, the outage probability achieved by the near users with different choice of density and path-loss coefficients for the three user selection schemes is demonstrated. Note that the same user selection criteria is applied for the nears users of NNNF and NNFF; we use NNN(F)F to represent these two selection schemes in Figures 6.2 and 6.4.

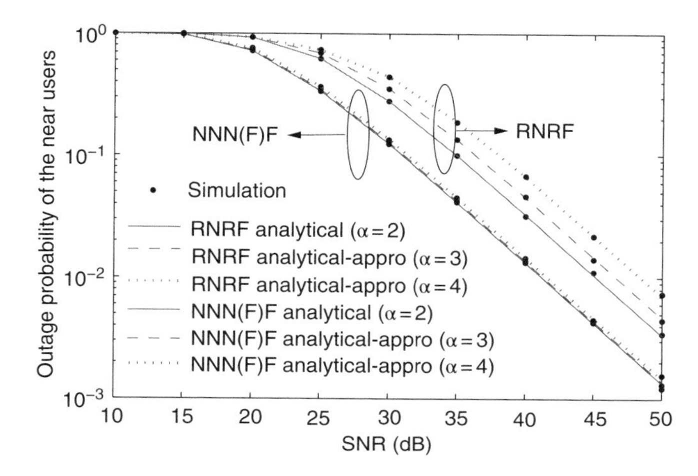

Figure 6.2 Outage probability of the near users versus SNR with different α, $R_1 = 0.3$, $R_2 = 1$, $R_{D_A} = 10$, $R_{D_B} = 2$, $R_{D_C} = 8$, $\lambda_{\Phi_A} = 1$, and $\lambda_{\Phi_B} = 1$.

Figure 6.2 plots the outage probability of the near users versus SNR with different path-loss coefficients for both RNRN and NNN(F)F. One can observe that by performing NNNF and NNFF (which we refer to as NNN(F)F in the figure), lower outage probability is achieved than RNRF since the shorter distance makes less path loss and leads to better performance. The figure also demonstrates that as α increases, outage will occur more because of higher path loss. It is worth noting that for NNNF and NNFF, the performance is very close for different values of α. This is because we use the bounded path-loss model (i.e., $1 + d_i^{\alpha} > 1$) to ensure that the path loss is always larger than one. When selecting the nearest near user, d_i will approach zero and the path loss will approach one, which makes the performance difference of the three selection schemes insignificant.

Figure 6.3 plots the outage probability of the near users versus the density with different path-loss coefficients of NNN(F)F. One can observe that the outage probability decreases as the density of the near users increases. This is because the multi-user diversity gain is improved with the increasing number of near users. The figure also demonstrates that the outage probability decreases with decreasing α because there is smaller path loss. It is worth noting that an outage floor exits as the density of the near users increases for different values of α. The reason is, once again, because we use the bounded path-loss model. When the number of the near users exceeds a threshold, the selected near users will be very close to the source and make the path-loss approach one.

Figure 6.4 plots the outage probability of the near users versus the rate of the far users with different rates of the near users for both RNRN and NNN(F)F. One can observe that the outage of the near users occurs more frequently as

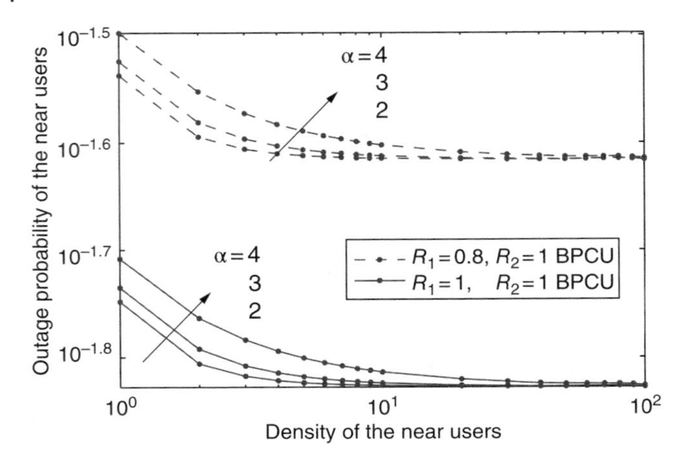

Figure 6.3 Outage probability of the near users versus density with different α, $R_{D_A} = 10$, $R_{D_B} = 2$, $R_{D_C} = 8$, $\lambda_{\Phi_B} = 1$, and $SNR = 30$ dB.

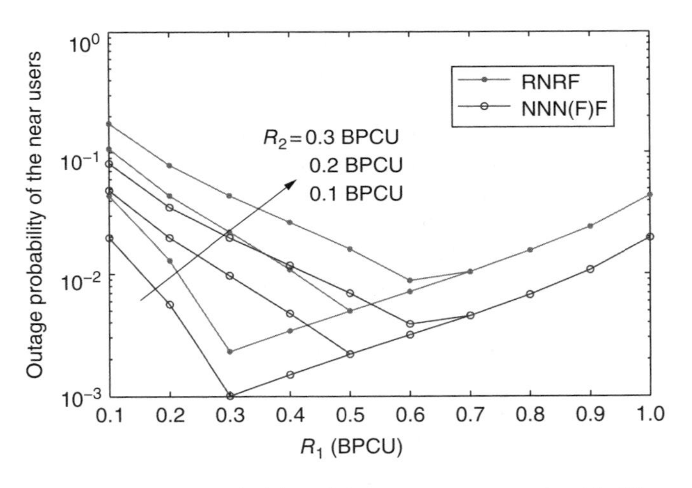

Figure 6.4 Outage probability of the near users versus R_1 with different R_2, $\alpha = 2$, $R_{D_A} = 10$, $R_{D_B} = 2$, $R_{D_C} = 8$, and $SNR = 30$ dB.

the rate of the far user R_2 increases from 0.1 to 0.3 bit per channel use (BPCU). An interesting observation is that as the rate of the near users R_1 increases, the outage probability first decreases then increases. This indicates that there exists an optimal choice of R_1 that can minimize the outage probability. This behavior can be explained as follows. On the one hand, in the NOMA system, the near user B_i needs to first decode x_{i1} intended for the far user A_i then decode its own. Therefore increasing R_1 makes it harder to decode x_{i1}, which will lead

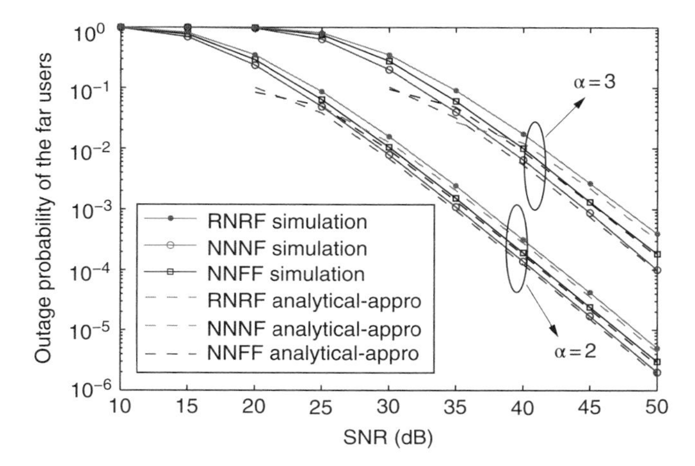

Figure 6.5 Outage probability of the far users with different α, $R_1 = 0.3$, $R_{D_A} = 10$, $R_{D_B} = 2$, $R_{D_C} = 9$, $\lambda_{\Phi_A} = 1$, and $\lambda_{\Phi_B} = 1$.

to more outage. On the other hand, since B_i also acts as an energy harvesting relay and uses the harvested energy to decode x_{i1} and x_{i2}, with a larger R_1, more observation flow will be distributed to the decoding circuit, which in turn decreases the outage probability.

6.4.2 Outage Probability of the Far Users

In this subsection, we demonstrate the outage probability of the far users with different choices of the density and path-loss coefficients of the three user selection schemes.

Figure 6.5 plots the outage probability of the far users versus SNR with different path-loss coefficients of RNRN, NNNF, and NNFF. Several observations are drawn: (i) NNNF achieves the lowest outage probability among the three selection schemes since both the near and far users have the smallest path loss, (ii) NNFF achieves lower outage than RNRF, which indicates that the distance of the near users has more impact than that of the far users, and (iii) it is clear that all of the curves in Figure 6.5 have the same slopes, which indicates that their diversity gains are the same. In the diversity analysis part, we derive the diversity gain of the three selection schemes to be two. The simulation validates the analytical results and indicates that the achievable diversity gain is the same for different user selection schemes. It is demonstrated that the use of SWIPT will not jeopardize the diversity gain of the proposed cooperative NOMA protocol compared to the conventional NOMA.

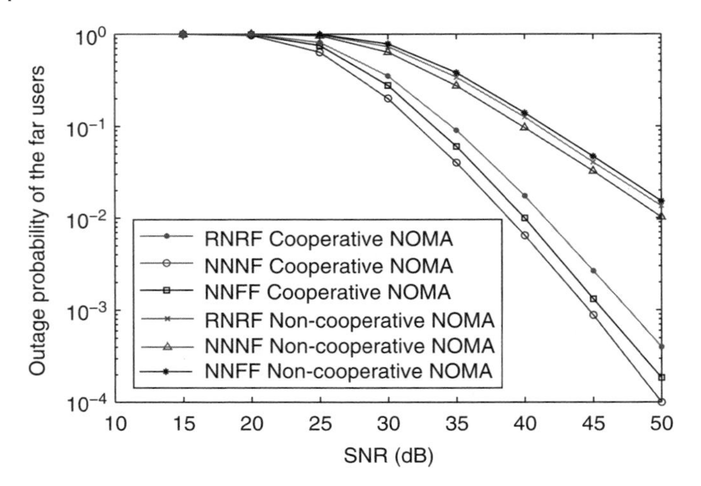

Figure 6.6 Comparison of outage probability with non-cooperative NOMA, $\alpha = 3$, $R_1 = 1$, $R_{D_A} = 10$, $R_{D_B} = 2$, $R_{D_C} = 8$, $\lambda_{\Phi_A} = 1$, and $\lambda_{\Phi_B} = 1$.

Figure 6.6 plots the outage probability of the far users versus SNR for both cooperative NOMA and non-cooperative NOMA. Several observations are drawn: (i) by applying an energy constrained relay to perform cooperative NOMA transmissions, the outage probability of the far users has a larger slope than the non-cooperative NOMA for all of these three user selection schemes. This is due to the fact that the cooperative NOMA can achieve a larger diversity gain and guarantee more reliable reception for the far users, (ii) NNNF achieves the lowest outage probability among these three selection schemes both for cooperative NOMA and non-cooperative NOMA because of its smallest path loss, and (iii) it is worth noting that NNFF has higher outage probability than RNRF in non-cooperative NOMA, but it achieves lower outage probability than RNRF in cooperative NOMA. This phenomenon indicates that it is very helpful and necessary to apply cooperative NOMA in NNFF due to the largest performance gain over non-cooperative NOMA.

6.4.3 Throughput in Delay-Sensitive Transmission Mode

Figure 6.7 plots the delay-sensitive throughput versus SNR with different rates. We use solid marked lines to represent the low rate case with $R_1 = 1, R_2 = 1$ and use dashed marked lines to represent the high rate case with $R_1 = 1, R_2 = 2$. One can observe that NNNF achieves the highest throughput since it has the lowest outage probability among these three selection schemes. The figure also demonstrates that in the low SNR region the low rate case performs better while in the high SNR region the high rate case performs better. This is because the

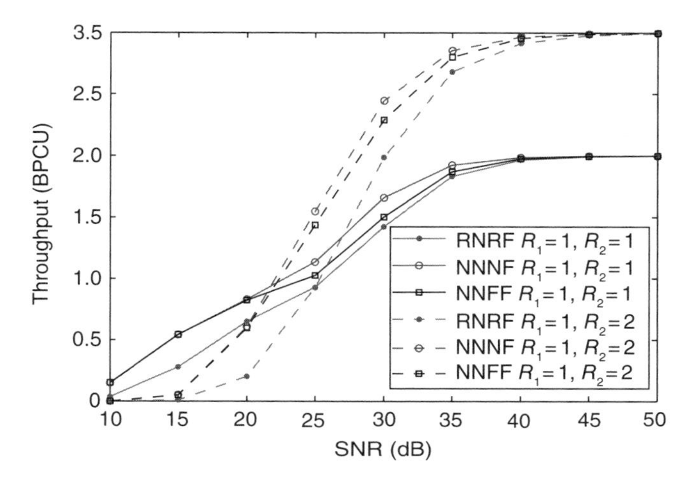

Figure 6.7 Throughput in delay-sensitive mode versus SNR with different rate, $\alpha = 2$, $R_{D_A} = 10$, $R_{D_B} = 2$, $R_{D_C} = 8$, $\lambda_{\Phi_A} = 1$, and $\lambda_{\Phi_B} = 1$.

outage probability plays a more important role in the low SNR region while the data rate plays a more important role in the high SNR region. It is also worth noting that a throughput ceiling exits in the high SNR region. This is due to the fact that the outage probability is approaching zero and the throughput is determined only by the transmission rate.

6.5 Conclusions

In this chapter the application of SWIPT to NOMA has been considered. A novel cooperative SWIPT NOMA protocol with three different user selection criteria has been proposed. We have used the stochastic geometric approach to provide a complete framework to model the locations of users and evaluate the performance of the proposed user selection schemes. Closed-form results have been derived in terms of outage probability and delay-sensitive throughput to determine the system performance. The diversity gain of the three user selection schemes has also been characterized and proved to be the same as that of a conventional cooperative network. For the proposed protocol, the decreasing rate of the outage probability of far users is $\frac{\ln SNR}{SNR^2}$ while it is $\frac{1}{SNR^2}$ for a conventional cooperative network. Numerical results have been presented to validate our analysis. We conclude that by carefully choosing the parameters of the network (e.g., transmission rate or power splitting coefficient), acceptable system performance can be guaranteed even if the users do not use their own batteries to power the relay transmission.

Bibliography

1 Z. Qin, J. Fan, Y. Liu, Y. Gao and G.Y. Li (2018) Sparse representation for wireless communications. *IEEE Signal Process. Mag.*, **36** (3): 40–58.
2 Z. Qin, Y. Gao, M. Plumbley and C. Parini (2016) Wideband spectrum sensing on real-time signals at sub-Nyquist sampling rates in single and cooperative multiple nodes. *IEEE Trans. Signal Process.*, **64** (12): 3106–3117.
3 Z. Ding, Y. Liu, J. Choi, Q. Sun, M. Elkashlan, C.-L. I, and H.V. Poor (2017) Application of non-orthogonal multiple access in LTE and 5G networks. *IEEE Commun. Mag.*, **55** (2): 185–191.
4 Y. Cai, Z. Qin, F. Cui, G.Y. Li, and J.A. McCann (2018) Modulation and multiple access for 5G networks. *IEEE Commun. Surveys & Tutorials*, **20** (1): 629–646.
5 Y. Liu, Z. Qin, M. Elkashlan, Z. Ding, A. Nallanathan, and L. Hanzo (2017) Non-orthogonal multiple access for 5G and beyond. *Proceedings of the IEEE*, **105** (12): 2347–2381.
6 Y. Saito, A. Benjebbour, Y. Kishiyama, and T. Nakamura (2013) System-level performance evaluation of downlink non-orthogonal multiple access (NOMA). In *Proceedings of the IEEE Annual Symposium on Personal, Indoor and Mobile Radio Communications (PIMRC)*, London.
7 Z. Qin, X. Yue, Y. Liu, Z. Ding, and A. Nallanathan (2017) User association and resource allocation in unified non-orthogonal multiple access enabled heterogeneous ultra dense networks. *IEEE Commun. Mag.*, **56** (6): 86–92.
8 Z. Ding, P. Fan, and H.V. Poor (2016) Impact of user pairing on 5Gnon-orthogonal multiple access. *IEEE Trans. Veh. Technol.*, **65** (8): 6010–6023.
9 Z. Ding, Z. Yang, P. Fan, and H.V. Poor (2014) On the performance of non-orthogonal multiple access in 5G systems with randomly deployed users. *IEEE Signal Process. Lett.*, **21** (12): 1501–1505.
10 M. Al-Imari, P. Xiao, M.A. Imran, and R. Tafazolli (2014) Uplink non-orthogonal multiple access for 5G wireless networks. In *Proceedings of the 11th International Symposium on Wireless Communications Systems (ISWCS)*, Barcelona, pp. 781–785.
11 Z. Ding, M. Peng, and H.V. Poor (2015) Cooperative non-orthogonal multiple access in 5G systems. *IEEE Commun. Lett.*, **19** (8): 1462–1465.
12 Y. Liu, Z. Qin, M. Elkashlan, Y. Gao, and L. Hanzo (2017) Enhancing the physical layer security of non-orthogonal multiple access in large-scale networks. *IEEE Trans. Wireless Commun.*, **16** (3): 1656–1672.
13 Y. Liu, Z. Qin, M. Elkashlan, A. Nallanathan, and J.A. McCann (2017) Non-orthogonal multiple access in large-scale heterogeneous networks. *IEEE J. Selected Areas Commun.*, **35** (12): 2667–2680.

14 Y. Liu, Z. Ding, M. Elkashlan, and J. Yuan (2016) Non-orthogonal multiple access in large-scale underlay cognitive radio networks. *IEEE Trans. Veh. Technol.*, **65** (12): 10152–10157.

15 Y. Liu, M. Elkashlan, Z. Ding, and G.K. Karagiannidis (2016) Fairness of user clustering in MIMO non-orthogonal multiple access systems. *IEEE Commun. Lett.*, **20** (7): 1465–1468.

16 L. Varshney (2008) Transporting information and energy simultaneously. In *Proceedings of the IEEE International Symposium on Information Theory (ISIT)*, Toronto, ON, pp. 1612–1616.

17 Z. Ding et al. (2015) Application of smart antenna technologies in simultaneous wireless information and power transfer. In *IEEE Commun. Mag.*, **53** (4): 86–93.

18 Z. Qin, Y. Liu, Y. Gao, M. Elkashlan, and A. Nallanathan (2017) Wireless powered cognitive radio networks with compressive sensing and matrix completion. *IEEE Trans. Commun.*, **65** (4): 1464–1476.

19 R. Zhang and C.K. Ho (2013) MIMO broadcasting for simultaneous wireless information and power transfer. *IEEE Trans. Commun.*, **12** (5); 1989–2001.

20 A.A. Nasir, X. Zhou, S. Durrani, and R.A. Kennedy (2013) Relaying protocols for wireless energy harvesting and information processing. *IEEE Trans. Wireless Commun.*, **12** (7): 3622–3636.

21 I. Krikidis, S. Sasaki, S. Timotheou, and Z. Ding (2014) A low complexity antenna switching for joint wireless information and energy transfer in MIMO relay channels. *IEEE Trans. Commun.*, **62** (5): 1577–1587.

22 Y. Liu, S.A. Mousavifar, Y. Deng, C. Leung, and M. Elkashlan (2016) Wireless energy harvesting in a cognitive relay network. *IEEE Trans. Wireless Commun.*, **15** (4): 2498–2508.

23 Z. Ding, I. Krikidis, B. Sharif, and H.V. Poor (2014) Wireless information and power transfer in cooperative networks with spatially random relays. *IEEE Trans. Wireless Commun.*, **13** (8): 4440–4453.

24 Z. Ding and H. Poor (2013) Cooperative energy harvesting networks with spatially random users. *IEEE Signal Process. Lett.*, **20** (12): 1211–1214.

25 Y. Liu, L. Wang, S. Raza Zaidi, M. Elkashlan, and T. Duong (2016) Secure d2d communication in large-scale cognitive cellular networks: A wireless power transfer model. *IEEE Trans. Commun.*, **64** (1): 329–342.

26 T.M. Cover and J.A. Thomas (1991) *Elements of information theory*, 6th edn. Wiley and Sons, New York.

27 E. Hildebrand (1987) *Introduction to Numerical Analysis*. Dover, New York.

28 J.N. Laneman, D.N. Tse, and G.W. Wornell (2004) Cooperative diversity in wireless networks: Efficient protocols and outage behavior. *IEEE Trans. Inf. Theory*, **50** (12): 3062–3080.

29 Y. Liu, Z. Ding, M. Elkashlan, and H.V. Poor (2016) Cooperative non-orthogonal multiple access with simultaneous wireless information and power transfer. *IEEE J Selected Areas Commun.*, **34** (4): 938–953.

30 I.S. Gradshteyn and I.M. Ryzhik (2000) *Table of Integrals, Series and Products*, 6th edn. Academic Press, New York.

7

Fairness-Aware Wireless Powered Communications with Processing Cost

Zoran Hadzi-Velkov[1], Slavche Pejoski[1], and Nikola Zlatanov[2]*

[1] *Faculty of Electrical Engineering and Information Technologies, Ss. Cyril and Methodius University, Skopje 1000, Macedonia*
[2] *Department of Electrical and Computer Systems Engineering, Monash University, Clayton, VIC 3800, Australia*

7.1 Introduction

Wireless powered communication (WPC) is an essential paradigm for future energy-efficient communication systems, such as the Internet of Things (IoT), [1]–[3]. The WPC promises autonomy and self-sufficiency in terms of energy of emerging low-power systems, such as wireless sensors and IoT devices [4, 5]. As a result, the lifetime of these systems can be extensively prolonged while maintaining their performance comparable to their conventional systems counterparts. The research hype for WPC has been facilitated by recent technological advances in the efficiency of the radio-frequency (RF) energy harvesting (EH) circuits [5], and of the energy storage circuits with rapid recharging capabilities [6, 7]. However, the RF EH circuitry is far from perfect. On one hand, the circuitry consumes additional power on top of the radiated power at the RF antenna, [8, 9]. On the other hand, the efficiency curve of the RF-to-direct power conversion is not linear [10, 11], resulting in a nonlinear end-to-end wireless power transfer.

Background. An important WPC research direction is the design and optimization of wireless networks with RF EH, also referred to as *wireless powered communication networks* (WPCNs) [12, 13]. Typically, a WPCN is composed of a RF power source, a base station (BS), and multiple EH users (EHUs). Different protocols are devised for maximizing the sum rate of the network subject to various power constraints at the RF power source [14]. However, the system design based upon the sum rate maximization may lead to extremely unfair

*Corresponding author: Zoran Hadzi-Velkov; zoranhv@feit.ukim.edu.mk

Wireless Information and Power Transfer: Theory and Practice, First Edition.
Edited by Derrick Wing Kwan Ng, Trung Q. Duong, Caijun Zhong, and Robert Schober.

resource sharing among the EHUs due to the *double near-far effect* [12]. When the BS and the RF power source are collocated (as a single hybrid BS), the EHUs close to the BS receive more energy from the BS due to the better average channel quality, and as a result are able to achieve higher data rates compared to EHUs that are more distant to the BS. In the conventional (i.e., non-EH) cellular networks, the resource allocation mechanisms typically guarantee fair utilization of the system resources (time and/or bandwidth) by applying criteria such as max–min, harmonic fairness, and *proportional fairness* (PF) [15]–[18]. Here we apply the PF criterion [16] in order to achieve rate fairness among the WPCN nodes.

Apart from the fairness requirements, the optimization of WPCN is fundamentally affected by the harvested energy storage strategy employed at the EHUs. Specifically, depending on the type of energy storage, EHUs may employ either short-term energy storage or long-term energy storage [6, 7]. The short-term energy storage strategy implies that, regardless of the fading state, the EHU transmits information by completely depleting its energy storage, thus giving rise to the *harvest-then-transmit* policy [12]. The long-term storage strategy implies that, depending on the fading state, an EHU transmits information by spending only a part of its stored energy while saving the rest for future use [21].

In this chapter, we tackle the issue of fair resource allocation in WPCNs for both types of energy storage strategies [22]–[24]. For the purpose of analytical tractability, we consider only the additional power consumed by the circuitry and assume linear RF-to-direct power conversion. Specifically, we devise policies for optimal power allocation at the BS and optimal time sharing among the EHUs, such that the resource sharing is proportionally fair. Both proposed policies satisfy a constraint for an average power consumption at the BS, which necessitates imposing silence at the BS in the extremely poor channel conditions, and therefore facilitates high energy efficiency of the WPCN. Unlike related works [18], our proposed power and time allocations for fairness-aware WPCNs are provided in closed forms, and the practical implementation is computationally simple.

7.2 System Model

We consider a WPCN with a single BS and K EHUs that operate in a wireless fading environment (Figure 7.1). Each network node is equipped with a single antenna and operates in a half-duplex mode. The fading channel between the BS and EHU k ($1 \leq k \leq K$) is assumed to be a stationary and ergodic random process that follows the quasi-static block fading model, i.e. the channel is constant during a single fading epoch, but changes independently from one epoch to the next. The duration of the fading epochs is assumed to be fixed to T. The channel

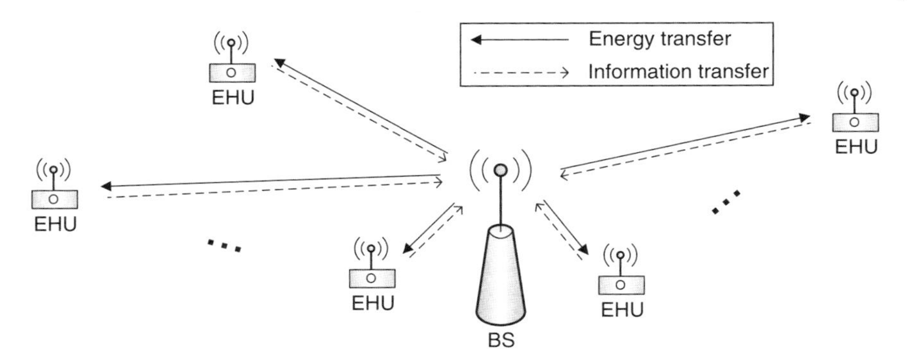

Figure 7.1 Wireless powered communication network with downlink RF energy broadcast and uplink information transmission.

between the BS and kth EHU is assumed to be reciprocal and its fading power gain in epoch i is denoted by $x'_k(i)$. For convenience, the fading power gains are normalized with respect to the power of the additive white Gaussian noise (AWGN), denoted by N_0, such that $x_k(i) = x'_k(i)/N_0$. The average value of $x_k(i)$ is denoted by Ω_k and given by $\Omega_k = E[x_k(i)] = E[x'_k(i)]/N_0$, where $E[\cdot]$ denotes expectation. The proposed protocols necessitate the following requirements for the channel state information (CSI) available at the network nodes. In each fading epoch, the BS should have instantaneous CSI of all fading links, whereas each EHU should know only its own fading channel.

We assume that the transmission time is divided into M orthogonal epochs, where $M \to \infty$. Each epoch contains a single TDMA frame, which is divided into $K + 1$ phases: an EH phase and K successive information transmission (IT) phases. The EH phase is the time period during which the BS broadcasts RF energy to the EHUs, and the IT phase of the kth EHU is the time period during which it transmits information to the BS. The durations of the EH and the IT phases are dynamically adjusted in each epoch. The duration of the EH phase in epoch i is denoted by $\tau_0(i)T$, and the duration of the IT phase of the kth EHU is denoted by $\tau_k(i)T$, where $\tau_k(i)$ are the time-sharing parameters that satisfy

$$\tau_0(i) + \sum_{k=1}^{K} \tau_k(i) = 1, \tag{7.1}$$

where $0 \leq \tau_k(i) \leq 1$. During the EH phase of epoch i, the BS broadcasts RF energy with an output power $p_0(i)$, which satisfies constraints on both the maximum power, P_{max}, and the average power, P_{avg}, given by $0 \leq p_0(i) \leq P_{max}$ and $E[p_0(i)\tau_0(i)] \leq P_{avg}$, respectively.

7.2.1 Energy Storage Strategies

During the EH phase, each EHU harvests the radiated RF energy from the BS and stores it in the EHU's energy storage unit. Depending on the method for the spending of energy by the EHU during its IT, we differentiate between the *short-term energy storage* (STES) strategy and the *long-term energy storage* (LTES) strategy. Each strategy necessitates a different type of energy storage unit at the EHUs.

When employing the STES strategy, whenever an EHU transmits information in a given epoch it spends the total amount of energy available in its energy storage for that transmission. In this case, an EHU harvests energy during a short period of time before completely depleting its stored energy. The energy storage employed at the EHU for this type of operation should be fully recharged during the EH phase, but also should be capable of rapid consumption of its stored energy during the IT phase. The energy storage with such properties applies to the *supercapacitor*, which has high recharge rate yet relatively low storage capacity [6, 7]. The WPCN whose EHUs store their energy in supercapacitors typically employ the harvest-then-transmit protocol [12].

When employing the LTES strategy, whenever an EHU transmits information in a given epoch, it spends either a part or the total amount of energy available in its storage, depending on the channel fading conditions. Thus, EHU typically harvests energy over multiple epochs before transmitting information in favorable channel conditions. The energy storage that facilitates such a mode of operation should have high storage capacity, such as *rechargeable batteries*. Examples of rechargeable batteries include Li-ion batteries and thin-film batteries [6, 7].

7.2.2 Circuit Power Consumption

In a practical EH transmitter, besides the RF power radiated from the RF antenna, an additional power is consumed by the non-ideal electric circuitry, due to the AC/DC converters, the analog RF amplifiers, and the processing units. Hence, apart from the radiated RF power, a realistic power consumption model should also incorporate the circuit power consumption at the EHU. To this end, we employ a similar model to [9] and [19], where the total power consumed by the kth EHU in epoch i is expressed as

$$p_{T,k}(i) = \begin{cases} P_k(i) + p_c, & \text{if } P_k(i) > 0 \\ 0, & \text{if } P_k(i) = 0, \end{cases} \tag{7.2}$$

where $P_k(i)$ is the power of the transmitted signal by the kth EHU. The term p_c in (7.2) denotes the constant circuit power consumed during the EHU's operation, which is referred to as the *processing cost* [9]. In this work, the processing cost is assumed to be fixed and equal for all EHUs. From (7.2), $P_k(i)$ can be expressed as a function of the total consumed power $p_{T,k}(i)$ as $P_k(i) = (p_{T,k}(i) - pc)^+$, where $(x)^+ = \max\{0, x\}$.

7.3 Proportionally Fair Resource Allocation

Depending on the EH strategy employed at the EHUs, we consider two resource allocation schemes that provide proportional fairness. Each scheme implies an optimal power allocation at the BS, $p_0(i)$, and an optimal duration of the EH and IT phases, $\tau_k(i)$, $0 \le k \le K$, in order to maximize the sum of the logarithmic achievable rates of all EHUs, given by

$$\sum_{k=1}^{K} \log \overline{R}_k, \tag{7.3}$$

where the achievable rate of the kth EHU, $\overline{R}_k$, is given by

$$\overline{R}_k = \lim_{M\to\infty} \frac{1}{M} \sum_{i=1}^{M} \tau_k(i) \log(1 + P_k(i)x_k(i)). \tag{7.4}$$

As a measure for the level of system fairness, we use the Jain's fairness index, J, defined by [20]

$$J = \frac{\left(\sum_{k=1}^{K} \overline{R}_k\right)^2}{K \sum_{k=1}^{K} \overline{R}_k^2}. \tag{7.5}$$

7.3.1 Short-term Energy Storage Strategy

In this subsection, we consider the STES strategy and the harvest-then-transmit protocol. For the purpose of computational simplicity of the proposed policies, we assume a linear efficiency curve for the RF-to-direct power conversion at the EHUs. For the kth EHU, the conversion efficiency coefficient is denoted by η_k $(0 < \eta_k < 1)$, which determines the fraction of the incident RF power that is stored in the energy storage. In this case, the amount of energy harvested by the kth EHU during the EH phase of epoch i is determined by

$$E_k(i) = \eta_k x_k(i) N_0 p_0(i) \tau_0(i) T. \tag{7.6}$$

Based upon (7.6), we can obtain the power harvested by this EHU in epoch i as $E_k(i)/(\tau_k(i)T)$. If the harvested power by an EHU is less than p_c, it is completely depleted by the processing units during the IT phase, thus averting transmission of the useful signal. Thereby, the transmit power of the kth EHU during its IT phase is compactly written as

$$P_k(i) = \left(\frac{E_k(i)}{\tau_k(i)T} - p_c\right)^+. \tag{7.7}$$

Thus, the optimal power allocation at the BS, $p_0(i)$, and the optimal time sharing among the EHUs, $\tau_k(i), \forall k$, is devised by solving the following maximization problem for $M \to \infty$,

$$\begin{aligned}
\underset{\tau_k(i),\tau_0(i),p_0(i)}{\text{maximize}} \quad & \sum_{k=1}^{K} \log\left(\frac{1}{M}\sum\nolimits_{i=1}^{M} \tau_k(i)\log(1 + x_k(i)P_k(i))\right) \\
\text{s.t.} \quad & C1 : P_k(i) = \left(\frac{E_k(i)}{\tau_k(i)T} - p_c\right)^{+}, \forall i, \ \ \forall k, \\
& C2 : \frac{1}{M}\sum\nolimits_{i=1}^{M} p_0(i)\tau_0(i) \le P_{avg}, \\
& C3 : 0 \le p_0(i) \le P_{max}, \forall i, \\
& C4 : \sum\nolimits_{k=0}^{K} \tau_k(i) \le 1, \forall i, \\
& C5 : \tau_0(i) > 0, \ \ \tau_k(i) > 0, \forall i, \forall k,
\end{aligned} \tag{7.8}$$

where $C1$ represents the STES strategy, given by (7.7), $C2$ and $C3$ are due to the average and the maximum power constraints at the BS, respectively, and $C4$ is due to (7.1). The solution of (7.8) is given by the following theorem.

Theorem 7.1 The optimal transmit power of the BS is given by

$$p_0^*(i) = \begin{cases} P_{max}, & \text{if } \sum_{k=1}^{K} \frac{1}{\overline{R}_k} \frac{N_0 \eta_k x_k^2(i)}{1 - p_c x_k(i) + z_k(i)} > \lambda_0 \\ 0, & \text{otherwise.} \end{cases} \tag{7.9}$$

The optimal durations of EH and IT phases are determined by

$$\tau_0^*(i) = \frac{1}{1 + \sum_{k=1}^{K} \frac{\eta_k x_k^2(i) N_0 P_{max}}{z_k(i)}}, \tag{7.10}$$

$$\tau_k^*(i) = \frac{\eta_k x_k^2(i) N_0 P_{max} \tau_0^*(i)}{z_k(i)}, \quad 1 \le k \le K, \tag{7.11}$$

respectively. In (7.9), (7.10), and (7.11), $z_k(i)$ is an auxiliary variable, determined by

$$z_k(i) = -(1 - p_c x_k(i)) \left[1 + \frac{1}{W(-(1 - p_c x_k(i))\, e^{-1-\beta(i)P_{max}\overline{R}_k})}\right], \tag{7.12}$$

where $W(\cdot)$ is the Lambert W function, and the parameter $\beta(i)$ is found as the root of the following transcendental equation

$$\lambda_0 + \beta(i) + \sum_{k=1}^{K} \frac{N_0 \eta_k x_k^2(i)}{\overline{R}_k (1 - p_c x_k(i))} W(-(1 - p_c x_k(i))\, e^{-1-\beta(i)P_{max}\overline{R}_k}) = 0. \tag{7.13}$$

The constant λ_0 is selected such that $C2$ in (7.8) is satisfied with equality.

Proof: Please see the appendix of [23].

7.3.2 Long-term Energy Storage Strategy

In order to tackle the LTES strategy, we apply the framework devised in [21], which leads to asymptotically optimal solutions to the corresponding optimization problem (i.e., as $M \to \infty$). In this case, it is necessary to define the notion of a desired transmit power of kth EHU in epoch i, $P_{d,k}(i)$, which may be different than the actual transmit power of the kth EHU, i.e. $P_k(i) \leq P_{d,k}(i)$. Let $B_k(i-1)$ denote the available energy in the battery of the kth EHU at the beginning of epoch i, and let the storage capacity of the battery be unlimited (i.e., $B_{max} \to \infty$). Therefore, the amount of energy available after the EH phase at the kth EHU is found as $B_k(i-1) + E_k(i)$, where $E_k(i)$ is given by (7.6). Considering the power model (7.2), in order for the kth EHU to transmit a codeword with power $P_{d,k}(i)$, the actual extracted power from this EHU's battery should be $P_{d,k}(i) + p_c$ if $P_{d,k}(i) > 0$, and zero if $P_{d,k}(i) = 0$. However, this amount of power may not necessarily be available in the battery. In this case, the kth EHU transmits its codeword with the power actually available in the battery, $(B_k(i-1) + E_k(i))/(\tau_k(i)T)$, reduced by p_c. Thereby, the actual transmit power of the kth EHU can be compactly written as

$$P_k(i) = \left(\min\left\{\frac{B_k(i-1) + E_k(i)}{\tau_k(i)T} - p_c, P_{d,k}(i)\right\}\right)^+. \tag{7.14}$$

Given the harvested energy, $E_k(i)$, and consumed power, $P_k(i) + p_c$, the energy available in the battery of the kth EHU at the end of epoch i is given by

$$B_k(i) = B_k(i-1) + E_k(i) - (P_k(i) + p_c)I_k(i)\tau_k(i)T, \tag{7.15}$$

where $E_k(i)$ is given by (7.6) and $I_k(i)$ is an indicator function, defined by

$$I_k(i) = \begin{cases} 1, & \text{if } P_k(i) > 0 \\ 0, & \text{if } P_k(i) = 0, \end{cases} \tag{7.16}$$

which makes sure that the kth EHU consumes energy only when it is transmitting (c.f. (7.2)).

The optimal allocations for $\tau_k(i)$, $P_{d,k}(i)$, $\tau_0(i)$, and $p_0(i)$ are obtained as the solution of the following maximization problem, which guarantees proportional fairness in WPCNs employing the LTES strategy:

$$\begin{aligned} \underset{\tau_k(i),\tau_0(i),p_0(i),P_{d,k}(i)}{\text{maximize}} \quad & \sum_{k=1}^{K} \log\left(\frac{1}{M}\sum\nolimits_{i=1}^{M} \tau_k(i)\log(1 + x_k(i)P_k(i))\right) \\ \text{s.t.} \quad C1: \; & B_k(i) = B_k(i-1) + E_k(i) - (P_k(i) + p_c)I_k(i)\tau_k(i)T, \forall i, \forall k, \end{aligned}$$

$$
\begin{aligned}
&C2: P_k(i)=\left(\min\left\{\frac{B_k(i-1)+E_k(i)}{\tau_k(i)T}-p_c, P_{d,k}(i)\right\}\right)^+, \forall i, \forall k,\\
&C3: P_{d,k}(i)\geq 0, \forall i, \forall k,\\
&C4: \frac{1}{M}\sum\nolimits_{i=1}^{M} p_0(i)\tau_0(i)\leq P_{avg},\\
&C5: 0\leq p_0(i)\leq P_{max}, \forall i,\\
&C6: \sum_{k=0}^{K}\tau_k(i)\leq 1, \forall i,\\
&C7: \tau_0(i)\geq 0,\ \tau_k(i)\geq 0, \forall i, \forall k.
\end{aligned}
\tag{7.17}
$$

In (7.17), $C1$ and $C2$ are due to the LTES strategy, i.e. (7.15) and (7.14), respectively, whereas $C4$ and $C5$ are due to the average and maximum power constraints at the BS, respectively. In order to be able to tackle (7.17) analytically, we relax $C1$ into $\overline{C}1$ by omitting the indicator function $I_k(i)$, which yields the following optimization problem:

$$
\begin{aligned}
\underset{\tau_k(i),\tau_0(i),p_0(i),P_{d,k}(i)}{\text{maximize}} \quad & \sum_{k=1}^{K}\log\left(\frac{1}{M}\sum_{i=1}^{M}\tau_k(i)\log(1+x_k(i)P_k(i))\right)\\
\text{s.t.} \quad & \overline{C}1: B_k(i)=B_k(i-1)+E_k(i)-(P_k(i)+p_c)\tau_k(i)T, \forall i, \forall k,\\
& C2,\ C3,\ C4,\ C5,\ C6, \text{ and } C7 \text{ as in (7.17)}.
\end{aligned}
\tag{7.18}
$$

Note that the proposed optimal allocations in the forthcoming Theorem 7.2 satisfy $\tau_k^*(i)I_k(i)=\tau_k^*(i)$, in which case $C1$ and $\overline{C}1$ are identical, and the solution of (7.17) is therefore identical to the solution of (7.18).

For finite M, solving (7.18) is very difficult and may require non-causal CSI knowledge. However, based upon [21], (7.18) can be simplified significantly when $M\to\infty$, as shown by the following lemma.

Lemma 7.1 For $M\to\infty$ and $B_{max}\to\infty$, (7.18) is equivalently written as

$$
\begin{aligned}
\underset{\tau_k(i),\tau_0(i),p_0(i),P_{d,k}(i)}{\text{maximize}} \quad & \sum_{k=1}^{K}\log\left(\frac{1}{M}\sum_{i=1}^{M}\tau_k(i)\log(1+P_{d,k}(i)x_k(i))\right)\\
\text{s.t.} \quad & \overline{C}2: \frac{1}{M}\sum\nolimits_{i=1}^{M}(P_{d,k}(i)+p_c)\tau_k(i)\\
& \leq \frac{1}{M}\sum\nolimits_{i=1}^{M}N_0\eta_k x_k(i)p_0(i)\tau_0(i), \forall k,\\
& C3,\ C4,\ C5,\ C6, \text{ and } C7 \text{ as in (7.17)}.
\end{aligned}
\tag{7.19}
$$

Proof: In [21], it is proven that, given $M\to\infty$, if an EHU with unlimited energy storage capacity has a desired power allocation policy for which the average harvested energy in its battery is larger than or equal to the average extracted energy from its battery, then this EHU can transmit with its desired transmit power in almost all epochs. In this case, the number of epochs in which the battery cannot supply its EHU with its desired output power is negligible compared to the number of epochs in which the battery can provide for the desired output power as $M \to \infty$. Moreover, if we add constraint $\overline{C2}$ in (7.19) as an additional constraint into the optimization problem (7.18), we will find that the feasible power allocation policies of this new optimization problem are only those for which the average harvested energy in the battery of each EHU is larger than or equal to the average depleted energy from that battery. Thus, constraint $C2$ in (7.18) is reduced to $P_k(i) = P_{d,k}(i)$, since it is satisfied in almost all epochs. Consequently, the constraint $\overline{C1}$ in (7.18) can be removed, yielding (7.19). The solution of (7.19) is given by the following theorem.

Theorem 7.2 Optimally, the TDMA frame in a given epoch i should contain either an EH phase ($\tau_0^*(i) = 1$) or an IT phase ($\tau_0^*(i) = 0$), where

$$\tau_0^*(i) = \begin{cases} 1, \text{ if } \lambda_0 < \sum_{k=1}^{K} N_0 \eta_k x_k(i)\ \lambda_k / \overline{R}_k \\ 0, \textit{ otherwise.} \end{cases} \tag{7.20}$$

The optimal BS transmit power is given by

$$p_0^*(i) = \begin{cases} P_{max}, & \text{if } \lambda_0 < \sum_{k=1}^{K} N_0 \eta_k x_k(i)\ \lambda_k / \overline{R}_k \\ 0, & \textit{otherwise.} \end{cases} \tag{7.21}$$

If $\tau_0^*(i) = 0$, only a single EHU with an index s_i $(1 \le s_i \le K)$ is scheduled to transmit information during the IT phase. The optimal time-sharing parameter for the kth EHU $(1 \le k \le K)$ is thus determined by

$$\tau_k^*(i) = \begin{cases} 1, \text{ if } k = s_i^* \text{and } \lambda_0 \ge \sum_{k=1}^{K} N_0 \eta_k x_k(i)\ \lambda_k / \overline{R}_k \\ 0, \textit{ otherwise.} \end{cases} \tag{7.22}$$

In (7.22), the index of the optimally scheduled EHU is determined by

$$s_i^* = \arg\max_{k \in \mathcal{D}(i)} \frac{1}{\overline{R}_k} \left(\log(1 + P_{d,k}(i) x_k(i)) - \frac{P_{d,k}(i) x_k(i)}{1 + P_{d,k}(i) x_k(i)} - \lambda_k p_c \right), \tag{7.23}$$

where $P_{d,k}(i)$ is the desired transmit power of the kth EHU, determined by

$$P_{d,k}(i) = \begin{cases} \frac{1}{\lambda_k} - \frac{1}{x_k(i)}, & \text{if } x_k(i) > a_k \lambda_k \\ 0, & \text{otherwise,} \end{cases} \tag{7.24}$$

whereas $\mathcal{D}(i)$ is the subset from which the scheduled EHU is selected,

$$\mathcal{D}(i) = \{1 \le k \le K : x_k(i) > a_k \lambda_k\} . \tag{7.25}$$

Note that if $\mathcal{D}(i)$ is an empty set in epoch i, then none of the EHUs is scheduled in that epoch. When the EHU with index s_i is scheduled for transmission, its desired transmit power is given by $P^*_{d,s_i}(i) = (1/\lambda_{s_i} - 1/x_{s_i}(i))^+$, whereas the rest of the EHUs are silent. In (7.24) and (7.25), the parameter a_k is the non-negative root of the transcendental equation

$$\log(a_k) + \frac{1}{a_k} = 1 + \lambda_k p_c, \tag{7.26}$$

such that $a_k > 1$. The constants $\{\lambda_k\}_{k=1}^K$ and λ_0 are determined by solving the equation set of constraints $\overline{C}2$ and $C4$ in (7.19), with "$\leq$" replaced by "$=$".

Proof: Please refer to Section 7.5.1 of the Appendix.

Note that the value of a_k, determined by (7.26), guarantees that the argument of (7.23) is positive for any $k \in \mathcal{D}(i)$. If $p_c = 0$, then $a_k = 1$, and (7.24) reduces to the conventional "water-filling" power allocation.

7.3.3 Practical Online Implementation

In practice, the values of the constants λ, λ_k, and $\overline{R}_k$ that appear in Theorem 7.1 and Theorem 7.2 may not be available in advance. For an online estimation of λ values, we apply the stochastic gradient descent method ([26], Sec. III.C), and therefore the constant λ_0 in Theorem 7.1 and Theorem 7.2 is estimated as

$$\hat{\lambda}_0(i) = \hat{\lambda}_0(i-1) + \beta_0 \left(\frac{1}{i-1} \sum_{n=1}^{i-1} p_0(n)\tau_0(n) - P_{avg} \right), \tag{7.27}$$

whereas the constants λ_k of Theorem 7.2 are estimated as

$$\hat{\lambda}_k(i) = \hat{\lambda}_k(i-1) + \frac{\beta_k}{i-1} \sum_{n=1}^{i-1} (P_{d,k}(n) + p_c)\tau_k(n) - N_0\eta_k\, x_k(n)p_0(n)\tau_0(n), \tag{7.28}$$

where β_0 and β_k are some small step sizes, and averages are estimated from the previous $i-1$ epochs. The rate $\overline{R}_k$ can be estimated according to a simple iterative rule

$$\hat{R}_k(i) = \frac{i-1}{i}\, \hat{R}_k(i-1) + \frac{1}{i}\, r_k(i). \tag{7.29}$$

Actually, each iteration of $\hat{R}_k(i)$ is based upon an ever-increasing window size, equal to the elapsed session time i, which guarantees the maximization of $\sum_{k=1}^K \log \overline{R}_k$, ([15], Lemma 4). The algorithmic implementation of the policy proposed by Theorem 7.1 is available in [23].

7.3.4 Numerical Results

We now illustrate the effect of the two proposed policies over the performance of a WPCN with employed STES strategy and a WPCN with employed LTES strategy. We consider a WPCN that operates in a Rayleigh fading environment with a predefined path loss of 30 dB at a reference distance of 1 m and a path-loss exponent $\alpha = 3$. The average channel gains are thus determined by $E[x'_k(i)] = 10^{-3}D_k^{-\alpha}$, where D_k is the distance between the BS and the kth EHU. The EHUs are located at different distances around the BS in order to verify if the proposed policies assure proportional fairness. The maximum and the average BS transmit power satisfy $P_{max} = 5P_{avg}$, whereas the AWGN power is fixed to $N_0 = 10^{-12}$W. The sum rate over the uplink is calculated as $\sum_{k=1}^{K} \overline{R}_k$, and the system fairness is expressed in terms of the Jain's fairness index, calculated by (7.5). The constants λ_0, λ_k, and the rates $\overline{R}_k$ are iteratively estimated according to (7.27), (7.28), and (7.29), respectively.

Figure 7.2 depicts the sum rate vs. P_{avg} for negligible ($p_c = 0$) and non-negligible processing cost ($p_c = 10^{-5}$W), and $K = 3$. The sum rate steadily increases as P_{avg} increases from 1 W to 10 W. These sum rates are compared against the sum rate of a benchmark protocol studied in [12], which maximizes the uplink sum rate in each epoch by adjusting only the time-sharing parameters, $\tau_k(i)$, whereas the BS transmit power is fixed to P_0 in all epochs, such that $E[P_0\tau_0(i)] = P_{avg}$. Note that the benchmark scheme actually utilizes the STES strategy and is applicable only to the case when $p_c = 0$, but not when $p_c > 0$. For $p_c = 0$, the sum rates of the proposed policies

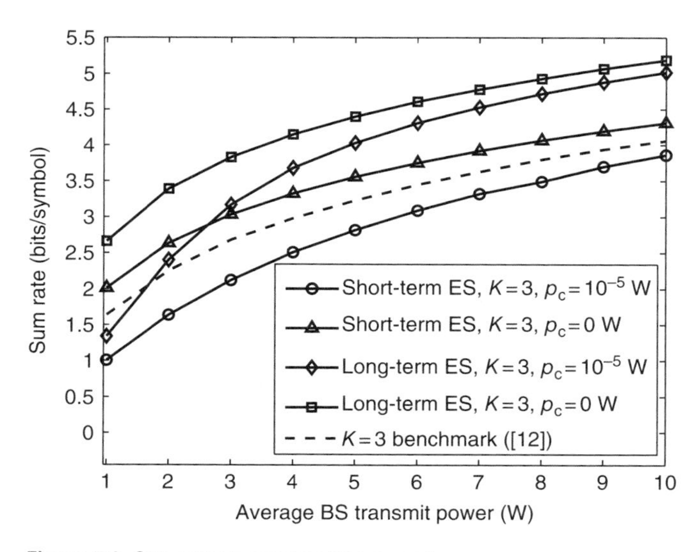

Figure 7.2 Sum rate vs. average BS transmit power.

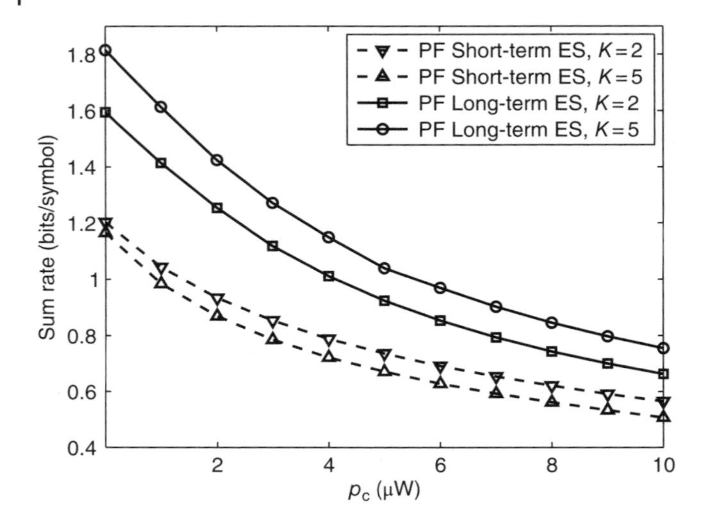

Figure 7.3 Sum rate vs. processing cost.

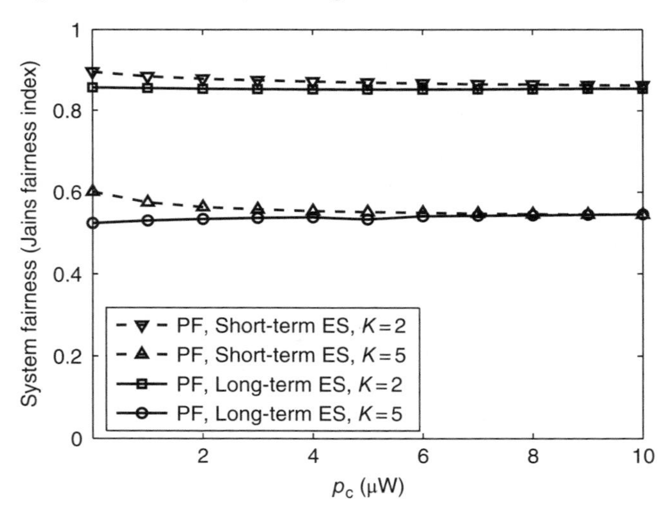

Figure 7.4 System fairness vs. processing cost.

exceed that of the benchmark protocol, but their sum rates decrease as the processing cost attains a realistic value of $p_c = 10^{-5}$ W. Moreover, the policy employing STES performs worse than the benchmark curve.

Figures 7.3 and 7.4 illustrate the impact of the processing cost, p_c, over the sum rate and the system fairness, when employing both the STES and the LTES strategies. For the setup of $K = 2$, the EHU distances from the BS are set as $D_1 = 10$ m and $D_2 = 19$ m, respectively. For the setup of $K = 5$, the EHU distances from the BS are set as $D_1 = 10$ m, $D_2 = 12.5$ m, $D_3 = 15$ m, $D_4 = 17$ m, and $D_5 = 19$ m, respectively. For both proposed policies, as p_c increases, the sum rates steadily decrease. Given p_c, the sum rate of the WPCN under the LTES strategy exceeds that of the WPCN under STES strategy because, depending on the channel fading state, the long-term EHUs control how much power is spent during the IT phase. The performance mismatch is more pronounced for lower p_c, but diminishes as p_c increases. On the other hand, both policies lead to nearly the same level of system fairness, which is kept nearly independent of p_c (Figure 7.4).

7.4 Conclusion

In this chapter we studied WPCS with non-negligible circuit power consumption at the EH nodes. We developed the optimal power and time allocations for WPCNs employing both STES and LTES strategies, which tackle the double near-far problem by providing proportionally fair resource sharing in these systems. It turns out that the processing cost has a significant effect on the EHU achievable rates and the sum rate, but only a minor effect on the system fairness. The two proposed operational policies are prone to simple online implementations, and therefore are suitable for the emerging fairness-aware WPCNs.

Acknowledgements. This work was supported in part by the Ministry of Education and Science of Macedonia under Grant 16-4644.

7.5 Appendix

7.5.1 Proof of Theorem 7.2

After introducing the change of variables $e(i) = p_0(i)\tau_0(i)$ and $p_{d,k}(i) = P_{d,k}(i)\tau_k(i)$, (7.19) is transformed as

$$\begin{aligned}
\underset{\tau_k(i),\tau_0(i),e(i),p_{d,k}(i)}{\text{maximize}} \quad & \sum_{k=1}^{K} \log\left(\frac{1}{M}\sum_{i=1}^{M}\tau_k(i)\log\left(1+\frac{x_k(i)p_{d,k}(i)}{\tau_k(i)}\right)\right) \\
\text{s.t.} \quad & \overline{C}2: \frac{1}{M}\sum\nolimits_{i=1}^{M} p_{d,k}(i) + p_c\tau_k(i) \leq \frac{1}{M}\sum\nolimits_{i=1}^{M} N_0\eta_k x_k(i)e(i), \forall k, \\
& C3: p_{d,k}(i) \geq 0, \forall i, \forall k, \\
& C4: \frac{1}{M}\sum\nolimits_{i=1}^{M} e(i) \leq P_{avg}, \\
& C5: 0 \leq e(i) \leq P_{max}\tau_0(i), \forall i, \\
& C6: \sum\nolimits_{k=1}^{K} \tau_k(i) \leq 1-\tau_0(i), \forall i, \\
& C7: \tau_0(i) \geq 0,\ \tau_k(i) \geq 0,\ \forall i, \forall k.
\end{aligned} \tag{7.30}$$

The objective function of (7.30) is jointly concave in $p_{d,k}(i)$ and $\tau_k(i)$, since the ith summation term is the perspective of the function $\log(1 + p_{d,k}(i)x_k(i))$. Furthermore, all the constraints are affine in $p_{d,k}(i)$ and $\tau_k(i), \forall k, \forall i$. Therefore, (7.30) is a convex optimization problem and can be solved by applying the Lagrangian dual method. Its Lagrangian is given by

$$\begin{aligned}
\mathcal{L} = &\sum_{k=1}^{K} \log\left(\frac{1}{M}\sum_{i=1}^{M} \tau_k(i)\log\left(1 + \frac{x_k(i)p_{d,k}(i)}{\tau_k(i)}\right)\right) + \sum_{k=1}^{K}\sum_{i=1}^{M} \delta_k(i)p_{d,k}(i) \\
&- \lambda_0\left(\frac{1}{M}\sum_{i=1}^{M} e(i) - P_{avg}\right) - \sum_{k=1}^{K} \lambda_k \sum_{i=1}^{M}(p_{d,k}(i) + p_c\tau_k(i) - N_0\eta_k x_k(i)e(i)) \\
&+ \sum_{i=1}^{M} \alpha(i)e(i) - \sum_{i=1}^{M} \beta(i)(e(i) - P_{max}\tau_0(i)) \\
&- \sum_{i=1}^{M} \mu(i)\left(\sum_{k=0}^{K} \tau_k(i) - 1\right) + \sum_{k=0}^{K}\sum_{i=1}^{M} \nu_k(i)\tau_k(i).
\end{aligned} \tag{7.31}$$

In (7.31), the Lagrange multipliers λ_k are associated with $\overline{C}2$, $\delta_k(i)$ are associated with $C3$, λ_0 is associated with $C4$, $\alpha(i)$, and $\beta(i)$ are associated with the left-hand side and right-hand side of $C5$, respectively. $\mu(i)$ are associated with $C6$ and $\nu_k(i)$ are associated with $C7$. The Lagrange multipliers are non-negative and according to the KKT conditions should satisfy the following complementary slackness conditions [25]:

$$\begin{aligned}
\alpha(i)e(i) =& \beta(i)(e(i) - P_{max}\tau_0(i)) = \delta_k(i)p_{d,k}(i) \\
=& \nu_k(i)\tau_k(i) = \mu(i)\left(\sum_{k=0}^{K} \tau_k(i) - 1\right) = 0.
\end{aligned} \tag{7.32}$$

The derivatives of (7.31) with respect to $e(i)$, $p_{d,k}(i)$, $\tau_k(i)$, and $\tau_0(i)$ are set equal to zero, which respectively yields

$$-\lambda_0 + \sum_{k=1}^{K} \lambda_k N_0\eta_k x_k(i) + \alpha(i) - \beta(i) = 0, \tag{7.33}$$

$$\frac{1}{\overline{R}_k}\frac{x_k(i)}{1 + \frac{p_{d,k}(i)x_k(i)}{\tau_k(i)}} - \lambda_k + \delta_k(i) = 0, \tag{7.34}$$

$$\frac{1}{\overline{R}_k}\left[\log\left(1 + \frac{p_{d,k}(i)x_k(i)}{\tau_k(i)}\right) - \frac{p_{d,k}(i)x_k(i)}{p_{d,k}(i)x_k(i) + \tau_k(i)}\right] - \lambda_k p_c - \mu(i) + \nu_k(i) = 0, \tag{7.35}$$

$$P_{max}\beta(i) - \mu(i) + \nu_0(i) = 0. \tag{7.36}$$

Case 1: Let $e(i) = P_{max}\tau_0(i)$, i.e. the BS is assumed to broadcast RF energy. In this case, due to the complementary slackness conditions (7.32), $\alpha(i) = 0$, $\beta(i) > 0$. From (7.33), we thus obtain the condition for the occurrence of this case, as $\beta(i) = \sum_{k=1}^{K} \lambda_k N_0 \eta_k x_k(i) - \lambda_0 > 0$. Since the BS is active, $\tau_0(i) > 0$, and, therefore, according to (7.32), $\nu_0(i) = 0$. From (7.36), we thus obtain $\mu(i) = P_{max}\beta(i) > 0$, since $\beta(i) > 0$. Using (7.35), we obtain

$$0 < \left(\sum_{k=1}^{K} \lambda_k N_0 \eta_k x_k(i) - \lambda_0\right) P_{max} = \nu_k(i) - \lambda_k p_c + \frac{1}{\overline{R}_k}\left[\log\left(1 + \frac{p_{d,k}(i)x_k(i)}{\tau_k(i)}\right) - \frac{p_{d,k}(i)x_k(i)}{p_{d,k}(i)x_k(i) + \tau_k(i)}\right]. \tag{7.37}$$

Note, $\nu_k(i)$ must be adjusted such that (7.37) is satisfied for each k $(1 \leq k \leq K)$. Therefore, $\nu_k(i) > 0, \forall k$, which implies $\tau_k(i) = 0, \forall k$. As a result, all EHUs must be silent in this case, i.e. $P_{d,k}(i) = 0$.

Case 2: Let $e(i) = 0$, i.e. the BS is assumed to be silent. In this case, due to the complementary slackness conditions (7.32), $\alpha(i) > 0$, $\beta(i) = 0$. From (7.33), we thus obtain the condition for the occurrence of this case, as $\alpha(i) = \lambda_0 - \sum_{k=1}^{K} \lambda_k N_0 \eta_k x_k(i) > 0$. Since the BS is silent, $\tau_0(i) = 0$, and, therefore, according to (7.32), $\nu_0(i) > 0$. From (7.36), we obtain $\mu(i) = \nu_0(i) > 0$. From (7.35), we obtain

$$\nu_0(i) = \frac{1}{\overline{R}_k}\left[\log\left(1 + \frac{p_{d,k}(i)x_k(i)}{\tau_k(i)}\right) - \frac{p_{d,k}(i)x_k(i)}{p_{d,k}(i)x_k(i) + \tau_k(i)}\right] - \lambda_k p_c + \nu_k(i) > 0. \tag{7.38}$$

In order to see that only a single EHU must be active, let us assume for a moment that $K = 2$ and denote the two EHUs by A and B. In this case, from (7.38), we obtain

$$\begin{aligned}
&\frac{1}{\overline{R}_A}\left[\log\left(1 + \frac{p_{d,A}(i)x_A(i)}{\tau_A(i)}\right) - \frac{p_{d,A}(i)x_A(i)}{p_{d,A}(i)x_A(i) + \tau_A(i)}\right] - \lambda_A p_c + \nu_A(i) = \\
&\frac{1}{\overline{R}_B}\left[\log\left(1 + \frac{p_{d,B}(i)x_B(i)}{\tau_B(i)}\right) - \frac{p_{d,B}(i)x_B(i)}{p_{d,B}(i)x_B(i) + \tau_B(i)}\right] - \lambda_B p_c + \nu_B(i).
\end{aligned} \tag{7.39}$$

The assumptions $p_{d,A}(i) > 0$ and $p_{d,B}(i) > 0$ imply $\nu_A(i) = \nu_B(i) = 0$, in which case (7.39) cannot be satisfied since $x_A(i)$ and $x_B(i)$ are continuous random variables. The case $p_{d,A}(i) > 0$ and $p_{d,B}(i) = 0$ implies $\nu_A(i) = 0$ and $\nu_B(i) > 0$, in which case (7.39) are satisfied and the optimal scheduling policy is given by (7.23). According to (7.34), the scheduled EHU A must satisfy

$$p_{d,A}(i) = \tau_A(i)\left(\frac{1}{\lambda_A \overline{R}_A} - \frac{1}{x_A(i)}\right) > 0, \tag{7.40}$$

which leads to (7.24). The combination of (7.38) and (7.40) implies that the scheduled EHU A must also satisfy

$$\frac{1}{\overline{R}_A}\left[\log\left(\frac{x_A(i)}{\lambda_A\overline{R}_A}\right)-\frac{\frac{x_A(i)}{\lambda_A\overline{R}_A}-1}{\frac{x_A(i)}{\lambda_A\overline{R}_A}}\right]-\lambda_A p_c>0, \tag{7.41}$$

which is a more stringent condition then (7.40) and leads to (7.25) and (7.26).

Bibliography

1 L.R. Varshney (2008) Transporting information and energy simultaneously. *Proceedings of the IEEE International Symposium on Information Theory*, pp. 1612–1616.

2 P. Grover and A. Sahai (2010) Shannon meets Tesla: wireless information and power transfer. *Proceedings of the IEEE International Symposium on Information Theory*, pp. 2363–2367.

3 P. Kamalinejad et al. (2015) Wireless energy harvesting for the Internet of Things. *IEEE Commun. Mag.*, **53**: 102–108.

4 S. Bi, C.K. Ho, and R. Zhang (20150 Wireless powered communication: Opportunities and challenges. *IEEE Commun. Mag.*, **53**: 117–125.

5 Y. Zeng, B. Clerckx, and R. Zhang (2017) Communications and signals design for wireless power transmission. *IEEE Trans. Commun.*, **65**: 2264–2290.

6 R.J.M. Vullers et al. (2009) Micropower energy harvesting. *Solid-State Electronics*. **53**: 684–693.

7 P. Simon, Y. Gogotsi, and B. Dunn (2014) Where do batteries end and supercapacitors begin? *Science Magazine*, **343**: 1210–1211.

8 H. Kim and G. Veciana (2010) Leveraging dynamic spare Ccpacity in wireless system to conserve mobile terminals energy. *IEEE/ACM Trans. Network.*, **18**: 802–815.

9 O. Orhan, D. Gunduz, and E. Erkip (2012) Throughput maximization for an energy harvesting communication system with processing cost. *Proceedings of the IEEE Information Theory Workshop*, pp. 84–88.

10 T. Le, K. Mayaram, and T. Fiez (2008) Efficient far-field radio frequency energy harvesting for passively powered sensor networks. *IEEE J. Solid-State Circuits*, **43**: 1287–1302.

11 E. Boshkovska, D.W.K. Ng, N. Zlatanov, and R. Schober (2015) Practical non-linear energy harvesting model and resource allocation for SWIPT systems. *IEEE Commun. Lett.*, **19**: 2082–2085.

12 H. Ju and R. Zhang (2014) Throughput maximization in wireless powered communication networks. *IEEE Trans. Wireless Commun.*, **13**: 418–428.

13 Z. Hadzi-Velkov, I. Nikoloska, G.K. Karagiannidis, and T.Q. Duong (2016) Wireless networks with energy harvesting and power transfer: joint power and time allocation. *IEEE Signal Process. Lett.*, **23**: 50–54.

14 X. Lu et al. (2015) Wireless networks with RF energy harvesting: A contemporary survey. *IEEE Commun. Surveys & Tutorials*, 757–789.

15 P. Viswanath, D.N.C. Tse, and R. Laroia (2002) Opportunistic beamforming using dumb antennas. *IEEE Trans. Inf. Theory*, **48**: 1277–1294.

16 T.-D. Nguyen and Y. Han (2006) A proportional fairness algorithm with QoS provision in downlink OFDMA systems. *IEEE Commun. Lett.*, **10**: 760–762.

17 Z.-Q. Luo and S. Zhang (2008) Dynamic spectrum management: Complexity and duality. *IEEE J. Selected Topics Signal Process.*, **2**: 57–73.

18 C. Guo et al. (2016) Convexity of fairness-aware resource allocation in wireless powered communication networks. *IEEE Commun. Lett.*, **20**: 474–477.

19 J. Xu and R. Zhang (2014) Throughput optimal policies for energy harvesting wireless transmitters with non-ideal circuit power. *IEEE J Selected Areas Commun.*, **32**: 322–332.

20 R. Jain, D. Chiu, and W. Hawe (1984) A quantitative measure of fairness and discrimination for resource allocation in shared computer systems. Technical Report TR-301, Dec.

21 N. Zlatanov, R. Schober, and Z. Hadzi-Velkov (2017) Asymptotically optimal power allocation for energy harvesting communication networks. *IEEE Trans. Veh. Technol.*, **66**: 7286–7301.

22 Z. Hadzi-Velkov, I. Nikoloska, H. Chingoska, and N. Zlatanov (2016) Proportional fair scheduling in wireless networks with RF energy harvesting and processing cost. *IEEE Commun. Lett.*, **20**: 2107–2110.

23 S. Pejoski, Z. Hadzi-Velkov, T.Q. Duong, and C. Zhong (2017) Wireless powered communication networks with non-ideal circuit power consumption. *IEEE Commun. Lett.*, **21**: 1429–1432.

24 Z. Hadzi-Velkov, I. Nikoloska, H. Chingoska, and N. Zlatanov (2017) Opportunistic scheduling in wireless powered communication networks. *IEEE Trans. Wireless Commun.*, **16**: 4106–4119.

25 S. Boyd and L. Vandenberghe (2004) *Convex Optimization.* Cambridge University Press.

26 X. Wang, and N. Gao (2010) Stochastic resource allocation over fading multiple access and broadcast channels. *IEEE Trans. Inf. Theory*, **56**: 2382–2391.

8

Wireless Power Transfer in Millimeter Wave

Talha Ahmed Khan and Robert W. Heath Jr.*

Department of Electrical and Computer Engineering, The University of Texas at Austin, USA

8.1 Introduction

Background The millimeter wave (mmWave) band is a key enabler of future fifth generation (5G) networks. This is due to the availability of large spectrum resources at mmWave frequencies promising high data rates [1]. It is also attractive for wireless power transfer as it allows large antenna arrays to be deployed in small form-factors. With a large antenna array, a wireless power transmitter can focus energy towards a desired location, thus improving the power transfer efficiency. Unfortunately, mmWave signals suffer from poor penetration and diffraction characteristics, making them susceptible to blockage by many common materials such as buildings and human bodies [1]. In this chapter we review a case study illustrating how mmWave can be used for wireless power transfer despite the presence of blockages.

Literature Survey MmWave power transfer is an active area of research in the circuits community [2–4]. The focus is on designing an efficient mmWave energy harvesting circuit. It has also gained traction among communication theorists thanks to the rise of mmWave communications. Their research is about developing system-level analytical models to gain insights into the performance of a mmWave energy and/or information transfer system. We now summarize some of the major contributions in this regard. Different propagation characteristics and directional antenna arrays are the main distinguishing features of mmWave compared to prior analyses on lower frequency systems. The work in [5–7] was first to investigate mmWave wireless information and power transfer. In [5], an outdoor urban scenario with mmWave transmitters and harvesters was considered. Using stochastic geometry, the energy coverage probability at an energy harvester was characterized

*Corresponding author: Talha Ahmed Khan; talhakhan@utexas.edu

Wireless Information and Power Transfer: Theory and Practice, First Edition.
Edited by Derrick Wing Kwan Ng, Trung Q. Duong, Caijun Zhong, and Robert Schober.

while incorporating blockages due to buildings. This work was extended in [6] to include simultaneous wireless information and energy transfer, and the joint energy-and-information coverage probability was characterized. The main takeaway from [5, 6] is that mmWave can leverage directional transmission to boost the harvested power despite the presence of blockages. In [7], a mmWave wireless-powered communication system with power transfer in the downlink and information transfer in the uplink was analyzed in terms of the average uplink achievable rate. In [8], the joint distribution of the harvested power and data rate was derived for a large-scale cellular network with simultaneous information and power transfer. In [9], wireless energy transfer in a mmWave massive multiple-input multiple-output (MIMO) system was analyzed while accounting for rainfall effects on signal propagation.

Wireless power transfer has also been studied in other operating scenarios, such as mmWave ad-hoc networks [10, 11], tactical networks [12], and relay-aided networks [13]. Specifically, in [10], a directional ad-hoc network with joint information and energy transfer was investigated in the presence of building blockages. In [11], the aggregate received power at an energy harvester in a mmWave ad-hoc network was characterized. In [12], a large-scale mmWave tactical network consisting of human nodes equipped with mmWave power beacons and energy harvesters was considered. A feasibility study for using mmWave power transfer was presented while accounting for human body blockages. It was shown that the system benefits from the large beamforming gains offered by the mmWave antenna arrays. In [13], the coverage performance in a mmWave network with wireless-powered relays was presented.

Overview In this chapter, we briefly review our work [5, 6] on characterizing the performance of wireless power transfer in a large-scale mmWave cellular network. Our analysis accounts for the key distinguishing features of mmWave systems, namely the sensitivity to blockage and the use of potentially large antenna arrays at the transmitter/receiver. We model two operating scenarios, one where a mmWave BS has its beam aligned to that of an energy harvesting device and the other where no such beam alignment is assumed. For both operating modes we provide analytical expressions for metrics such as the energy coverage probability and the average harvested power using tools from stochastic geometry.

To obtain design insights, we examine the network level performance trends in terms of key parameters such as the mmWave network density and the antenna geometry parameters for both operating modes of the energy harvesting devices. Numerical results suggest that narrower antenna beams are preferred when the BS and user beams are already aligned, whereas wider beams are favorable when no beam alignment is assumed. Our findings also suggest that there typically exists an optimum transmit antenna beamwidth that maximizes the network-wide energy coverage for a given user distribution. This implies that the mmWave BSs will need to adapt the antenna beam patterns depending on the fraction of the active users in each mode.

The rest of this chapter is organized as follows. In Section 8.2, we describe the system model for the considered mmWave wireless power transfer network. In Section 8.3, we summarize some analytical results that characterize the performance of wireless power transfer in the mmWave band. In Section 8.4, we present numerical examples to showcase the performance of mmWave wireless power transfer. In Section 8.5, we conclude this chapter.

8.2 System Model

In this section we describe the network model, the channel model, and the antenna model considered in this chapter.

Network Model. We consider a large-scale cellular network consisting of mmWave BSs and a population of wireless-powered devices (or users) that operate by extracting energy in the mmWave band. The mmWave BSs are located according to a homogeneous Poisson point process (PPP) $\Phi(\lambda)$ of density λ. The user population is drawn from another homogeneous PPP $\Phi_u(\lambda_u)$ of density λ_u, independently of Φ. In general, mmWave BSs and users may be located outdoors or indoors. Empirical evidence suggests that mmWave signals exhibit high penetration losses for many common building materials [14–16]. Assuming the building blockages to be impenetrable, we focus on the case where the BSs and users are located outdoors. We say that a BS-user link is line-of-sight (LOS) or non-line-of-sight (NLOS) depending on whether or not it is intersected by a building blockage. Channel measurement campaigns have reported markedly different propagation characteristics for LOS/NLOS links [1, 16]. To model blockage due to buildings, we leverage the results in [14] where the buildings are drawn from a boolean stochastic point process. We define a LOS probability function $p(r) = e^{-\beta r}$ for a link of length r, where β is a constant that depends on the geometry and density of the building blockage process: a BS-receiver link of length r is declared LOS with a probability $p(r)$, independently of other links. While conducting stochastic geometry analysis, we will apply this result to split the BS PPP into two independent but non-homogeneous PPPs consisting of LOS and NLOS BSs.

We allow the user population to consist of two types of users, namely *connected* and *nonconnected*. A connected user is assumed to be tagged with the BS, either LOS or NLOS, that maximizes the average received power at that user. Moreover, for the connected case, we assume perfect beam alignment between a BS and its tagged user, i.e. the BS and user point their beams towards each other so as to have the maximum directivity gain. Furthermore, we assume that a BS serves only one connected user at a given time. For a nonconnected user, we do not assume any prior beam alignment with a BS, i.e. it is not tagged with any BS. This allows us to model a wide range of scenarios. For instance, due to limited resources, the mmWave network may (directly) serve only a fraction of the user population as connected users, leaving the rest in the nonconnected mode. Another interpretation could be that due to the challenges associated with channel acquisition, not all the users could be simultaneously served in the connected mode. We let ϵ be the probability that a randomly selected node is a connected user, independently of other nodes. With this assumption, we can thin the user PPP Φ_u into two independent PPPs, $\Phi_{u,\text{con}}$ and $\Phi_{u,\text{ncon}}$, with densities $\epsilon\lambda_u$ and $(1-\epsilon)\lambda_u$, respectively. Note that an arbitrary user, either connected or nonconnected, may experience an energy outage if the received power falls short of a required threshold ψ. This threshold would depend on the power consumption requirements of the receiver. To capture the sensitivity requirements of the energy harvesting circuit, we define $\psi_{\min}$ to be the harvester activation threshold, i.e. the minimum received energy needed to activate the harvesting circuit (the energy outage threshold ψ would typically be greater than $\psi_{\min}$). We use ξ to denote the rectifier efficiency. We define $\text{P}_{\text{con}}(\lambda, \psi_{\text{con}})$ to be the energy coverage probability given an outage threshold ψ_{con} for a connected user, while $\text{P}_{\text{ncon}}(\lambda, \psi_{\text{ncon}})$ denotes the same for the nonconnected case. With these definitions, we can define the overall energy coverage probability $\Lambda(\epsilon, \lambda, \psi_{\text{con}}, \psi_{\text{ncon}})$ of the network as

$$\Lambda(\epsilon, \lambda, \psi_{\text{con}}, \psi_{\text{ncon}}) = \epsilon \text{P}_{\text{con}}(\lambda, \psi_{\text{con}}) + (1-\epsilon)\text{P}_{\text{ncon}}(\lambda, \psi_{\text{ncon}}), \tag{8.1}$$

where the energy coverage probability is a function of several parameters such as the BS density, the channel propagation parameters, as well as the antenna beam patterns at the transmitter/receiver. For cleaner exposition, we drop the subscript in ψ_{con} or ψ_{ncon}, using the notation $\Lambda(\epsilon, \lambda, \psi)$ when the context is clear. In Section 8.3, we provide analytical expressions to compute the energy coverage probability in a mmWave network.

Channel Model. We now describe the channel model for an arbitrary user in the network. Empirical evidence suggests that mmWave frequencies exhibit different propagation characteristics for the LOS/NLOS links [16]. While the LOS mmWave signals propagate as if in free space, the NLOS mmWave signals typically exhibit a higher path-loss exponent (and additional shadowing) [16]. We let α_{L} and α_{N} be the path-loss exponents for the LOS and NLOS links, respectively. We define the distance-dependent path loss for a user located a distance r_ℓ from the ℓth BS: $g_\ell(r_\ell) = C_{\text{L}} r_\ell^{-\alpha_{\text{L}}}$ when the link is LOS, where the constant C_{L} is the path-loss intercept and $g_\ell(r_\ell) = C_{\text{N}} r_\ell^{-\alpha_{\text{N}}}$ for the NLOS case. Note that by including blockages in our model (Section 8.2), we capture the distance-dependent signal attenuation due to buildings. To simplify the analysis, we do not include additional forms of shadowing in our model. We further define h_ℓ to be the small-scale fading coefficient corresponding to a BS $\ell \in \Phi$. Assuming independent Nakagami fading for each link, the small-scale

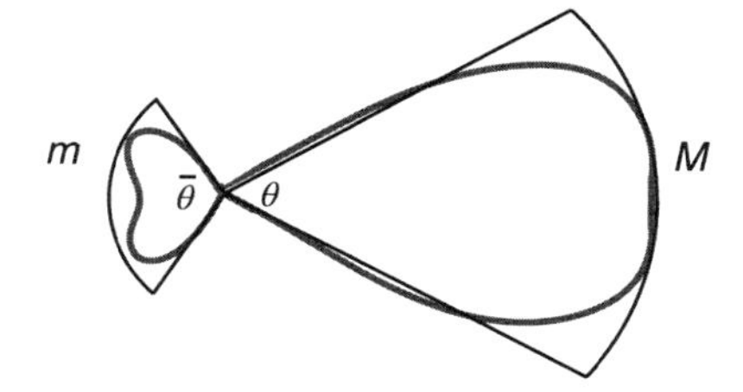

Figure 8.1 Sectored antenna model. The antenna beam pattern is parameterized by the directivity gains for the main lobe (M) and side lobe (m), and the half power beamwidths for the main lobe (θ) and side lobe ($\overline{\theta}$).

fading power $H_\ell = |h_\ell|^2$ can be modeled as a normalized Gamma random variable, i.e. $H_\ell \sim \Gamma(N_\mathrm{L}, 1/N_\mathrm{L})$ when the link is LOS and $H_\ell \sim \Gamma(N_\mathrm{N}, 1/N_\mathrm{N})$ for the NLOS case, where the fading parameters N_L and N_N are assumed to be integers for simplicity.

Antenna Model. To compensate for higher propagation losses, mmWave BSs will use large directional antennas arrays. We assume that the BSs and users are equipped with N_t and N_r antenna elements each. To simplify the analysis while capturing the key antenna characteristics, we use the sectored antenna model of Figure 8.1, similar to the one considered in [15]. We use $A_{M,m,\theta,\overline{\theta}}(\phi)$ to characterize the antenna beam pattern, where ϕ gives the angle from the boresight direction, M denotes the directivity gain, and θ is the half power beamwidth for the main lobe, while m and $\overline{\theta}$ give the corresponding parameters for the side lobe. With this notation, $A_{M_\mathrm{t},m_\mathrm{t},\theta_\mathrm{t},\overline{\theta}_\mathrm{t}}(\cdot)$ denotes the antenna beam pattern at an arbitrary BS in Φ and $A_{M_\mathrm{r},m_\mathrm{r},\theta_\mathrm{r},\overline{\theta}_\mathrm{r}}(\cdot)$ denotes the same for an energy harvesting user in Φ_u. We further define $\delta_\ell = A_{M_\mathrm{t},m_\mathrm{t},\theta_\mathrm{t},\overline{\theta}_\mathrm{t}}(\phi_\mathrm{t}^\ell) A_{M_\mathrm{r},m_\mathrm{r},\theta_\mathrm{r},\overline{\theta}_\mathrm{r}}(\phi_\mathrm{r}^\ell)$, the total directivity gain for the link between the ℓth BS and the typical user; ϕ_t^ℓ and ϕ_r^ℓ give the angle-of-arrival and angle-of-departure of the signal. Without any further assumptions about the beam alignment between a user and its BS, we model the directivity gain δ_ℓ as a random variable. We assume the angles ϕ_t^ℓ and ϕ_r^ℓ are uniformly distributed in $[0, 2\pi)$. Due to the sectored antenna model, the random variable $\delta_\ell = D_i$ with a probability p_i ($i \in \{1, 2, 3, 4, 5\}$), where $D_i \in \{M_\mathrm{t}M_\mathrm{r}, M_\mathrm{t}m_\mathrm{r}, m_\mathrm{t}M_\mathrm{r}, m_\mathrm{t}m_\mathrm{r}, 0\}$ with corresponding probabilities $p_i \in \{q_\mathrm{t}q_\mathrm{r}, q_\mathrm{t}\overline{q}_\mathrm{r}, \overline{q}_\mathrm{t}q_\mathrm{r}, \overline{q}_\mathrm{t}\overline{q}_\mathrm{r}, q_o\}$; the constants $q_\mathrm{t} = \frac{\theta_t}{2\pi}$, $\overline{q}_\mathrm{t} = \frac{\overline{\theta}_t}{2\pi}$, $q_\mathrm{r} = \frac{\theta_r}{2\pi}$, $\overline{q}_\mathrm{r} = \frac{\overline{\theta}_r}{2\pi}$, and $q_o = 2 - q_\mathrm{t} - \overline{q}_\mathrm{t} - q_\mathrm{r} - \overline{q}_\mathrm{r}$. Note that $D_5 = 0$ models the extreme case where the BS and user beams have no alignment at all. Note that for the connected mode, since we assume perfect beam alignment between the typical user and its serving BS (hereby denoted by subscript 0), the directivity gain $\delta_0 = M_\mathrm{t}M_\mathrm{r}$ due to the sectored antenna model.

8.3 Analytical Results

In this section we provide an analytical framework to characterize the performance of mmWave wireless power transfer using metrics such as the average harvested power and the energy coverage probability.

We assume that each user is equipped with an energy harvesting circuit and attempts to extract energy from the incident mmWave signals. We conduct the analysis at a typical energy harvesting user located at the origin without loss of generality. We let P_t be the BS transmit power and $Y = \sum_{\ell \in \Phi(\lambda)} P_t \delta_\ell H_\ell g_\ell(r_\ell)$ be the power received at the user. Recall that $\psi_{\min}$ denotes the harvester activation threshold defined in Section 8.2. The energy harvested at a typical receiver (in unit time) can be expressed as

$$\gamma = \xi Y \mathbb{1}_{\{Y > \psi_{\min}\}} \tag{8.2}$$

where $\xi \in (0, 1]$ is the rectifier efficiency and $\mathbb{1}_{\{\cdot\}}$ is the indicator function. Note that we have neglected the noise term since it is extremely small relative to the aggregate received signal. Recall that given a BS $\ell \in \Phi(\lambda)$, the corresponding fading parameters will be distinct depending on whether the link is LOS or NLOS, which in turn depends on the LOS probability function (Section 8.2). Further note that for the connected case, it follows from Section 8.2 that $\delta_0 = M_t M_r$ for the link from serving BS (denoted by subscript 0).

Connected Case. The following theorem provides an analytical expression for the energy coverage probability $\mathrm{P}_{\mathrm{con}}(\lambda, \psi) = \Pr\{\gamma > \psi\}$ at a connected user, where the random variable γ is given in (8.2) and ψ is the energy outage threshold. Note that $\mathrm{P}_{\mathrm{con}}(\lambda, \psi)$ can also be interpreted as the complementary cumulative distribution function (CCDF) of the harvested energy.

Theorem 8.1 In a mmWave network with density λ, the energy coverage probability $\mathrm{P}_{\mathrm{con}}(\lambda, \psi)$ for the connected case given an energy outage threshold ψ can be evaluated as

$$\mathrm{P}_{\mathrm{con}}(\lambda, \psi) = \mathrm{P}_{\mathrm{con,L}}(\lambda, \hat{\psi})\varrho_{\mathrm{L}} + \mathrm{P}_{\mathrm{con,N}}(\lambda, \hat{\psi})\varrho_{\mathrm{N}}, \tag{8.3}$$

where $\hat{\psi} = \max\left(\frac{\psi}{\xi}, \psi_{\min}\right)$, $\varrho_{\mathrm{L}} = 1 - \varrho_{\mathrm{N}}$ is given in Lemma 2 (Section 8.6), while $\mathrm{P}_{\mathrm{con,L}}(\cdot)$ and $\mathrm{P}_{\mathrm{con,N}}(\cdot)$ are the conditional energy coverage probabilities given the serving BS is LOS or NLOS. These terms can be tightly approximated as

$$\mathrm{P}_{\mathrm{con,L}}(\lambda, \psi) \approx \sum_{k=0}^{N} (-1)^k \binom{N}{k} \int_{r_g}^{\infty} \zeta_k^{\mathrm{L}}(r) e^{-\Upsilon_{k,1}(\lambda,\psi,r) - \Upsilon_{k,2}(\lambda,\psi,\rho_{\mathrm{L}}(r))} \tilde{\tau}_{\mathrm{L}}(r) \mathrm{d}r, \tag{8.4}$$

where $\zeta_k^{\mathrm{L}}(x) = \left(1 + \frac{akP_t M_t M_r C_{\mathrm{L}}}{\psi N_{\mathrm{L}} x^{\alpha_{\mathrm{L}}}}\right)^{-N_{\mathrm{L}}}$, the approximation constant $a = N(N!)^{-\frac{1}{N}}$ where N denotes the number of terms in the approximation, while r_g defines the minimum link distance and is included to avoid unbounded path loss at the receiver. Similarly,

$$\mathrm{P}_{\mathrm{con,N}}(\lambda, \psi) \approx \sum_{k=0}^{N} (-1)^k \binom{N}{k} \int_{r_g}^{\infty} \zeta_k^{\mathrm{N}}(r) e^{-\Upsilon_{k,1}(\lambda,\psi,\rho_{\mathrm{N}}(r)) - \Upsilon_{k,2}(\lambda,\psi,r)} \tilde{\tau}_{\mathrm{N}}(r) \mathrm{d}r, \tag{8.5}$$

where $\zeta_k^{\mathrm{N}}(x) = \left(1 + \frac{akP_{\mathrm{t}}M_{\mathrm{t}}M_{\mathrm{r}}C_{\mathrm{N}}}{\psi N_{\mathrm{N}}x^{\alpha_{\mathrm{N}}}}\right)^{-N_{\mathrm{N}}}$,

$$\Upsilon_{k,1}(\lambda,\psi,x) = 2\pi\lambda\sum_{i=1}^{4}p_i\int_x^{\infty}\left(1-\left[1+\frac{aP_{\mathrm{t}}kD_iC_{\mathrm{L}}}{\psi N_{\mathrm{L}}t^{\alpha_{\mathrm{L}}}}\right]^{-N_{\mathrm{L}}}\right)p(t)t\mathrm{d}t, \quad (8.6)$$

$$\Upsilon_{k,2}(\lambda,\psi,x) = 2\pi\lambda\sum_{i=1}^{4}p_i\int_x^{\infty}\left(1-\left[1+\frac{aP_{\mathrm{t}}kD_iC_{\mathrm{N}}}{\psi N_{\mathrm{N}}t^{\alpha_{\mathrm{N}}}}\right]^{-N_{\mathrm{N}}}\right)(1-p(t))t\mathrm{d}t, \quad (8.7)$$

and the distance distributions $\tilde{\tau}_{\mathrm{L}}(\cdot)$ and $\tilde{\tau}_{\mathrm{N}}(\cdot)$ follow from Lemma 8.3 (Section 8.6).

Proof: Please refer to [6] for a detailed proof. Recall that $p(t) = e^{-\beta t}$ is the LOS probability function defined in Section 8.2 and captures the effect of building blockages. In (8.4), the term $\zeta_k^{\mathrm{L}}(\cdot)$ models the contribution from the LOS serving link, $\Upsilon_{k,1}(\cdot)$ accounts for other LOS links, and $\Upsilon_{k,2}(\cdot)$ captures the effect of the NLOS links. Note that the ith term in (8.6) and (8.7) corresponds to the contributions from the BS-user links having directivity gain D_i. Similarly, $\zeta_k^{\mathrm{N}}(\cdot)$ in (8.5) models the case where the serving BS is NLOS. Note that these terms further depend on the channel propagation conditions ($\alpha_{\mathrm{L}}, \alpha_{\mathrm{N}}, N_{\mathrm{L}}, N_{\mathrm{N}}, C_{\mathrm{L}}, C_{\mathrm{N}}$), the network density λ as well as the antenna geometry parameters (via D_i, p_i). Furthermore, the outage threshold ψ will depend on the power requirements at a particular user, and would typically be greater than the sensitivity of the harvesting circuit (i.e., $\psi \geq \psi_{\min}$ such that $\hat{\psi} = \frac{\psi}{\xi}$).

Proposition 8.1 The average harvested power for the connected case $\overline{P}_{\mathrm{con}}(\lambda,\psi)$ for an energy outage threshold $\psi \in [\psi_{\min}, \infty)$ is given by

$$\overline{P}_{\mathrm{con}}(\lambda,\psi) = \int_{\psi}^{\infty} P_{\mathrm{con}}(\lambda,x)\mathrm{d}x + \psi P_{\mathrm{con}}(\lambda,\psi). \quad (8.8)$$

Proof: Please refer to [6] for a detailed proof.

Here, $P_{\mathrm{con}}(\cdot,\cdot)$ follows from Theorem 8.1. $\overline{P}_{\mathrm{con}}(\lambda,\psi)$ can be interpreted as the *useful* average harvested power. This is because only those incident signals that meet the activation threshold can be harvested. To get further insights, we now analyze the limiting case $\psi \to 0$. This provides an upper bound on the average harvested power.

Corollary 8.1 The average harvested power for the limiting case $\lim_{\psi\to 0}\overline{P}_{\mathrm{con}}(\lambda,\psi) = \overline{P}_{\mathrm{con}}(\lambda,0) = \xi(\rho_{\mathrm{L}}\overline{P}_{\mathrm{L}} + \rho_{\mathrm{N}}\overline{P}_{\mathrm{N}})$, where

$$\overline{P}_{\mathrm{L}} = \int_{r_g}^{\infty}(P_{\mathrm{t}}M_{\mathrm{t}}M_{\mathrm{r}}C_{\mathrm{L}}r^{-\alpha_{\mathrm{L}}} + \Psi_{\mathrm{L}}(r) + \Psi_{\mathrm{N}}(\rho_{\mathrm{L}}(r)))\tilde{\tau}_{\mathrm{L}}(r)\mathrm{d}r, \quad (8.9)$$

$$\overline{P}_{\mathrm{N}} = \int_{r_g}^{\infty} (P_{\mathrm{t}} M_{\mathrm{t}} M_{\mathrm{r}} C_{\mathrm{N}} r^{-\alpha_{\mathrm{N}}} + \Psi_{\mathrm{L}}(\rho_{\mathrm{N}}(r)) + \Psi_{\mathrm{N}}(r))\tilde{\tau}_{\mathrm{N}}(r)\mathrm{d}r, \tag{8.10}$$

$$\Psi_{\mathrm{L}}(x) = \kappa C_{\mathrm{L}} \sum_{i=1}^{4} D_i p_i \int_{x}^{\infty} t^{-(\alpha_{\mathrm{L}}-1)} p(t)\mathrm{d}t, \tag{8.11}$$

$$\Psi_{\mathrm{N}}(x) = \kappa C_{\mathrm{N}} \sum_{i=1}^{4} D_i p_i \left(\frac{x^{-(\alpha_{\mathrm{N}}-2)}}{\alpha_{\mathrm{N}} - 2} - \int_{x}^{\infty} t^{-(\alpha_{\mathrm{N}}-1)} p(t)\mathrm{d}t \right), \tag{8.12}$$

and $\kappa = 2\pi\lambda P_{\mathrm{t}}$.

Proof: Please refer to [6] for a detailed proof.

$\overline{P}_{\mathrm{L}}$ and $\overline{P}_{\mathrm{N}}$ denote the average harvested power given the user is tagged to an LOS or an NLOS BS, respectively. Note that the average harvested power is independent of the small-scale fading parameters. To reveal further insights, we provide the following approximation for the average harvested power.

Corollary 8.2 The average harvested power for the limiting case, $\overline{P}_{\mathrm{con}}(\lambda, 0)$, can be further approximated as

$$\begin{aligned} \overline{P}_{\mathrm{con}}(\lambda, 0) &\overset{(a)}{\approx} \kappa M_{\mathrm{t}} M_{\mathrm{r}} C_L \int_{r_g}^{R_B} \frac{e^{-\lambda\pi t^2}}{t^{\alpha-1}} \mathrm{d}t \\ &= \frac{\Gamma(1 - 0.5\alpha_{\mathrm{L}}; \lambda\pi r_g^2, \infty) - \Gamma(1 - 0.5\alpha_{\mathrm{L}}; \lambda\pi R_B^2, \infty)}{2(\kappa M_{\mathrm{t}} M_{\mathrm{r}} C_L)^{-1} (\lambda\pi)^{1-0.5\alpha_{\mathrm{L}}}}. \end{aligned} \tag{8.13}$$

Proof: Please refer to [6] for a detailed proof.

This approximation suggests that the average power is mainly determined by the LOS serving link. Note that $\overline{P}_{\mathrm{con}}(\cdot, \cdot)$ grows linearly with the transmit power P_{t}. Depending on the path-loss exponent α_{L}, it may exhibit a sublinear to approximately-linear scaling with the BS density λ. When α_{L} is large, the denominator $t^{\alpha_{\mathrm{L}}-1}$ in (a) overshadows the impact of λ on the numerator $e^{-\pi\lambda t^2}$. Therefore, the scaling behavior is essentially determined by $\kappa = 2\pi\lambda P_{\mathrm{t}}$, which is linear in λ. This suggests that increasing the transmit power or BS density has almost the same effect on the average harvested power when α_{L} is large. Also note that (8.13) is relatively simple as it is expressed in terms of the incomplete Gamma function only.

Nonconnected Case. Having discussed the connected case, we now consider the case where a user operates in the nonconnected mode. The following theorem characterizes the energy coverage probability at a typical user for the nonconnected case.

Theorem 8.2 In a mmWave network of density λ, the energy coverage probability for the nonconnected case $\mathrm{P}_{\mathrm{ncon}}(\lambda, \psi)$ given an outage threshold ψ can be evaluated using

$$\mathrm{P}_{\mathrm{ncon}}(\lambda, \psi) \approx \sum_{k=0}^{N} (-1)^k \binom{N}{k} e^{-\Upsilon_{k,1}(\lambda,\hat{\psi},r_g)-\Upsilon_{k,2}(\lambda,\hat{\psi},r_g)}, \tag{8.14}$$

where $\Upsilon_{k,1}(\cdot)$ and $\Upsilon_{k,2}(\cdot)$ are given by (8.6) and (8.7), respectively, $\hat{\psi} = \max\left(\frac{\psi}{\xi}, \psi_{\min}\right)$, and r_g is the minimum link distance.

Proof: See [6].

Similar to the connected case, the energy coverage probability for this case is also a function of the propagation conditions, the network density, and the antenna geometry parameters. We note that the expressions in Theorem 8.2 are efficient to compute, obviating the need for further simplification. We now consider the average harvested power for the nonconnected case.

Proposition 8.2 The average harvested power for the nonconnected case $\overline{P}_{\mathrm{ncon}}(\lambda, \psi)$ for an energy outage threshold $\psi \in [\psi_{\min}, \infty)$ is given by $\overline{P}_{\mathrm{ncon}}(\lambda, \psi) = \int_{\psi}^{\infty} P_{\mathrm{ncon}}(\lambda, x)\mathrm{d}x + \psi P_{\mathrm{ncon}}(\lambda, \psi)$.

Proof: Please refer to [6] for a detailed proof.

Corollary 8.3 The average harvested power for the limiting case $\lim_{\psi\to 0} \overline{P}_{\mathrm{ncon}}(\lambda, \psi) = \overline{P}_{\mathrm{ncon}}(\lambda, 0)$ is given by

$$\overline{P}_{\mathrm{ncon}}(\lambda, 0) = \xi(\Psi_{\mathrm{L}}(r_g) + \Psi_{\mathrm{N}}(r_g)), \tag{8.15}$$

where $\Psi_{\mathrm{L}}(\cdot)$ and $\Psi_{\mathrm{N}}(\cdot)$ are given in (8.11) and (8.12), respectively.

Proof: Please refer to [6] for a detailed proof.

The average harvested power for the nonconnected case scales *linearly* with the transmit power and the BS density. This follows from (8.15) as both $\Psi_{\mathrm{L}}(\cdot)$ and $\Psi_{\mathrm{N}}(\cdot)$ relate linearly with the transmit power and density via the term $\kappa = 2\pi\lambda P_{\mathrm{t}}$. This also suggests that increasing the transmit power or density has the same effect on the average harvested power. Note that this is different from the connected case where the path-loss exponent affects how average harvested power scales with the BS density.

8.4 Key Insights

In this section we present some simulation results to illustrate the performance of the considered mmWave power transfer system. In the following plots, the users are assumed to be equipped with a single omnidirectional receive

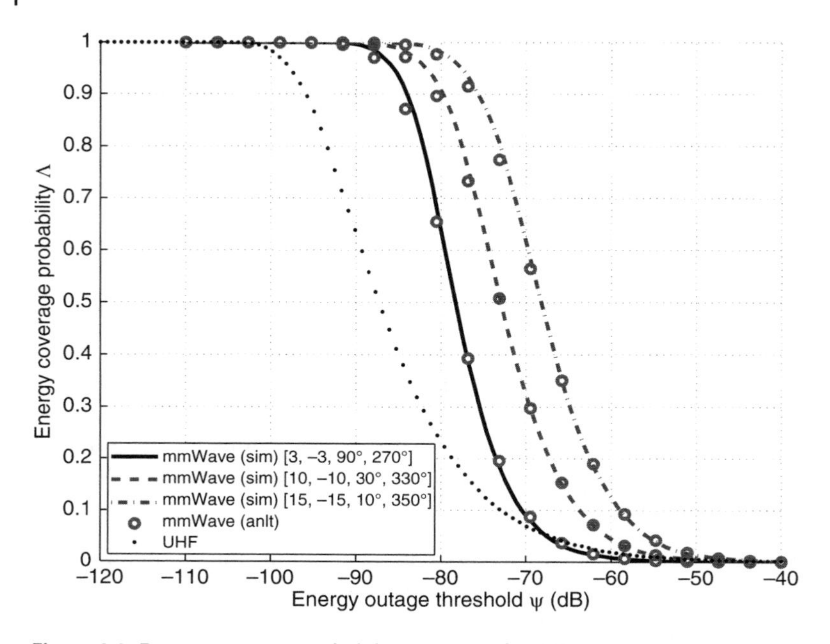

Figure 8.2 Energy coverage probability $\Lambda(\epsilon, \psi, \lambda)$ for different transmit antenna beam patterns parameterized by $[M_\mathrm{t}, m_\mathrm{t}, \theta_\mathrm{t}, \overline{\theta}_\mathrm{t}]$ in a purely connected network ($\epsilon = 1$, $\lambda = 100/\mathrm{km}^2$). The performance improves with narrower beams for this case. $P_\mathrm{t} = 13$ dB, $W = 100$ MHz, $\alpha_\mathrm{L} = 2$, $\alpha_\mathrm{N} = 4$, $N_\mathrm{L} = 2$, $N_\mathrm{N} = 3$, and $r_g = 1$ m. There is a nice agreement between Monte Carlo simulation (sim) results and the analytical (anlt) results obtained using Theorem 8.1 with $N = 5$ terms.

antenna, the mmWave carrier frequency is set to 28 GHz, the blockage constant $\beta = 0.0071$ [15], and $\psi > \psi_{\min}$. In other words, for a given ψ, the plots are valid for any $\psi_{\min} < \psi$. Note that for the less relevant case when $\psi < \psi_{\min}$, the energy coverage probability flattens out and is specified by $\mathrm{P}_{\mathrm{con}}(\lambda, \psi_{\min})$ or $\mathrm{P}_{\mathrm{ncon}}(\lambda, \psi_{\min})$. Without loss of generality, we set the rectifier efficiency $\xi = 1$ since this parameter does not impact the shape of the results, i.e. setting $\xi < 1$ results in shifting all the curves to the left by the same amount. We assume the rectifier efficiency to be the same when comparing mmWave and UHF. Note that there are no standard values for ξ since prior work has reported widely varying values [3, 17] depending on the device technology, operating frequency, etc. For example, [3, 4] suggest that a mmWave energy harvesting circuit may have better overall performance than its lower frequency counterparts.

Connected case ($\epsilon \to 1$). In Figure 8.2, we plot the energy coverage probability with three distinct transmit beam patterns for a given network density. We observe that the energy harvesting performance improves with narrower beams, i.e. smaller beamwidths and larger directivity gains. As the beamwidth decreases, relatively fewer beams from the neighboring BSs would be incident on a typical user. However, the beams that do reach will have larger directivity gains, resulting in an overall performance improvement. This is possible due to the use of potentially large antenna arrays at the mmWave BSs. Note that this performance boost will possibly be limited due to the ensuing equivalent isotropically radiated power (EIRP) or other safety regulations on future mmWave systems [18].

For the purpose of comparison, we also plot the energy coverage probability for UHF energy harvesting under realistic assumptions. Given the current state-of-the-art [1, 19], the UHF BSs are assumed to have eight transmit antennas each. Furthermore, they are assumed to employ maximal ratio transmit beamforming to serve a connected user. For the channel model, we assume an IID Rayleigh fading environment and a path-loss exponent of 3.6 (no blockage is considered). The network density is set to 25 nodes/km^2, which corresponds to an *average* distance of about 113 m to the closest UHF BS. The carrier frequency is set to 2.1 GHz and the transmission bandwidth is 100 MHz. As can be seen from Figure 8.2, mmWave energy harvesting could provide considerable performance gain over its lower frequency counterpart. Moreover, the anticipated dense deployments of mmWave networks would further enlarge this performance gap. This effect is illustrated in Figure 8.3, where we plot the energy coverage probability for different mmWave network densities for a given transmit antenna beam pattern. In Figure 8.4 we use Proposition 8.2 to plot the average harvested power at a typical mmWave user against the transmit array size. The plots based on Corollary 8.1 are also included. This figure confirms our earlier intuition that mmWave energy harvesting can benefit from (i) potentially large antenna arrays at the BSs and (ii) high BS density, which would be the key ingredients of future mmWave cellular systems. Figure 8.4 shows how the path-loss exponent impacts the scaling behavior of average harvested power with BS density (see discussion following Corollary 8.2).

Nonconnected Case ($\epsilon \rightarrow 0$). We now analyze the energy harvesting performance when the harvesting devices operate in the nonconnected mode. In a stark contrast to the connected case, Figure 8.5 shows that for the nonconnected case, mmWave energy harvesting could benefit from using wider beams. This is because BS connectivity (alignment) is critical for the nonconnected case. With wider beams, it is more likely that a mmWave BS gets aligned with a receiver, albeit at the expense of the beamforming gain. Furthermore, a comparison with UHF energy harvesting shows that mmWave energy harvesting gives a comparable performance to its UHF counterpart. Similarly, Figure 8.6 plots the energy coverage probability for different deployment densities. We note that performance can be substantially improved with denser deployments, which would be a key feature of future mmWave cellular systems.

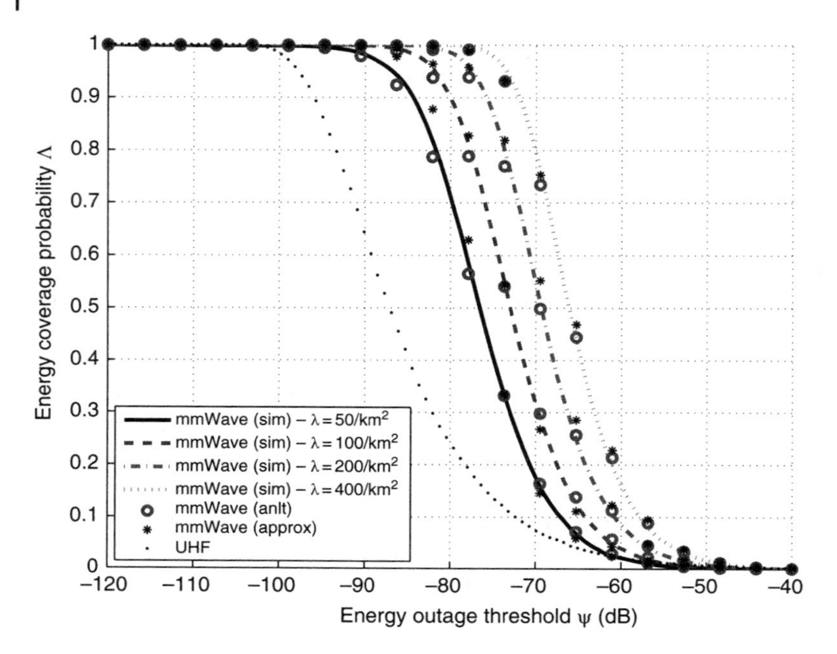

Figure 8.3 Energy coverage probability $\Lambda(1, \psi, \lambda)$ for different network densities for connected users. Transmit beam pattern is fixed to $[10, -10, 30^\circ, 330^\circ]$. Other parameters are the same as given in Figure 8.2. There is a nice agreement between Monte Carlo simulation (sim) results and the analytical (anlt) results obtained using Theorem 8.1 with $N = 5$ terms. Also included are the results based on the analytical approximation (approx) in [6], Proposition 1].

General Case $(0 < \epsilon < 1)$. Having presented the energy coverage trends for the two extreme network scenarios, we now consider the general case where the user population consists of both connected and nonconnected users. We expect this to be the likely scenario for the reasons explained in the network model (Section 8.2). As described in Section 8.2, an antenna beam pattern can be characterized by the half power beamwidth and directivity gain for both the main and side lobes. By tuning these parameters, the beam pattern can be particularized to a given antenna array. As an example, we assume that uniform linear arrays (ULA) are deployed at the mmWave BSs. We use the following relations to approximate the main and side lobe beamwidths as a function of the transmit array size: $\theta_{\rm t} \approx \frac{360}{\pi} \arcsin\left(\frac{0.892}{N_{\rm t}}\right)$ and $\overline{\theta}_{\rm t} \approx \frac{720}{\pi} \left|\arcsin\left(\frac{2}{N_{\rm t}}\right)\right|$ [20]. We use $M_{\rm t} = 10V \log(N_{\rm t})$ and $m_{\rm t} = V(M_{\rm t} - 12)$ for the directivity gains of the main and side lobes [20]. To ensure the power normalization, the constant V is chosen to satisfy $\frac{\theta_{\rm t}}{2\pi} M_{\rm t} + \frac{\overline{\theta}_{\rm t}}{2\pi} m_{\rm t} = 1$.

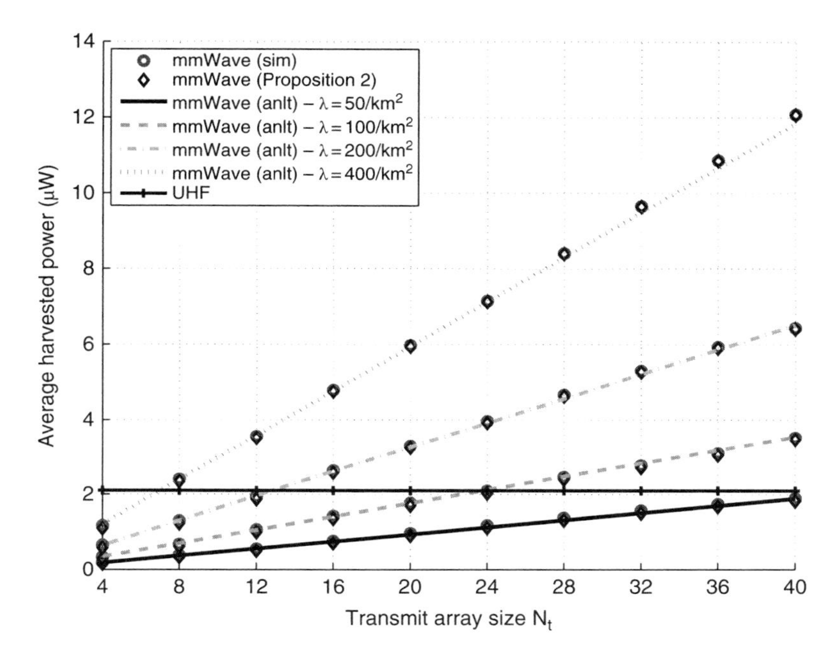

Figure 8.4 The average harvested power in a connected mmWave network for different number of BS antennas N_t and deployment densities λ. Results based on Proposition 8.2 are obtained for $\psi = -35$ dBm. The analytical (anlt) results based on Corollary 8.1 ($\psi \rightarrow 0$) are validated using Monte Carlo simulations (sim) and closely approximate the average harvested energy. The transmit antenna beam patterns are calculated using the approximations used for obtaining Figure 8.7. Other simulation parameters are the same as used in Figure 8.2. For comparison, a plot for a UHF system is also included [6].

In Figure 8.7 we plot the overall energy coverage probability $\Lambda(\epsilon, \psi, \lambda)$ against transmit array size N_t for different values of parameter ϵ. We find that the optimal transmit array size depends on the type of user population. For example, when ϵ is large, it is desirable to use large antenna arrays at the BSs. When ϵ is small, it is favorable to use small antenna arrays to improve the overall energy coverage probability. Depending on the network load (or the user population *mix*) captured via ϵ, the energy coverage probability can be substantially improved by intelligent antenna switching schemes. Since the parameter ϵ would typically vary over large time-scales, such schemes would be practically feasible.

8.5 Conclusions

In this chapter we discussed the viability of using mmWave for wireless power transfer. As a case study, we investigated an outdoor urban scenario where a mmWave cellular network delivers energy to RF energy harvesters.

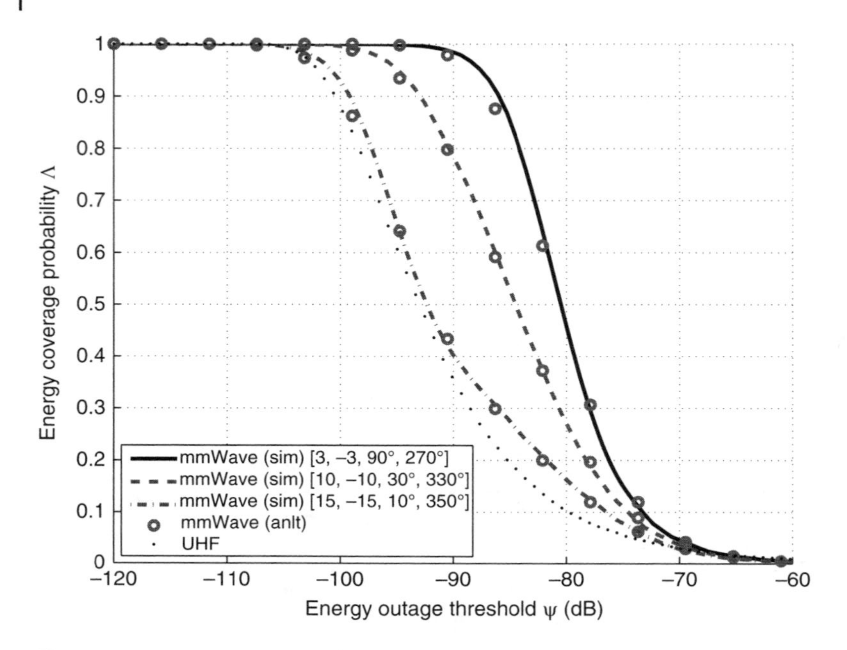

Figure 8.5 Energy coverage probability $\Lambda(0, \psi, \lambda)$ for different network densities for nonconnected users. The transmit beam pattern is fixed to $[10, -10, 30^\circ, 330^\circ]$. Other parameters are the same as given in Figure 8.2. Monte Carlo simulation (sim) results validate the analytical (anlt) results obtained using Theorem 2 with $N = 5$ terms [6].

We presented analytical expressions for the energy coverage probability and the average harvested power to characterize performance. Our analytical model captures the salient features of mmWave such as sensitivity to building blockages and the use of directional antenna arrays. We briefly review some of the key takeaways from this chapter. First, a sufficiently dense mmWave cellular network with directional beamforming may enable remotely powered operation of energy harvesting devices. Second, mmWave may outperform lower frequency solutions by leveraging directional antenna arrays. Third, the energy coverage probability benefits from narrower beams for connected and wider beams for nonconnected users. This tradeoff is evident in the more general scenario having both types of users, where there typically exists an optimal beamforming beamwidth that maximizes the network-wide energy coverage.

Acknowledgements. This work was supported in part by the Army Research Office under grant W911NF-14-1-0460, and a gift from Mitsubishi Electric Research Labs, Cambridge.

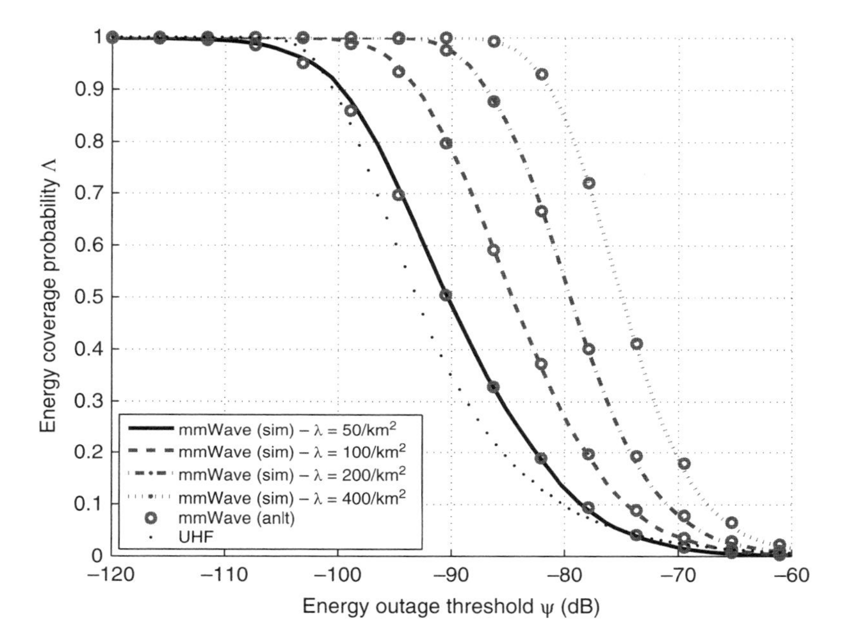

Figure 8.6 Energy coverage probability $\Lambda(0, \psi, \lambda)$ for different network densities for nonconnected users. Transmit beam pattern is fixed to $[10, -10, 30°, 330°]$. Other parameters are same as given in Figure 8.2. Monte Carlo simulation (sim) results validate the analytical (anlt) results obtained using Theorem 2 with $N = 5$ terms [6].

8.6 Appendix

We provide some lemmas here which play a key role in deriving the main analytical results.

Lemma 8.1 ***(Modified from [14, Theorem 8])*:** The probability density function (PDF) of the distance from an energy harvesting user to its nearest LOS BS, given that the user observes at least one LOS BS, is given by $\tau_{\mathrm{L}}(x) = 2\pi\lambda B_{\mathrm{L}}{}^{-1} x p(x) e^{-2\pi\lambda\int_0^x vp(v)\mathrm{d}v}$, where $x > 0$ and $B_{\mathrm{L}} = 1 - e^{-2\pi\lambda\int_0^\infty vp(v)\mathrm{d}v}$ is the probability that the receiver observes at least one LOS BS. Similarly, the distance distribution of the link between the user and its nearest NLOS BS, given that the user observes at least one NLOS BS, is given by $\tau_{\mathrm{N}}(x) = 2\pi\lambda B_{\mathrm{N}}{}^{-1} x(1 - p(x)) e^{-2\pi\lambda\int_0^x v(1-p(v))\mathrm{d}v}$, where $x > 0$ and $B_{\mathrm{N}} = 1 - e^{-2\pi\lambda\int_0^\infty v(1-p(v))\mathrm{d}v}$ is the probability that the user observes at least one NLOS BS.

Lemma 8.2 ***([Modified from [15, Lemma 2])*:** Let ϱ_{L} and ϱ_{N} denote the probability that the energy harvesting user is connected to a LOS and a NLOS

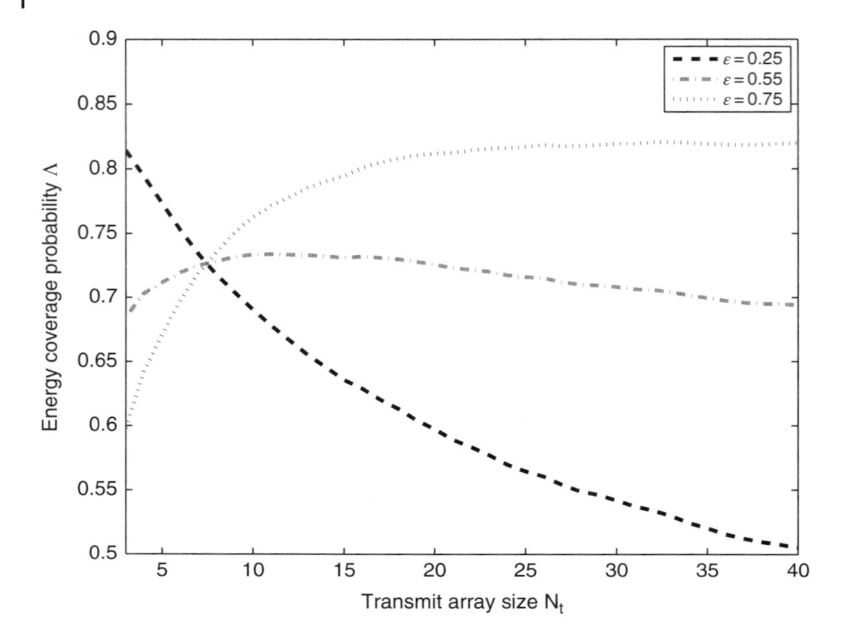

Figure 8.7 The overall energy coverage probability $\Lambda(\epsilon, \psi, \lambda)$ for different values of ϵ. Depending on the fraction of users operating in connected/nonconnnected modes, the transmit array size (which controls the beamforming beamwidth in this example) can be optimized to maximize the network-wide energy coverage. This could translate into massive gains given that the number of served devices would be potentially large. The users are assumed to be equipped with a single omnidirectional receive antenna. The energy outage threshold ψ is −70 dB for $\Phi_{u,\text{con}}$ and −85 dB for $\Phi_{u,\text{ncon}}$. $P_t = 13$ dB, $\lambda = 200/\text{km}^2$. Channel parameters are the same as used in Figure 8.2.

BS, respectively, then ϱ_{L} is given by $\varrho_{\text{L}} = B_{\text{L}} \int_0^{\infty} e^{-2\pi\lambda \int_0^{\rho_{\text{L}}(x)} (1-p(v))v\text{d}v} \tau_{\text{L}}(x)\text{d}x$, where $\rho_{\text{L}}(x) = \left(\frac{C_{\text{N}}}{C_{\text{L}}}\right)^{\frac{1}{\alpha_{\text{N}}}} x^{\frac{\alpha_{\text{L}}}{\alpha_{\text{N}}}}$ and $\varrho_{\text{N}} = 1 - \varrho_{\text{L}}$.

Lemma 8.3 ***(Modified from [15, Lemma 3])***: Given that the energy harvesting user is connected to a LOS mmWave BS, the PDF of the link distance is given by the expression $\tilde{\tau}_{\text{L}}(x) = \frac{B_{\text{L}}\tau_{\text{L}}(x)}{\varrho_{\text{L}}} e^{-2\pi\lambda \int_0^{\rho_{\text{L}}(x)} (1-p(v))v\text{d}v}$, where $x > 0$. Given that the user is connected to a NLOS mmWave BS, the PDF of the link distance is given by $\tilde{\tau}_{\text{N}}(x) = \frac{B_{\text{N}}\tau_{\text{N}}(x)}{\varrho_{\text{N}}} e^{-2\pi\lambda \int_0^{\rho_{\text{N}}(x)} p(v)v\text{d}v}$ for $x > 0$ and $\rho_{\text{N}}(x) = \left(\frac{C_{\text{L}}}{C_{\text{N}}}\right)^{\frac{1}{\alpha_{\text{L}}}} x^{\frac{\alpha_{\text{N}}}{\alpha_{\text{L}}}}$.

Bibliography

1 T.S. Rappaport, R.W. Heath Jr, R.C. Daniels, and J.N. Murdock (2014) *Millimeter Wave Wireless Communications*. Pearson Education.

2 M. Tabesh et al. (2014) A power-harvesting pad-less mm-sized 24/60 GHz passive radio with on-chip antennas. In *Proceedings of the Symposium on VLSI Circuits and Digital Technical Papers*, pp. 1–2, IEEE.

3 M. Tabesh et al. (2015) A power-harvesting pad-less millimeter-sized radio. *IEEE J. Solid-State Circuits*, **50**: 962–977.

4 J. Charthad, N. Dolatsha, A. Rekhi, and A. Arbabian (2016) System-level analysis of far-field radio frequency power delivery for mm-sized sensor nodes. *IEEE Trans. Circuits Syst. I: Reg. Papers*, **63**: 300–311.

5 T.A. Khan, A. Alkhateeb, and R.W. Heath (2015) Energy coverage in millimeter wave energy harvesting networks. In *Proceedings of the 2015 IEEE Globecom Workshops*, pp. 1–6.

6 T.A. Khan, A. Alkhateeb, and R. Heath (2016) Millimeter wave energy harvesting. *IEEE Trans. Wireless Commun.*, **15**: 6048–6062.

7 L. Wang et al. (2015) Millimeter wave power transfer and information transmission. In *Proceedings of the IEEE Global Communications Conference*, pp. 1–6.

8 M. Di Renzo and W. Lu (2016) System-level analysis and optimization of cellular networks with simultaneous wireless information and power transfer: Stochastic geometry modeling. *IEEE Trans. Veh. Tech.*, **66**: 2251–2275.

9 G. Kamga and S. Aissa (2017) Scaling laws for wireless energy transmission in mmWave massive MIMO systems. *2017 IEEE International Conference Communucations (ICC)*, Paris, pp. 1–6.

10 C. Psomas and I. Krikidis (2016) Blockage effects on joint information energy transfer in directional ad-hoc networks. In *Proceedings of the 24th European Signal Processing Conference*, pp. 808–812.

11 X. Zhou, S. Durrani, and J. Guo (2017) Characterization of aggregate received power from power beacons in millimeter wave ad hoc networks. *2017 IEEE International Conference on Communications (ICC)*, Paris, pp. 1–7.

12 T.A. Khan and R.W. Heath (2017) Wireless power transfer in millimeter wave tactical networks. *IEEE Signal Process. Lett.*, **24**: 1284–1287.

13 S. Biswas, S. Vuppala, and T. Ratnarajah (2016) On the performance of mmWave networks aided by wirelessly powered relays. *IEEE J. Selected Topics Signal Process.*, **10** (8): 1522–1537.

14 T. Bai, R. Vaze, and R. Heath (2014) Analysis of blockage effects on urban cellular networks. *IEEE Trans. Wireless Commun.*, **13**: 5070–5083.

15 T. Bai and R. Heath (2015) Coverage and rate analysis for millimeter-wave cellular networks. *IEEE Trans. Wireless Commun.*, **14**: 1100–1114.

16 S. Rangan, T. Rappaport, and E. Erkip (2014) Millimeter-wave cellular wireless networks: Potentials and challenges. *Proceedings of the IEEE*, **102**: 366–385.

17 C. Valenta and G. Durgin (2014) Harvesting wireless power: Survey of energy-harvester conversion efficiency in far-field, wireless power transfer systems. *IEEE Microwave Mag.*, **15**: 108–120.
18 T. Wu, T.S. Rappaport, and C.M. Collins (2015) The human body and millimeter-wave wireless communication systems: Interactions and implications. In *2015 IEEE International Conference on Communications (ICC)*, pp. 2423–2429.
19 A. Ghosh et al. (2010) *Fundamentals of LTE*. Pearson Education.
20 H. L. Van Trees (2004) *Detection, estimation, and modulation theory, optimum array processing*. John Wiley & Sons.

9

Wireless Information and Power Transfer in Relaying Systems

Panagiotis D. Diamantoulakis[1*], *Koralia N. Pappi*[2,3], *and George K. Karagiannidis*[3]

[1] *Institute for Digital Communications, Friedrich-Alexander University, Erlangen D-91058, Germany*
[2] *Intracom S. A. Telecom Solutions, GR-57001, Thessaloniki, Greece*
[3] *Electrical and Computer Engineering Department, Aristotle University of Thessaloniki, GR-54124 Thessaloniki, Greece*

9.1 Introduction

While the era of the fifth-generation (5G) communication systems and beyond is rapidly approaching, future networks pose new challenges, displaying a group of diverse characteristics. These networks are expected to accommodate a vast number of wireless connected devices with increased throughput, which will be low-cost, with extended lifetime, environmentally friendly, and with no need for hard-wiring. Furthermore, with the introduction of the Internet-of-Things (IoT), many objects will be connected to the internet, including devices such as smart meters, sensors, etc., which are remotely situated. These remote devices may not have access to energy resources, and they may also be situated at a great distance from the destination of their transmitted information [1].

In the light of the above considerations, energy harvesting (EH), as an emerging solution for prolonging the lifetime of the energy-constrained wireless devices, has gained significant research interest in recent years [1]. Conventional energy harvesting techniques which rely on external natural energy resources, such as wind or solar energy, may not be able to offer reliable and uninterrupted communication due to the intermittent nature of these resources. On the other hand, the abundance of radio frequency (RF) signals and its controllable nature render them a promising energy resource for powering communication devices [2]. Since RF signals are capable of transferring both information and energy, the concept of simultaneous wireless information and power transfer (SWIPT) has been proposed, where the receiving device can perform two functions, energy harvesting and information reception [3].

*Corresponding author: P. D. Diamantoulakis; padiamant@gmail.com

Wireless Information and Power Transfer: Theory and Practice, First Edition.
Edited by Derrick Wing Kwan Ng, Trung Q. Duong, Caijun Zhong, and Robert Schober.

Wireless power transfer (WPT), which is the remote powering of a device using RF signaling, has been demonstrated as a generic energy harvesting technique. However, its application to cooperative networks is of particular importance. A network with multiple energy-constrained nodes which may be remotely situated often relies on relaying the transmitted information between nodes in order to provide better coverage. The employment of relays is known to enhance the quality of service (QoS) and to increase the network coverage, especially when there is no line-of-sight (LOS) between the source and the destination. With simultaneous employment of energy harvesting and WPT, the relay nodes can be self-powered [4–6].

In practice, a node cannot harvest power and process the information from the received signal at the same time. Hence, two main practical receiver architectures have been proposed, namely, "time-switching", where the receiver switches between decoding information and harvesting energy, and "power-splitting", where the receiver splits the signal into two streams, one for information decoding and the other for energy harvesting [7, 8]. A number of works have appeared in the literature investigating different aspects of simultaneous information and energy transfer with practical receivers.

In this chapter, wireless cooperative networks are studied in which the available relays can harvest energy from the received RF signals. More specifically, two types of networks will be examined, those with a single source and destination, and those with multiple sources or destinations [4, 5, 9–12]. In both types, one or multiple relays will be considered, under the assumption that the power consumption at the relay due to information processing is negligible compared to the power used for signal transmission from the relay to the destination [7]. Furthermore, some interesting future research directions will be discussed.

9.2 Wireless-Powered Cooperative Networks with a Single Source–Destination Pair

9.2.1 System Model and Outline

In this section we examine the case of a single source–destination pair, the communication of which is assisted by one or multiple relays, which are wireless powered via RF signals.

The single source is denoted by S, the destination is denoted by D, and the relays are denoted by R_k, with $1 \leq k \leq K$, where K is the number of available relays. The operating relay(s) is considered to be powered by the RF signal of the source, by performing energy harvesting.

In the following subsections, the various relaying protocols employing wireless power transfer will be examined. Furthermore, the impact of multiple antennas and the impact of multiple relays and relay selection will also be

Figure 9.1 Energy-constrained relay-assisted communication.

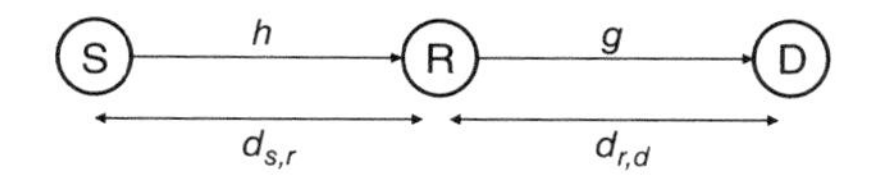

explored. Finally, in terms of the available energy at the relays, power allocation strategies at the source and relay will be presented, when multiple frequency bands are used.

9.2.2 Wireless Energy Harvesting Relaying Protocols

In this section the basic relaying protocols with energy harvesting at the relay are introduced, for a two-hop relaying network with one relay, as depicted in Figure 9.1, where h and g are the channel gains of the first and second hop, respectively, while $d_{s,r}$ and $d_{r,d}$ are the distances between S-R and R-D, respectively.

Two basic relaying protocols which facilitate SWIPT are described in [7], namely *time switching-based relaying protocol (TSR)* and *power splitting-based relaying protocol (PSR)*. These two protocols were examined for the delay-limited and the delay-tolerant transmission case, which refer to different code length. The results were presented for amplify-and-forward (AF) relaying, although they can be extended to decode-and-forward (DF) as well, in a similar manner.

The analysis was performed under the following assumptions:

1) The channel gains of the two hops are quasi-static and constant over the *block time* T, but change independently and identically distributed between blocks, following a Rayleigh distribution.
2) Perfect channel state information is only available at the destination.

Regarding the transmission modes, in the delay-limited mode the code length is no larger than the block time T, and each block is independently decoded. Thus, in the delay-limited case the throughput is determined by evaluating the outage probability at a fixed source transmission rate. On the other hand, in the delay-tolerant mode the code length can be kept very large compared to the transmission block time. Therefore, the source can transmit data at any rate less than or equal to the ergodic capacity, which is the metric used to evaluate the performance in the delay-tolerant case.

9.2.2.1 Time Switching-Based Relaying

If T is the block time in which a certain block of information is transmitted from the source node to the destination node, let a be the fraction of the block time in which the relay harvests energy from the source signal, where $0 \leq a \leq 1$. The remaining block time, $(1 - a)T$, is used for information transmission, such that half of that, $(1-a)\frac{T}{2}$, is used for the source to relay information transmission and the remaining half, $(1-a)\frac{T}{2}$, is used for the relay to destination information transmission. The choice of a affects the achievable throughput at the destination [7].

The harvested energy during the time fraction aT at the relay is given by

$$E_h = \frac{\eta P_s |h|^2}{d_{s,r}^{\alpha}} aT, \tag{9.1}$$

where P_s is the transmit power of the source, $0 < \eta < 1$ is the energy harvesting efficiency, and α is the path-loss exponent. During information transmission, the transmit power of the relay is given by

$$P_r = \frac{E_h}{(1-a)\frac{T}{2}} = \frac{2\eta P_s |h|^2 a}{(1-a) d_{s,r}^{\alpha}}. \tag{9.2}$$

Delay-limited mode: Let γ_d be the signal-to-noise ratio (SNR) at the destination. In the case of the delay-limited communication with fixed source transmission rate R, the outage probability for AF relaying is defined as

$$\mathcal{P} = \Pr(\gamma_d < 2^R - 1). \tag{9.3}$$

Given that the transmitter is communicating R bits/s/Hz and $(1-a)\frac{T}{2}$ is the effective communication time from the source node to the destination node in the block of time T seconds, then the throughput at the destination is given by

$$\tau = (1-\mathcal{P})R\frac{(1-a)\frac{T}{2}}{T} = \frac{(1-\mathcal{P})R(1-a)}{2}. \tag{9.4}$$

Delay-tolerant mode: In the case of delay-tolerant communication, the ergodic capacity can be calculated as

$$C = \mathbb{E}_{h,g}[\log_2(1+\gamma_d)]. \tag{9.5}$$

Given that the source is transmitting at a fixed rate equal to the ergodic capacity C, and that the code length is long enough so that C is an achievable rate, then the throughput is given by [7]

$$\tau = \frac{(1-a)\frac{T}{2}}{T} C = \frac{(1-a)C}{2}. \tag{9.6}$$

9.2.2.2 Power Splitting-Based Relaying

When power splitting is considered at the relay, the time block T is divided into two slots, where $\frac{T}{2}$ is used for the source to relay information transmission and the remaining half, $\frac{T}{2}$, is used for the relay to destination information transmission. During the first slot, a fraction θ of the total received power is used for energy harvesting, while the fraction $(1-\theta)$ is used for information detection, where $0 < \theta < 1$.

In this scenario, the energy harvested at the relay is given by

$$E_h = \frac{T\eta\theta P_s|h|^2}{2d_{s,r}^{\alpha}}. \tag{9.7}$$

Accordingly, in the second slot, the relay transmits with power

$$P_r = \frac{E_h}{\frac{T}{2}} = \frac{\eta P_s|h|^2\theta}{d_{s,r}^{\alpha}}. \tag{9.8}$$

Delay-limited mode: For PSR, the outage probability is calculated similar to the TSR protocol, where γ_d is a function of θ. However, the effective communication time is $\frac{T}{2}$, so the throughput at the destination is given by [7]

$$\tau = (1-\mathcal{P})R\frac{\frac{T}{2}}{T} = \frac{(1-\mathcal{P})R}{2}. \tag{9.9}$$

Delay-tolerant mode: Since $\frac{T}{2}$ is the effective communication time, the throughput is calculated as [7]

$$\tau = C\frac{\frac{T}{2}}{T} = \frac{C}{2}, \tag{9.10}$$

where C is the ergodic capacity, which is calculated similar to the TSR protocol, and it is a function of θ.

9.2.3 Multiple Antennas at the Relay

This section investigates the impact of deploying multiple antennas at the energy harvesting relay. Let N_r be the number of antennas equipped at the relay R, while the source S and the destination D are assumed to be equipped with a single antenna. In [12], the authors examined the outage probability, ergodic capacity, diversity gain, and optimal power splitting ratio for the above setup, when power splitting is assumed at the relay. More specifically, a common power splitting ratio θ was assumed at each antenna of the relay, therefore the received signal at the relay is [12]

$$\mathbf{y}_r = \sqrt{\frac{(1-\theta)P_s}{d_{s,r}^{\alpha}}}\mathbf{h}x + \mathbf{n}_r, \tag{9.11}$$

where $\mathbf{y}_r$, $\mathbf{h}$, and $\mathbf{n}_r$ are vectors of size $N_r \times 1$, x is the transmitted information symbol, and $\mathbf{n}_r$ is the additive white Gaussian noise (AWGN) vector at the relay.

Again, the time block T is split into two slots, as in the PSR protocol. After the first time slot of duration $\frac{T}{2}$, the harvested energy at the relay is

$$E_h = \frac{T\eta\theta P_s||\mathbf{h}||^2}{2d_{s,r}^{\alpha}}. \tag{9.12}$$

Similar to the AF protocol for a single-antenna relay, the relay transmits a transformed version of the received vector in the second time slot to the destination. Thus, the received signal at the destination can be written as [12]

$$y_d = \sqrt{\frac{1}{d_{r,d}^{\alpha}}}\mathbf{g}\mathbf{W}\mathbf{y}_r + n_d, \tag{9.13}$$

where $\mathbf{g}$ is the $1 \times N_r$ channel gain vector of the second hop, $\mathbf{W}$ is a $N_r \times N_r$ transformation matrix, and n_d is the AWGN noise at the destination.

In the case where there is no interference at relay, the optimal transformation matrix is given by [12]

$$\mathbf{W} = \omega\frac{\mathbf{g}^{\dagger}\mathbf{h}^{\dagger}}{||\mathbf{g}||||\mathbf{h}||}, \tag{9.14}$$

where $(\cdot)^{\dagger}$ denotes the conjugate transpose, while ω is defined by

$$\omega^2 = \frac{\frac{\eta\theta P_s}{d_{s,r}^{\alpha}}||\mathbf{h}||^2}{\frac{(1-\theta)P_s}{d_{s,r}^{\alpha}}||\mathbf{h}||^2 + N_0}, \tag{9.15}$$

where N_0 is the AWGN power spectral density.

The end-to-end SNR of the system can be expressed as follows

$$\gamma = \frac{\gamma_1\gamma_2}{\gamma_1 + \gamma_2 + 1}, \tag{9.16}$$

where

$$\gamma_1 = \frac{(1-\theta)P_s}{N_0 d_{s,r}^{\alpha}}||\mathbf{h}||^2 \tag{9.17}$$

and

$$\gamma_2 = \frac{\eta\theta P_s}{N_0 d_{s,r}^{\alpha} d_{r,d}^{\alpha}}||\mathbf{g}||^2||\mathbf{h}||^2. \tag{9.18}$$

Based on the above expressions, in [12] tight bounds and approximations were formulated for computing the outage probability and the ergodic capacity. Furthermore, a high SNR approximation of the outage probability showed that the diversity order of the network is equal to N_r, revealing that an energy harvesting MIMO relay can achieve full diversity.

Finally, based on the approximation of the ergodic capacity, an optimal power splitting ratio θ can be calculated, which is proven to be unique, for large values of SNR.

9.2.4 Multiple Relays and Relay Selection Strategies

The case of a network with one pair of source–destination and multiple relays, and various relay selection strategies, was investigated in [4]. More specifically, the authors in this work assume that the direct channel between the source and the destination is also available, while the relays perform DF relaying.

The following assumptions are made:

1) A time block T is divided into two equal slots of duration $\frac{T}{2}$.
2) During the first time slot, the destination and the relays receive the information transmitted by the source at rate R. At the relays, if the information is decoded successfully, the rest of the received power is harvested for transmission in the second time slot.
3) A selected relay, if it has enough harvested power, re-transmits the message to the destination during the second time slot, while the destination performs maximal ratio combining (MRC) for the signals received during both time slots.

Based on the assumptions above, if R is the transmission rate of the source, then the relays need to decode the transmitted message successfully, based on their received signal

$$y_{r,i} = \sqrt{\frac{(1-\theta_{r,i})P_s}{1+d_{s,r_i}^{\alpha}}} h_{r,i}x + n_{r,i}, \tag{9.19}$$

where $i = 1, 2, \ldots, K$ is the index of the relay, while in the expression above, the bounded path-loss model is used, i.e. $1 + d_{s,r_i}^{\alpha}$ instead of d_{s,r_i}^{α}, so that it is valid also for distances less than one. Note that each relay chooses its own power splitting factor θ_i, in order to ensure that a rate R can successfully be decoded, according to the following rule

$$\theta_{r,i} = \max\left(0, 1 - \frac{(1+\mathrm{d}_{s,r_i}^{\alpha})(2^{2R}-1)}{P_s|h_{r,i}|^2}\right). \tag{9.20}$$

If $\theta_i > 0$, then the harvested energy is used for transmission during the second time slot, with power

$$P_{r,i} = \eta\left(\frac{|h_{r,i}|^2}{1+d_{s,r_i}^{\alpha}}P_s - (2^{2R}-1)\right) \tag{9.21}$$

Three relay selection strategies were investigated, namely *random relay selection, relay selection based on second-order statistics*, and *distributed beamforming* [4], which are briefly described below.

9.2.4.1 Random Relay Selection

In this scenario, one random relay is selected by the source for retransmission. This scheme does not require any channel state information (CSI) [4]. Provided that $\theta_{r,i} > 0$, the relay retransmits the received information during the second time slot, utilizing the harvested energy. The received signal at the destination is then written as

$$y_d = \sqrt{\frac{P_{r,i}}{1 + d^{\alpha}_{r_i,d}}} g_{r,i} x + n_d, \tag{9.22}$$

while the SNR at the destination after combining, when the ith relay is selected, is given by

$$\gamma_{d,i} = \frac{|h_0|^2 P_s}{1 + d^{\alpha}_{s,d}} + \frac{\eta |g_{r,i}|^2}{1 + d^{\alpha}_{r_i,d}} \left(\frac{|h_{r,i}|^2}{1 + d^{\alpha}_{s,r_i}} P_s - (2^{2R} - 1) \right), \tag{9.23}$$

where h_0 is the channel gain and $d_{s,d}$ is the distance corresponding to the S-D path. The AWGN power is assumed to be unitary.

The outage probability, when the ith relay is selected, is formulated as

$$\begin{aligned} \mathcal{P}_i = \Pr & \left(\gamma_{d,i} < 2^{2R} - 1, \frac{P_s |h_{r,i}|^2}{1 + d^{\alpha}_{s,r_i}} > 2^{2R} - 1 \right) \\ & + \Pr \left(\frac{P_s |h_0|^2}{1 + d^{\alpha}_{s,d}} < 2^{2R} - 1, \frac{P_s |h_{r,i}|^2}{1 + d^{\alpha}_{s,r_i}} < 2^{2R} - 1 \right). \end{aligned} \tag{9.24}$$

Based on the above expression and its high SNR approximation, it was shown in [4] that, when the relays are closely situated to the source and the path-loss exponent is two, then the diversity gain of a randomly selected relay is two. Furthermore, it was also proved that the outage probability of a network with energy harvesting relays decays with a rate of $\ln SNR/SNR^2$, in contrast to the outage probability of a conventional network, which decays at a faster rate of $1/SNR^2$.

9.2.4.2 Relay Selection Based on Distance

In many practical scenarios, the second-order statistics of the channel can be obtained. These mainly depend on the distance between nodes, which changes more slowly than the fast fading channel coefficients.

Under the assumption that there are multiple available relays, closely situated to the source, and for $\alpha = 2$, in [4] it was proven that the relay which is closest to the source minimizes the outage probability at high SNR. Based on this, and by using a high-SNR approximation of the outage probability, it was also proven that the diversity order of this relay selection scheme is also two. Interestingly, the knowledge of the second-order statistics does not increase the diversity gain, compared to the random relay selection scheme. However, it improves the achieved outage probability.

9.2.4.3 Distributed Beamforming

Another relay selection scheme was also investigated in [4], namely *distributed beamforming*, which assumes global channel knowledge at the source and the relays.

In this case, the relays are divided into two sets, the set of qualified relays, denoted by $\mathcal{A}$, which are able to decode the source message, and the set of unqualified relays, denoted by $\tilde{\mathcal{A}}$, where $|\mathcal{A}| + |\tilde{\mathcal{A}}| = K$, with $|\cdot|$ being the cardinality of the set and K being the total number of available relays.

Then, the ith relay, with $i \in \mathcal{A}$, transmits

$$x_{r_i} = \frac{g_{r,i}^* P_{r,i}}{\sqrt{\xi(1 + d_{r_i,d}^{\alpha})}} x, \tag{9.25}$$

where x is the originally transmitted message by the source, $(\cdot)^*$ denotes the conjugate of a complex value, and ξ is the power normalization factor to ensure that the transmission power of the ith relay is less than $P_{r,i}$, which is given by

$$\xi = \sum_{i \in \mathcal{A}} \frac{|g_{r,i}|^2 P_{r,i}}{1 + d_{r_i,d}^{\alpha}}. \tag{9.26}$$

Then, the observation at the destination is given by

$$y_d = \sum_{i \in \mathcal{A}} \frac{g_{r,i}}{\sqrt{1 + d_{r_i,d}^{\alpha}}} \frac{g_{r,i}^* P_{r,i}}{\sqrt{\xi(1 + d_{r_i,d}^{\alpha})}} x + n_d. \tag{9.27}$$

Employing MRC over the two time slots and assuming unitary AWGN power, the SNR at the destination is given by

$$\gamma_{d,\mathcal{A}} = \frac{|h_0|^2 P_s}{1 + d_{s,d}^{\alpha}} + \sum_{i \in \mathcal{A}} \eta \frac{|g_{r,i}|^2}{1 + d_{r_i,d}^{\alpha}} \left(\frac{|h_{r,i}|^2}{1 + d_{s,r_i}^{\alpha}} P_s - (2^{2R} - 1) \right). \tag{9.28}$$

If Π is the set containing all possible partitions of the available relays into distinct pairs of $\mathcal{A}$ and $\tilde{\mathcal{A}}$, then the outage probability of distributed beamforming for K relays can be written as

$$\begin{aligned} \mathcal{P}_K = \sum_{\Pi} Pr \Bigg(& \gamma_{d,\mathcal{A}} < 2^{2R} - 1 \cap \frac{P_s |h_{r,i}|^2}{1 + d_{s,r_i}^{\alpha}} > 2^{2R} - 1, i \in \mathcal{A} \\ & \cap \frac{P_s |h_{r,j}|^2}{1 + d_{s,r_j}^{\alpha}} < 2^{2R} - 1, j \in \tilde{\mathcal{A}} \Bigg). \end{aligned} \tag{9.29}$$

Based on the above expression, and assuming multiple available relays, closely situated to the source, for $\alpha = 2$, it was proven in [4] that the system can achieve full diversity with distributed beamforming, i.e. diversity of $(K + 1)$.

This is a very important result, since it reveals that the system can still achieve full diversity, despite the fact that the relays are wirelessly powered.

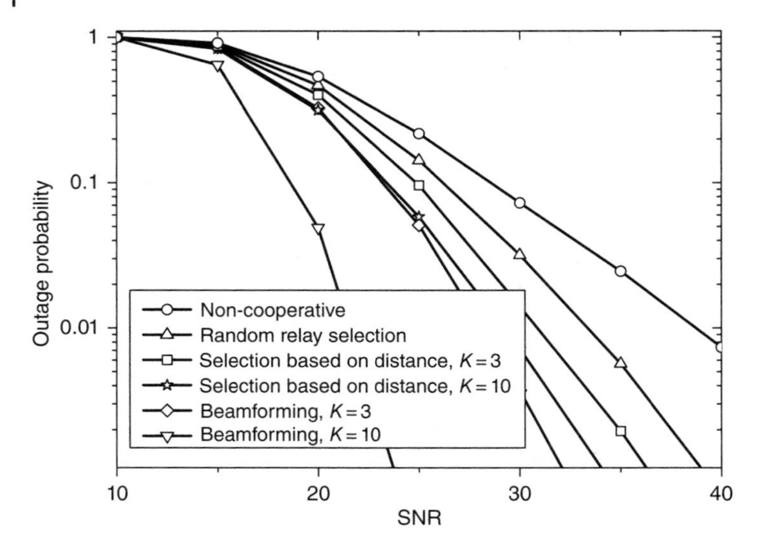

Figure 9.2 Outage probability achieved by different strategies [4] for $R = 1$ bits/channel and $\eta = 0.5$ [4].

9.2.4.4 Comparison of Relay Selection Strategies

In Figure 9.2 the three aforementioned relay selection strategies are compared in terms of outage probability, while, for the sake of comparison, the performance of non-cooperative direct transmission scheme is also included. It is assumed that the source is located at the origin of the disc and the radius of the disc is 5 m. It is further assumed that the K available relays are Poisson distributed. The distance between the source and the destination is 10 m. It is observed that distributed beamforming outperforms the other two schemes, while the random relay selection achieves the worst performance. Also, it is observed that the slope of the outage curve for the relay selection based on distance does not depend on the number of relays. On the other hand, it becomes apparent that when distributed beamforming is used the diversity gain increases proportionally to the number of relays.

9.2.5 Power Allocation Strategies for Multiple Carriers

In multicarrier cooperative networks, the efficient power utilization at both the source and the relay is critical. It should be noted that the optimal power allocation at both nodes is quite challenging, since it also depends on the level of the harvested energy and, thus, the corresponding parameter, i.e. the time-switching or power-splitting parameter, which should also be optimized [9]. For the sake of simplicity, in this subsection a single relay is assumed, while all nodes are assumed to have a single antenna [9]. Also, assuming that

N_c available channels, AF relaying, and power-splitting are used, the total achievable rate can be written as

$$\mathcal{R} = \sum_{i=1}^{N_c} \frac{1}{2}\log_2(1 + \gamma^{[i]}), \tag{9.30}$$

where $\gamma^{[i]}$ is the end-to-end SNR in each frequency band.

Moreover, the achievable rate maximization problem can be written as [9]

Problem 1: Achievable rate maximization [9]:

$$\begin{aligned}
&\underset{\mathcal{P}_s,\theta,\mathcal{P}_r}{\text{maximize}} \ \mathcal{R} \\
\text{s.t.} \quad & \text{C1}: \sum_{i=1}^{N_c} P_r^{[i]} \le P_{rt}, \text{C2}: \sum_{i=1}^{N_c} P_r^{[i]} \le P_{rm}, \\
& \text{C3}: \sum_{i=1}^{N_c} P_s^{[i]} \le P_{sm}, \text{C4}: 0 \le \theta \le 1, \\
& \text{C5}: P_s^{[i]} \ge 0, \ \forall i, \text{C6}: P_r^{[i]} \ge 0, \ \forall i,
\end{aligned} \tag{9.31}$$

where $\mathcal{P}_s = \{P_s^{[1]} \cdots P_s^{[N_c]}\}$ and $\mathcal{P}_r = \{P_r^{[1]} \cdots P_r^{[N_c]}\}$ are the sets of the allocated power, at the source and the relay, respectively. Constraint C_1 represents the limited harvested power which is available for retransmission. The total harvested power at the relay, denoted by P_{rt}, is a function of $\mathcal{P}_s$ and θ and is given by [9]

$$P_{rt} = \eta\theta \sum_{i=1}^{N_c} P_s^{[i]} \frac{|h^{[i]}|^2}{d_{s,r}^{\alpha_i}}, \tag{9.32}$$

where $h^{[i]}$ denotes the source to relay channel gain over the ith carrier and α_i is the path-loss exponent for the ith carrier. Constraints C_2 and C_3 include the hardware and regulations limitations P_{rm} and P_{sm} on the total transmitted power by R and S, respectively.

It is highlighted that this is not a convex optimization problem, but it can be solved by convex optimization tools, using one-dimensional search for θ of alternating optimization. The optimal method, i.e. the solution of the optimization problem in (9.31), achieves a significantly higher achievable rate compared to other simpler methods, e.g., (i) equal power allocation between the channels and (ii) selection of only the best channel, at the expense of complexity. For example, as shown in Figure 9.3, under the assumption of Rayleigh fading, bounded path-loss model [4, 9], $d_{s,r} = 0.9$ m, $d_{r,d} = 1.35$ m, and the path-loss exponent given by $\alpha_i = 2 + 0.5i$ for the ith carrier, the optimal

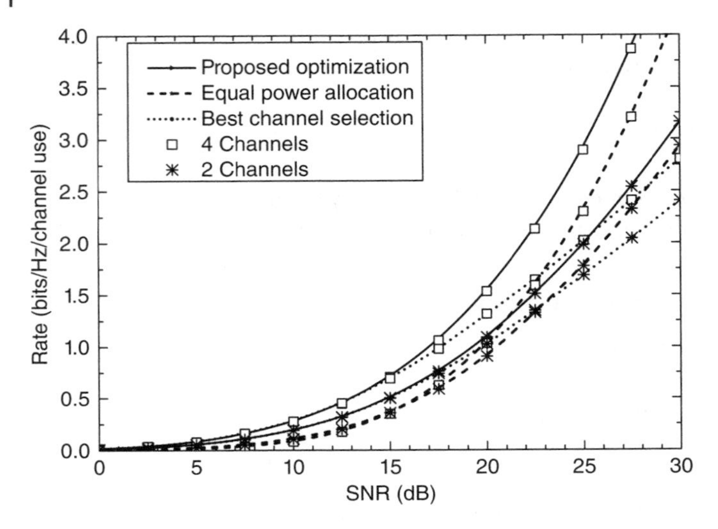

Figure 9.3 Rate against SNR for different number of channels [9].

method, for power ratio values over 15 dB, displays a performance gain gap of about 1 dB and 2 dB when two or four channels are used, respectively.

9.3 Wireless-Powered Cooperative Networks with Multiple Sources

9.3.1 System Model

In this section, a wireless cooperative network is considered, in which multiple sources communicate with multiple destinations through energy harvesting relays, using orthogonal channels, e.g., different time slots. This setup creates several challenges, such as the efficient power allocation at the relay, the relay selection, and the integration of two-way relaying capability. To give further insight on these challenges, in the next subsections, we will focus on some cases of these networks with special interest. More specifically, optimal power allocation will be revisited under the assumption of M source-destination pairs and a single relay, while relay selection will be discussed under the assumption of a single destination. Moreover, for the two-way relaying networks two sources and one relay will be assumed. It is noted that this section focuses solely on DF relaying, although the results can be extended to AF as well. Moreover, for the sake of simplicity, hereinafter the utilized channel gain parameters will represent both the small-scale fading and path loss, while all power values will be normalized to N_0.

9.3.2 Power Allocation Strategies

As has already been mentioned, this subsection focuses on the investigation of power allocation in energy harvesting relaying networks with multiple pairs of sources and destinations. The cooperative transmission consists of two time slots, each of which is of duration $\frac{T}{2}$, while the sources are assumed to transmit over orthogonal resources (e.g., different frequencies). The relay uses the power-splitting protocol. Let θ_i denote the power splitting coefficient for the ith user pair. The observation at the relay which is used for information decoding of the ith user's message is given by [5]

$$y_{r,i} = \sqrt{(1-\theta_i)P_{s,i}}h_i s_i + n_r, \tag{9.33}$$

where $P_{s,i}$ denotes the transmission power of the ith source, $h_{i,r}$ is the channel gain between the ith source and the relay, s_i is the source message with unit power, and n_r is the additive white Gaussian noise (AWGN).

The data rate at which the relay can decode the ith source's signal is

$$R_{r,i} = \frac{1}{2}\log_2(1 + (1-\theta_i)P_{s,i}|h_i|^2). \tag{9.34}$$

When the targeted rate is R, then the optimal selection of θ_i can be written as

$$\theta_i = 1 - \frac{2^{2R}-1}{P_{s,i}|h_i|^2}. \tag{9.35}$$

At the end of the first time slot, the energy that is harvested by the relay from the ith source is given by

$$E_{H,i} = \eta P_{s,i}|h_i|^2\theta_i\frac{T}{2}. \tag{9.36}$$

At the end of the second time slot, the observation at the ith destination is

$$y_{d,i} = \sqrt{P_i}g_i s_i + n_{d,i}, \tag{9.37}$$

where P_i is the relaying transmission power, g_i is the channel gain between the relay and the ith destination, and $n_{d,i}$ is the AWGN. Thus, the data rate at which the destination can decode the intended relaying transmission is

$$R_{d,i} = \frac{1}{2}\log_2(1 + P_i|g_i|^2). \tag{9.38}$$

Considering both time slots, the outage probability for the ith user pair transmission can be expressed as

$$\mathcal{P}_i = \Pr(R_{r,i} < R) + \Pr(R_{r,i} > R, R_{d,i} < R) \tag{9.39}$$

In the next subsections, four different power allocation schemes are revisited, namely the *non-cooperative individual transmission strategy*, *equal power allocation*, *opportunistic power allocation strategy*, and *auction based power allocation scheme* [5]. These strategies affect the fraction of $\sum_i^M E_{H,i}$ that is allocated to each pair's transmission and, thus, they have a nontrivial impact on the performance that each user can achieve.

9.3.2.1 Non-Cooperative Individual Transmission Strategy

When the users' pairs do not cooperate, then the relay only uses the harvested energy from the ith source for the second time slot of the cooperative transmission [5]. In this case, the relaying transmission power for the ith destination can be written as

$$P_i = \frac{E_{H,i}}{\frac{T}{2}} = \eta P_{s,i}|h_i|^2\theta_i. \tag{9.40}$$

Assuming that channels are identically and independently quasi-static Rayleigh fading, this strategy yields an outage probability at the ith destination of

$$\mathcal{P}_i = 1 - \frac{\exp(-\frac{2^{2R}-1}{P_{s,i}})}{P_{s,i}}\sqrt{\frac{4(2^{2R}-1)P_{s,i}}{\eta}}K_1\left(\sqrt{\frac{4(2^{2R}-1)}{\eta P_{s,i}}}\right). \tag{9.41}$$

$K_n(\cdot)$ denotes the modified Bessel function of the second kind with order n, where in the expression above it is $n = 1$.

It is easy to realize that this strategy might be inefficient, considering a simple case of two user pairs with $|h_1|^2 \gg |h_2|^2$ and $|g_1|^2 \gg |g_2|^2$. Then, the second pair has much higher probability of being in outage than the first one, since even if the relay manages to successfully decode the corresponding message, the capacity of the second link is much lower. This is due to the lower channel gain (g_i) and relaying transmission power, which lead to a double degradation of the second link.

9.3.2.2 Equal Power Allocation

In this strategy, the same amount of power is allocated by the relay to each user, i.e. $P_i = \frac{\sum_i^N E_{H,i}}{N}$, where N denotes the number of sources whose information can reliably be detected at the relay. Based on this and assuming that $P_{s,i} = P, \forall i$ the outage probability for the ith destination is given by [5]

$$\begin{aligned}\mathcal{P}_i = &\sum_{n=1}^{M}\frac{1}{(n-1)!}\left((n-1)! - 2\left(\frac{n(2^{2R}-1)}{\eta P}\right)^{\frac{n}{2}}K_n\left(2\sqrt{\frac{n(2^{2R}-1)}{\eta P}}\right)\right)\\ &\times\frac{(M-1)!}{(n-1)!(M-n)!}\exp\left(-\frac{n(2^{2R}-1)}{P}\right)\left(1-\exp\left(-\frac{2^{2R}-1}{P}\right)\right)^{M-n}\\ &+\left(1-\exp\left(-\frac{2^{2R}-1}{P}\right)\right)\end{aligned} \tag{9.42}$$

Using asymptotic analysis to compare this strategy with the non-cooperative individual transmission strategy, it is proven that equal power allocation achieves asymptotically lower average outage probability, as well as faster rate of decay, i.e. $\frac{1}{\text{SNR}}$ instead of $\frac{\log(\text{SNR})}{\text{SNR}}$.

9.3.2.3 Opportunistic Power Allocation Strategy

This strategy is based on sequential water filling according to which the relay orders the transmission power, such that the user with the higher channel gain value $g_{i,r}$ is allocated with the least power, provided that this user can be served. This depends on the sufficiency of harvested energy. More specifically, the relay first serves the destination with the strongest channel, e.g., destination j, by allocating the required amount of power, i.e. $\frac{2^{2R}-1}{|g_j|^2}$, such that the capacity of the second link is equal to the required rate R. Then, if there is sufficient energy left, the relay serves the sequential users by allocating $\frac{2^{2R}-1}{|g_i|^2}$ to each of them [5].

In contrast to former strategies, the opportunistic power allocation strategy increases the system complexity, since it requires the global channel state information at the relay. Also, the outage probability for any specific user cannot be calculated. Thus, in order to quantify performance, the outage probability of the user with the best and worst channel conditions can be calculated. However, it is proven that this strategy offers some important advantages compared to the former two. First, it manages to increase the average number of destinations that successfully receive the intended information. Moreover, it considerably reduces the outage probability for the destinations with the best and worst channel conditions. Furthermore, using asymptotic analysis to give further insight on the outage probability of the user with the worst channel conditions, the rate of decay of $\frac{1}{\text{SNR}}$ can be proved.

9.3.2.4 Game-Theoretic Formulation

Game-theoretic tools, e.g., the auction-based method [5, 13], can also be used to model the power allocation problem at the EH relay. Although there is an overlap between the game-theoretic and optimization-theoretic approaches, game theory tends to focus on the multi-user competitive nature of the problem and on the users' interaction. Also, game-theoretic strategies enable the development of distributed algorithms, which present particular advantages, such as scalability and, in some cases, reduction of the total complexity.

More specifically, when the auction-based method is used, the relay does not need to have access to the global CSI, while each destination only knows its own channel information. According to this method, each destination pays per amount of allocated power. To this end, each destination submits a bid to the relay, while higher bids increase both the allocated power and the required payment. A mathematic representation of the users' payoffs can be written as [5]

$$U_i = \frac{1}{2}\log_2(1 + P_i|g_i|^2) - cP_i \tag{9.43}$$

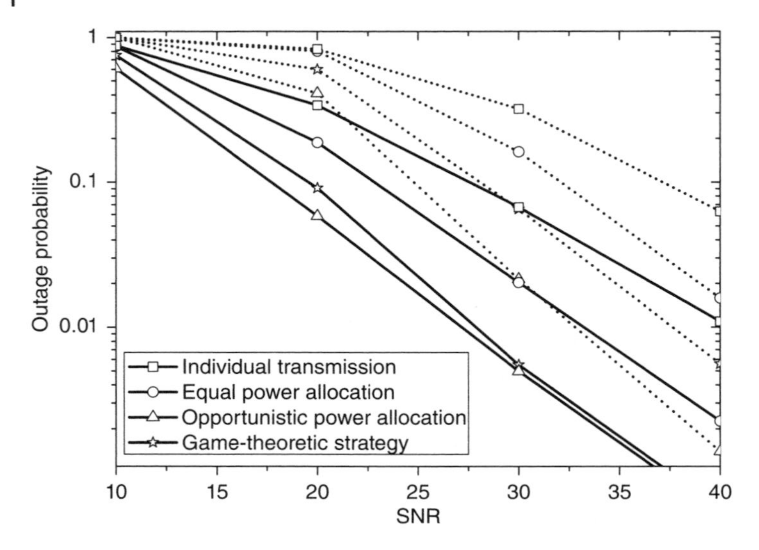

Figure 9.4 Average outage performance achieved by the studied transmission protocols for $R = \frac{1}{2}$ bits/channel use [5] for $\eta = 1$ (solid curves) and $\eta = 0.1$ (dotted curves) [5].

where

$$P_i = \frac{b_i}{\sum_{j=1}^{M} b_j + \psi} \sum_{j=1}^{M} \frac{E_{H,i}}{\frac{T}{2}}, \tag{9.44}$$

with ψ, b_i, and c being the factor related to the power reserved at the relay, the submitted bid, and the required payment per amount of utilized power, respectively. The optimal bidding profile is given by the Nash equilibrium of the game, which can be calculated iteratively. Specifically, the relay first announces the price to all destinations, which then update their bids. It is noted that, by using an appropriate updating function, each destination updates its bid without requiring the actions of other users.

9.3.2.5 Comparison of Power Allocation Strategies

As shown in Figure 9.4, where the four strategies are compared in terms of outage probability for $M = 20$, the outage performance of the auction-based method is worse than the water-filling strategy, but better than equal power allocation. Especially for high values of energy harvesting efficiency (η) it achieves performance close to that of water-filling, which does not happen as the value of η reduces.

9.3.3 Multiple Relays and Relay Selection Strategies

Game theory is also a useful tool in order to model the relay selection problem with multiple relays, which is the main focus of this section. More specifically, it is assumed that there are K available relays, which form M disjoint groups to assist the M source transmissions to a common destination by using a two-phase protocol [4]. The first time slot is dedicated to the sources' transmission, which occurs via orthogonal channels, while the second one is devoted to the relays' transmission. The destination combines the observation from first time slot with the received signal over the second time slot. For the sake of simplicity, a sole destination is considered. In order to achieve the maximum diversity gain, distributed beamforming can be used by the relays from the same group, which is denoted by S_m. Taking this into account, the SNR for the mth source message at the destination can be written as

$$\gamma_{S_m} = P_{s,m}|h_{0,m}|^2 + \sum_{i \in S_m} P_{r,i}|g_{r,i}|^2, \tag{9.45}$$

where $h_{0,m}$ is the channel gain from the mth source to the destination.

Next, two different strategies will be discussed, namely *opportunistic strategy* and *fairness-aware strategy* [4]. Both strategies are based on coalition formation games.

9.3.3.1 Opportunistic Strategy

In opportunistic strategy, when a relay joins group S_m, its payoff consists of two terms. The first term considers the acquired benefit, i.e. the increase in the SNR if the relay joins a specific group (coalition), while the second term is related to the size of the coalition and the increase in complexity. Let $S_m^{[c]} \subseteq S_m$ be the set of relays that correctly decode the mth source's message, that is, $S_m^{[c]} \subseteq S_m$. Then, mathematically, the following representation can be used for the payoff of each relay [4]:

$$\phi_i(S_m) = P_{r,i}|g_{r,i}|^2 - k|S_m^{[c]}|, \tag{9.46}$$

where k is the complexity cost of distributed beamforming per active relay. It is highlighted that a relay joins or leaves a coalition if and only if this action increases its payoff. In other words, between two groups with equal size, the relay will opt for the source with the higher channel gain. Thus, it is easy to realize that this strategy might be particularly unfair for users with weak channel conditions.

9.3.3.2 Fairness-Aware Strategy

In order to consider fairness, the payoff needs to include an extra constraint that will discourage the relays to join groups that already achieve much higher

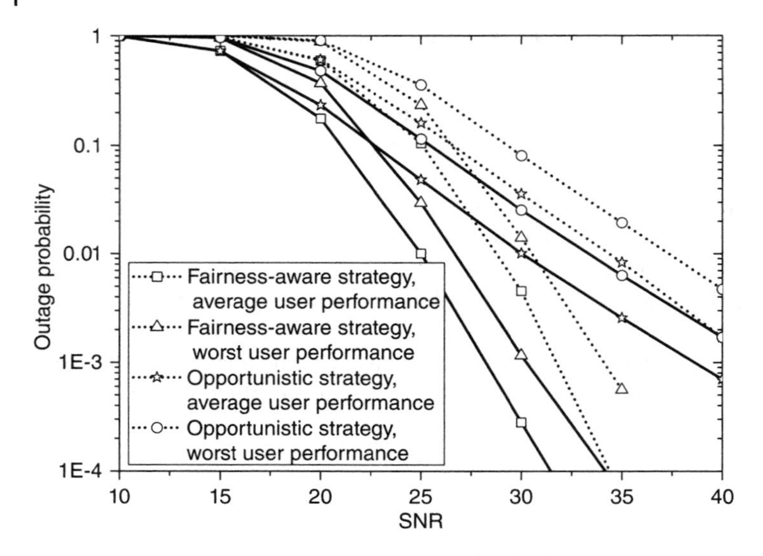

Figure 9.5 Outage probability achieved by different strategies for $k = 0.001$ for $R = 0.5$ bits/channel use (solid curves) and $R = 1$ bits/channel use (dotted curves) [4].

SNR compared to other groups. An example of such a payoff is the following [4]:

$$\phi_i(S_m) = \frac{P_{r,i}|g_{r,i}|^2}{\gamma_{S_m}} - k|S_m^{[c]}|. \tag{9.47}$$

Indeed, in (9.47) it is noted that as γ_{S_m} increases, the payoff of the relay reduces. It is highlighted that this strategy yields better outage performance compared to opportunistic relay selection. This is because the payoff in (9.47) captures better than (9.46) how significantly a relay contributes to a coalition.

9.3.3.3 Comparison of Relay Selection Strategies with Multiple Sources

In Figure 9.5 the two aforementioned relay selection strategies are compared in terms of outage probability [4]. It is assumed that there are four sources located at $(2.5, 5)$, $(5, 2.5)$, $(7.5, 5)$, and $(5, 7.5)$, while the $K = 8$ available relays are located inside the disk with origin located at $(5, 5)$ and radius 5 m. Also, it is assumed that the K available relays are Poisson distributed, while the distance between the source and the destination is 10 m. It is observed that the fairness-aware strategy outperforms the opportunistic one both in terms of average and worst user performance. Also, it is noted that by increasing the targeted data rate, the outage performance achieved by both coalition formation algorithms is reduced.

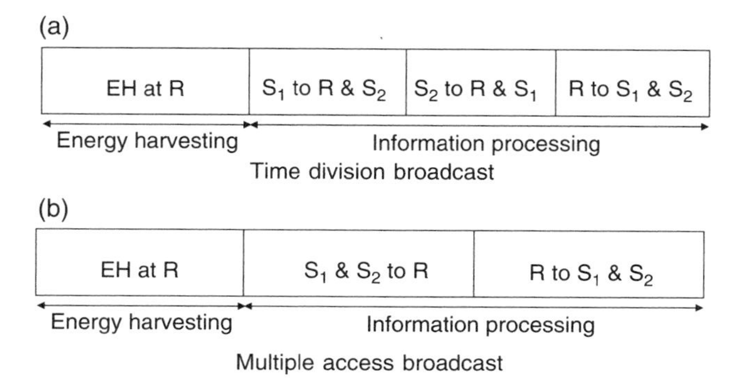

Figure 9.6 Frame structures of energy harvesting for TDBC and MABC.

9.3.4 Two-Way Relaying Networks

The communication between two sources in half-duplex two-way relaying networks can be facilitated by the utilization of one of two protocols, *multiple access broadcast (MABC)* and *time division broadcast (TDBC)* [10, 11], which are depicted in Figure 9.6. In both protocols, when power transfer to the relay is performed by time-switching, then the first time slot is dedicated to EH. Power transfer can be performed by three different policies, i.e. dual-source, single-fixed-source, and single-best-source. The main differences between multiple access and time division broadcast are the mechanism of information processing and the required number of distinct time slots. It is noted that these protocols can be slightly adapted to support the existence of a direct link between the sources. In this case, maximal ratio combining (MRC) can be used to combine the signals from the relay and the direct link, which will occur in the last time slot of information processing. The basic principles of two protocols are described in detail in the following subsections.

9.3.4.1 Time Division Broadcast

As shown in Figure 9.6, this protocol requires three distinct time slots for the information processing. More specifically, time slot 2 and time slot 3 are dedicated to information transmission by S_1 and S_2. Next, in time slot 4, the relay transmits the received observations by both sources back to them.

9.3.4.2 Multiple Access Broadcast

In this protocol the information processing is performed in solely two time slots, which take place after the slot that is dedicated to EH, according to 9.6. More specifically, in time slot 2, the two sources simultaneously transmit their information to the relay, which then forwards the received observation to the sources in time slot 3.

It is highlighted that this protocol outperforms time division broadcast, which happens because it requires a lower number of time slots to transmit information. Also, for both protocols, dual-source policy is the optimal one, which is reasonable since it transfers the largest power to the relay.

9.4 Future Research Challenges

9.4.1 Nonlinear Energy Harvesting Model and Hardware Impairments

It is noted that most existing schemes that investigate wireless information and power transfer in relaying systems rely on various idealizing assumptions and ignore the impact of practical constraints, such as hardware impairments and nonlinear energy harvesting characteristics, which have not been sufficiently investigated or jointly considered. For example, the impact of transceiver impairments on the specific case of two-way EH harvesting relaying has been investigated by [11] for a linear EH model. Also, regarding relaying with nonlinear energy harvesters, only a few recent works have been reported, such as [14].

9.4.2 NOMA-based Relaying

Non-orthogonal multiple access (NOMA), which has been proposed for the uplink of wireless powered networks with some promising results [15, 16], can also be used in relaying scenarios with EH. Rather than typical scenarios with multiple users that access the EH relay using NOMA [17], EH can be used to facilitate cooperative NOMA [18, 19]. The basic principle of cooperative NOMA is the employment of the user with better channel conditions as a relay to the weaker user. However, it becomes apparent that this protocol has a negative impact on the battery-life of the strong user, which discourages the establishment of cooperative NOMA. To overcome this, the strong user can harvest energy from the received signals and exploit the harvested energy to power the relay transmission. Also, it is noted that there is no work on NOMA-based relaying that considers the nonlinear EH model or hardware imperfections.

9.4.3 Large-Scale Networks

Most of existing works are based on specific and small-scale architectures (e.g., limited number of relays, sources, destinations, interferes, etc.), which do not give sufficient insight into the performance of general, large-scale networks [20]. In such types of networks, the most crucial issue is to control and minimize the total energy consumption and the generated interference, under EH constraints. To investigate the performance of these large-scale networks, stochastic geometric tools can be utilized.

9.4.4 Cognitive Relaying

Cognitive EH relaying is a quite interesting case, since it enables secondary transmission by using the primary users' resources, i.e. bandwidth and energy [21]. This is motivated as follows. The main constraint of secondary transmission is the source's transmit power, which is limited by primary user requirements and tolerance to interference. Thus, a relay can be used to provide a reliable link for the secondary transmission. As has already been mentioned, EH motivates existing nodes of the network to play the role of relays. Also, the RF signals from the primary users can be considered as constant energy sources, especially as the number of primary users increases. As a result, the investigation of cognitive EH relaying, especially in realistic large-scale networks, seems to be a promising direction and has the potential to create new opportunities in this area. Another interesting case is when both the secondary source and the relay are powered from the signals received by the primary users [22]. To this end, research focus has to be directed to exploring new relaying protocols and relay selection schemes.

Bibliography

1 S. Sudevalayam and P. Kulkarni (2011) Energy harvesting sensor nodes: survey and implications. *IEEE Commun. Surveys & Tutorials*, **13** (3): 443–461.

2 S. Bi, Y. Zeng, and R. Zhang (2016) Wireless powered communication networks: an overview. *IEEE Wireless Commun.*, **23** (2): 10–18.

3 I. Krikidis, S. Timotheou, S. Nikolaou, G. Zheng, D.W.K. Ng, and R. Schober (2014) Simultaneous wireless information and power transfer in modern communication systems. *IEEE Commun. Mag.*, **52** (11): 104–110.

4 Z. Ding, I. Krikidis, B. Sharif, and H.V. Poor (2014) Wireless information and power transfer in cooperative networks with spatially random relays. *IEEE Trans. Wireless Commun.*, **13** (8): 4440–4453.

5 Z. Ding, S.M. Perlaza, I. Esnaola, and H.V. Poor (2014) Power allocation strategies in energy harvesting wireless cooperative networks. *IEEE Trans. Wireless Commun.*, **13** (2): 846–860.

6 C. Zhong, G. Zheng, Z. Zhang, and G.K. Karagiannidis (2015) Optimum wirelessly powered relaying. *IEEE Signal Process. Lett.*, **22** (10): 1728–1732.

7 A.A. Nasir, X. Zhou, S. Durrani, and R.A. Kennedy (20130 Relaying protocols for wireless energy harvesting and information processing. *IEEE Trans. Wireless Commun.*, **12** (7): 3622–3636.
8 S. Nikoletseas, Y. Yang, and A. Georgiadis (2016) *Wireless Power Transfer Algorithms, Technologies and Applications in Ad Hoc Communication Networks*. Springer.
9 P.D. Diamantoulakis, G.D. Ntouni, K.N. Pappi, G.K. Karagiannidis, and B.S. Sharif (2015) Throughput maximization in multicarrier wireless powered relaying networks. *IEEE Wireless Commun. Lett.*, **4** (4): 385–388.
10 Y. Liu, L. Wang, M. Elkashlan, T.Q. Duong, and A. Nallanathan (2014) Two-way relaying networks with wireless power transfer: policies design and throughput analysis. In *Proceedings of the IEEE Global Communications Conference*, pp. 4030–4035.
11 D.K. Nguyen, M. Matthaiou, T.Q. Duong, and H. Ochi (2015) RF Energy harvesting two-way cognitive DF relaying with transceiver impairments. In *Proceedings of the IEEE International Conference on Communication Workshop (ICCW)*, pp. 1970–1975.
12 G. Zhu, C. Zhong, H.A. Suraweera, G.K. Karagiannidis, Z. Zhang, and T.A. Tsiftsis (2015) Wireless information and power transfer in relay systems with multiple antennas and interference. *IEEE Trans. Commun.*, **63** (4): 1400–1418.
13 J. Huang, Z. Han, M. Chiang, and H.V. Poor (2008) Auction-based resource allocation for cooperative communications. *IEEE J. Selected Commun.*, **26** (7): 1226–1237.
14 Y. Dong, M.J. Hossain, and J. Cheng (2016) Performance of wireless powered amplify and forward relaying over Nakagami-*m* fading channels with nonlinear energy harvester. *IEEE Commun. Lett.*, **20** (4): 672–675.
15 P.D. Diamantoulakis, K.N. Pappi, Z. Ding, and G.K. Karagiannidis (2016) Wireless-powered communications with non-orthogonal multiple access. *IEEE Trans. Wireless Commun.*, **15** (12): 8422–8436.
16 P.D. Diamantoulakis, K.N. Pappi, G.K. Karagiannidis, H. Xing, and A. Nallanathan (2017) Joint downlink/uplink design for wireless powered networks with interference. *IEEE Access*, **5**: 1534–1547.
17 Z. Yang, Z. Ding, P. Fan, and N. Al-Dhahir (2017) The impact of power allocation on cooperative non-orthogonal multiple access networks with SWIPT. *IEEE Trans. Wireless Commun.*, **16** (7): 4332–4343.
18 Z. Ding, X. Lei, G.K. Karagiannidis, R. Schober, J. Yuan, and V.K. Bhargava (2017) A survey on non-orthogonal multiple access for 5G networks: research challenges and future trends. *IEEE J. Selected Areas Commun.*, **35** (10): 2181–2195.
19 Y. Liu, Z. Ding, M. Elkashlan, and H.V. Poor (2016) Cooperative non-orthogonal multiple access with simultaneous wireless information and power transfer. *IEEE J. Selected Areas Commun.*, **34** (4): 938–953.

20 I. Krikidis (2014) Simultaneous information and energy transfer in large-scale networks with/without relaying. *IEEE Trans. Commun.*, **62** (3): 900–912.

21 Y. Liu, S.A. Mousavifar, Y. Deng, C. Leung, and M. Elkashlan (2016) Wireless energy harvesting in a cognitive relay network. *IEEE Trans. Wireless Commun.*, **15** (4): 2498–2508.

22 L. Mohjazi, M. Dianati, G.K. Karagiannidis, S. Muhaidat, and M. Al-Qutayri (20150 RF-powered cognitive radio networks: technical challenges and limitations. *IEEE Commun. Mag.*, **53** (4): 94–100.

10

Harnessing Interference in SWIPT Systems

Stelios Timotheou[1], Gan Zheng[2], Christos Masouros[3], and Ioannis Krikidis[1]*

[1] *KIOS Research and Innovation Center of Excellence and Department of Electrical and Computer Engineering, University of Cyprus, Cyprus*
[2] *Wolfson School of Mechanical, Electrical and Manufacturing Engineering, Loughborough University, Loughborough, United Kingdom*
[3] *Department of Electronic and Electrical Engineering, University College London, WC1E 7JE, London, United Kingdom*

10.1 Introduction

Recently, simultaneous wireless information and power transfer (SWIPT) via the radio frequency *energy harvesting* (EH) technology has emerged as a new solution for sustainable wireless network operation [1]. The first practical methods proposed for SWIPT were the *time switching* (TS) and *power splitting* (PS) methods that separate the received signal for decoding information and harvesting energy in the time and power domains [2], respectively. SWIPT in the spatial domain was introduced in [3] to decompose the multiple-input multiple-output (MIMO) channel and use the derived *eigenchannels* to either convey information or harness energy. The optimal precoding design for energy and information transfer in a multiple-input single-output (MISO) broadcast channel is studied in [4]. Joint information and power transfer is studied for a general K-user MISO interference channel in [5] based on PS receivers, where semi-definite relaxation is used to solve the precoding design. A more efficient and decentralized second-order cone programming (SOCP) relaxation is used in [6]. The work in [7] studies the optimal precoding for a MIMO relay channel where the relay has full-duplex radio capabilities. Finally, more recent works study the optimal waveform design when circuit nonlinearities associated with the RF harvesting process are taken into account, e.g., [8].

The above works consider conventional precoding design which involves a statistical view of interference and focuses on the optimal beamforming that maximizes the quality-of-service (QoS) – most commonly the signal to

*Corresponding author: Gan Zheng; g.zheng@lboro.ac.uk

Wireless Information and Power Transfer: Theory and Practice, First Edition.
Edited by Derrick Wing Kwan Ng, Trung Q. Duong, Caijun Zhong, and Robert Schober.

interference-plus-noise ratio (SINR) – and minimizes interference subject to a transmit power, or minimizes the transmit power subject to QoS constraints. A large body of literature exists on this topic that involves the solution of optimization problems that arise in different configurations [9, 10], objective functions [11, 12], and channel state information (CSI) knowledge [13, 14].

A new branch of the downlink beamforming optimization literature offers an alternative view of the interference, where as opposed to the above statistical approach interference is treated on an instantaneous basis, by *symbol-level precoding*. The relevant works focus on exploiting the constructive superposition of useful and interfering signals, to utilize interfering signals as first explored for closed-form precoders [15]–[18]. In [19] a symbol-level precoding is introduced where the conventional optimization constraints are adapted to accommodate *constructive interference* (CI) for phase shift keying modulation (PSK). Further work in [20, 21] focuses on a more relaxed optimization where the optimization constraints are designed based on the constructive interference regions in the PSK constellation.

In this chapter we introduce the adaptation of the conventional SWIPT beamforming [5] to the case of symbol-level precoding in order to harness interference for both information decoding and energy transfer. Explicitly, we aim to optimize data-aided precoding design in a MISO broadcast channel with PS receivers by exploiting constructive interference as a useful source for both signal and power transfer. We study the problem of transmit power minimization under both SINR and EH constraints. Our simulation results show that compared to the conventional precoding, the proposed data-aided precoding leads to the reduction of transmit power by an order of magnitude. We note that, while the analysis focuses on PSK modulation, the above concept and relevant optimizations can be extended to other modulation formats.

The rest of the chapter is organized as follows. Section 10.2 introduces the system model. Section 10.3 provides a review of the conventional precoding design problem and its solution. Section 10.4 formulates the considered optimization problem based on CI precoding and provides mathematical programming formulations offering lower and upper bounds to the solution of the problem; a *successive linear approximation* algorithm is also developed that provides excellent quality solutions. Section 10.5 examines the performance of the developed algorithms and investigates the benefits of harnessing interference signals. Finally, Section 10.6 concludes the chapter.

Notation. We use upper case boldface letters for matrices and lower case boldface letters for vectors. $(\cdot)^*$ and $(\cdot)^T$ denote the conjugate and transpose, respectively. $\|\cdot\|$ stands for the Frobenius norm. A complex Gaussian random vector variable $\mathbf{z}$ with mean $\boldsymbol{\mu}$ and variance Σ is represented as $\mathbf{z} \sim \mathcal{CN}(\boldsymbol{\mu}, \Sigma)$. A uniform random variable in the range $[a, b]$ is denoted by $z \sim U(a, b)$. $\mathbb{E}\{\cdot\}$ denotes

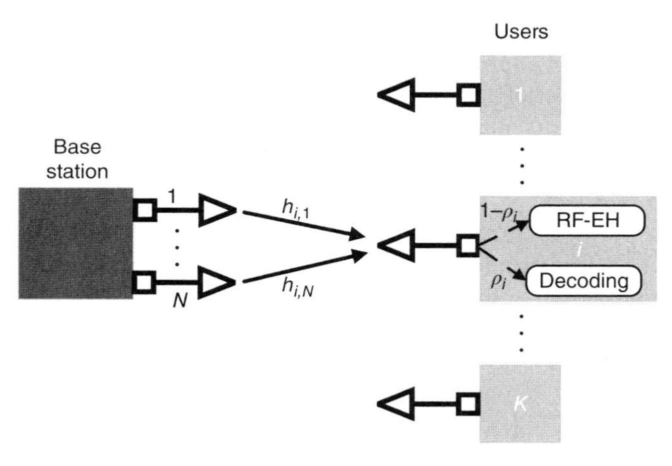

Figure 10.1 The MISO system model. Each receiver splits the received signal into two parts: information decoding and energy harvesting.

the expectation. Re(x) and Im(x) denote the real part and imaginary part of a complex number $x \in \mathbb{C}$, respectively.

10.2 System Model

Consider a MISO broadcast channel where an N-antenna base station (BS) transmits both signals and energy to K single-antenna users, as shown in Figure 10.1. For user i, its channel vector, precoding vector, received noise, data, SINR, and EH constraints are denoted as $\mathbf{h}_i^T$, $\mathbf{t}_i$, n_i, d_i, Γ_i, and E_i, respectively. The PSK modulated symbol can be expressed as $d_i = de^{j\phi_i}$ where d denotes the constant amplitude and ϕ_i is the phase. Without loss of generality, we assume $d = 1$. The average transmit power is

$$P_T = \mathbb{E}\left\{\left\|\sum_{k=1}^{K} \mathbf{t}_k d_k\right\|^2\right\}. \tag{10.1}$$

All wireless links exhibit independent frequency non-selective block fading. The received signal at user i is

$$y_i = \mathbf{h}_i^T \sum_{k=1}^{K} \mathbf{t}_k d_k + n_i, \tag{10.2}$$

where $n_i \sim \mathcal{CN}(0, N_0)$ is the additive white Gaussian noise (AWGN) with variance N_0. To decode the information and harvest RF energy at the receiver side, the practical PS technique [2] is adopted. Specifically, the receiver splits the RF signal into two parts: one for information decoding and the other for energy harvesting, with relative power of ρ_i and $1-\rho_i$, respectively.

The signal for information decoding is expressed as

$$\tilde{y}_i = \sqrt{\rho_i} y_i + \tilde{n}_i = \sqrt{\rho_i} \mathbf{h}_i^T \sum_{k=1}^{K} \mathbf{t}_k d_k + \sqrt{\rho_i} n_i + \tilde{n}_i, \tag{10.3}$$

where $\tilde{n}_i \sim \mathcal{CN}(0, N_C)$ is the complex AWGN with variance N_C introduced in the RF to baseband conversion in the decoding process, which is independent of n_i.

The signal for energy harvesting is

$$\bar{y}_i = \sqrt{1-\rho_i} y_i = \sqrt{1-\rho_i} \left(\mathbf{h}_i^T \sum_{k=1}^{K} \mathbf{t}_k d_k + n_i \right) \tag{10.4}$$

with average power

$$P_i = (1-\rho_i)\mathbb{E}\left\{ \left| \mathbf{h}_i^T \sum_{k=1}^{K} \mathbf{t}_k d_k + n_i \right|^2 \right\}.$$

The problem of interest is to minimize the total transmit power P_T in (10.1) subject to QoS (i.e., SINR) constraints $\{\Gamma_i\}$ and energy harvesting constraints $\{E_i\}$, respectively. This will be achieved by optimizing beamforming design, power allocation, and splitting by exploiting the CI concept.

In the following we first review the conventional precoding design then we introduce the proposed approach based on the CI.

10.3 Conventional Precoding Solution

In conventional MISO downlink precoding, users' data are independent of each other, i.e. $\mathbb{E}(d_i^* d_j) = 0, \forall j \neq i$. In this case, the transmit power in (10.1) becomes

$$P_T = \sum_{i=1}^{K} \| \mathbf{t}_i \|^2. \tag{10.5}$$

Based on the signal model (10.3) for information decoding, the received SINR for user i is given by

$$\Gamma_i^{\text{con}} = \frac{|\mathbf{h}_i^T \mathbf{t}_i|^2}{\sum_{j=1, j\neq i}^{K} |\mathbf{h}_i^T \mathbf{t}_j|^2 + N_0 + \frac{N_C}{\rho_i}}. \tag{10.6}$$

The harvested energy can be derived from (10.5) as

$$P_i^{\text{con}} = (1 - \rho_i)\left(\sum_{k=1}^{K} |\mathbf{h}_i^T \mathbf{t}_k|^2 + N_0\right). \tag{10.7}$$

Consequently, the power minimization problem with both QoS and EH constraints can be formulated as

Problem 2: Joint precoding design and power splitting without CI.

$$\underset{\{\mathbf{t}_i, \rho_i\}}{\text{minimize}} \sum_{i=1}^{K} \| \mathbf{t}_i \|^2$$
$$\text{s.t. } \Gamma_i^{\text{con}} \geq \Gamma_i, P_i^{\text{con}} \geq E_i, 0 < \rho_i < 1, \forall i.$$

It is easy to see that Problem 2 is non-convex and hence challenging to solve. In previous work [5], semi-definite programming (SDP) relaxation has been used to tackle Problem 2, showing that the SDP relaxation is tight for two-user and three-user MISO interference channels. This result was extended in [4], showing that the SDP relaxation is tight for the general MISO downlink case ($K \in \mathbb{N}$).

10.4 Joint Precoding and Power Splitting with Constructive Interference

Traditional precoding techniques refer to Gaussian information signals and treat users' data as totally random and independent information streams, while the interference between them is considered harmful. For the practical PSK signaling, however, interference can be constructive to the signal's detection on an instantaneous basis when it shifts the received constellation point away from the decision thresholds in the constellation [15]–[19]. Therefore, in this context one user's data do not always generate harmful interference to others. With the knowledge of both the instantaneous CSI and the data symbols at the BS, the received interference can be classified as constructive or destructive. In brief, while destructive interference deteriorates the system performance, CI moves the received symbols away from the decision thresholds of the constellation and thus improves the detection. Hence, the main idea is to jointly optimize the precoding design and the power split ratio to exploit the CI for both information decoding and energy harvesting.

After describing the problem formulation, we propose an upper bounding (UB) SOCP formulation [23] which provides an approximate solution to the considered problem. We further design a successive linear approximation

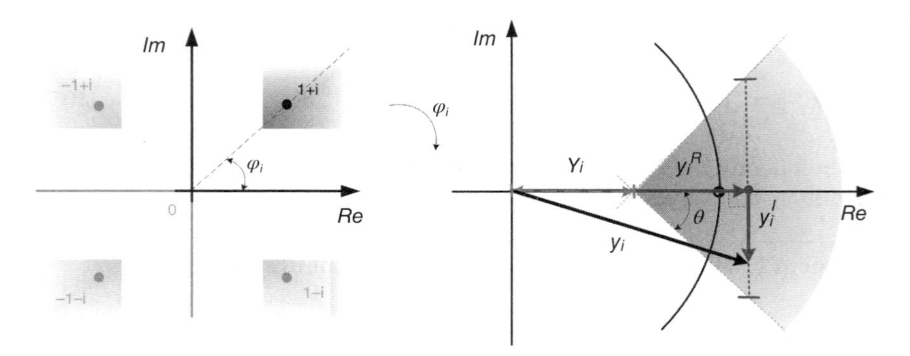

Figure 10.2 Illustration of constructive interference for decoding QPSK signaling as an example.

algorithm that starts from the UB provided by the first formulation and iterates towards better solutions. We also develop a lower bounding (LB) SOCP algorithm to have an indication of how far the obtained solution is from the optimal.

10.4.1 Problem Formulation

The received signal at user i in (10.2) can be rewritten as

$$y_i = \mathbf{h}_i^T \sum_{k=1}^{K} \mathbf{t}_k d_k + n_i = \mathbf{h}_i^T \sum_{k=1}^{K} \mathbf{t}_k e^{j(\phi_k - \phi_i)} d_i + n_i. \tag{10.8}$$

The information decoding part can be written as

$$\sqrt{\rho_i} y_i + \tilde{n}_i = \sqrt{\rho_i} \mathbf{h}_i^T \sum_{k=1}^{K} \mathbf{t}_k e^{j(\phi_k - \phi_i)} d_i + \sqrt{\rho_i} n_i + \tilde{n}_i. \tag{10.9}$$

We illustrate the derivation of the SINR constraint for the example of QPSK in Figure 10.2. Here, the shaded regions in the left-hand side of Figure 10.2 indicate the constructive interference regions of each QPSK symbol that move the received symbols away from the decision thresholds of the constellation. Notice that the constructive interference regions have a minimum distance from the decision thresholds as per the SNR constraints. The right-hand side of Figure 10.2 shows the rotation of the constructive interference region of symbol 1+i by ϕ_i and the decomposition of a received signal. In the figure we have used the definitions $y_i^R = \text{Re}\left(\mathbf{h}_i^T \sum_{k=1}^{K} \mathbf{t}_k e^{j(\phi_k - \phi_i)}\right)$, $y_i^I = \text{Im}\left(\mathbf{h}_i^T \sum_{k=1}^{K} \mathbf{t}_k e^{j(\phi_k - \phi_i)}\right)$, and $\gamma_i = \sqrt{\Gamma_i \left(N_0 + \frac{N_C}{\rho_i}\right)}$.

By means of their definition, y_i^R and y_i^I essentially shift the observation of the received symbol onto the axis from the origin of the constellation diagram to the constellation symbol of interest. Clearly, y_i^R provides a measure of the amplification of the received constellation point along the axis of the nominal constellation point due to constructive interference, and y_i^I provides a measure of the angle shift from the original constellation point, i.e. the deviation from the axis of the nominal constellation point with phase ϕ_i.

In conventional precoding optimization, y_i^R and y_i^I are constrained such that the received symbol is contained within a circle around the nominal constellation point, so that the interference caused by the other symbols is limited. In contrast to this, the concept of CI is exploited to allow a relaxation of y_i^R and y_i^I for all transmit symbols, under the condition that the interference caused is constructive, lying in the shaded regions of Figure 10.2. It can be seen that y_i^R and y_i^I are allowed to grow infinitely, as long as their ratio is kept such that the received symbol is contained within the constructive area of the constellation, i.e. the distances from the decision thresholds, as set by the SNR constraints γ_i, are not violated. It can be seen that the angle of interference need not be strictly aligned with the angle of the useful signal, as long as it falls within the constructive area of the constellation. For a given modulation order M the maximum angle shift in the CI area is given by $\theta = \pi/M$. By using basic geometry we arrive at the SINR constraint expressed as $|y_i^I| \leq (y_i^R - \gamma_i)\tan\theta$, [20], which is expanded to

$$\left|\mathrm{Im}\left(\mathbf{h}_i^T\sum_{k=1}^{K}\mathbf{t}_k e^{j(\phi_k-\phi_i)}\right)\right| \leq \left(\mathrm{Re}\left(\mathbf{h}_i^T\sum_{k=1}^{K}\mathbf{t}_k e^{j(\phi_k-\phi_i)}\right) - \sqrt{\Gamma_i\left(N_0+\frac{N_C}{\rho_i}\right)}\right)\tan\theta. \tag{10.10}$$

The harvested energy and the total transmit power can be derived based on (10.1) and (10.5), respectively, as $P_i = |\mathbf{h}_i^T\sum_{k=1}^{K}\mathbf{t}_k e^{j(\phi_k-\phi_i)}|^2$ and $P_T = \|\sum_{k=1}^{K}\mathbf{t}_k e^{j(\phi_k-\phi_i)}\|^2$. Therefore, the power minimization problem subjective to both SINR and EH constraints with the aid of the CI can be formulated as

$$\begin{aligned}
&\min_{\{\mathbf{t}_i,\rho_i,i\in\mathcal{K}\}} \left\|\sum_{k=1}^{K}\mathbf{t}_k e^{j(\phi_k-\phi_i)}\right\|^2 \\
&\text{s.t. constraint (10.10)},\\
&\left|\mathbf{h}_i^T\sum_{k=1}^{K}\mathbf{t}_k e^{j(\phi_k-\phi_i)}\right| \geq \sqrt{\frac{E_i}{1-\rho_i}}, i\in\mathcal{K}\\
&0<\rho_i<1, i\in\mathcal{K}.
\end{aligned} \tag{10.11}$$

By defining $\tilde{\mathbf{h}}_i = \mathbf{h}_i e^{j(\phi_1-\phi_i)}$ and $\mathbf{w} \triangleq \sum_{k=1}^{K}\mathbf{t}_k e^{j(\phi_k-\phi_1)}$, we can write (10.11) as

Problem 3: Joint precoding design and power splitting with CI.

$$\underset{\{\mathbf{w},\ \rho\}}{\text{minimize}} \parallel \mathbf{w}\|^2 \tag{10.12a}$$

$$\text{s.t.}|\text{Im}(\tilde{\mathbf{h}}_i^T\mathbf{w})| \leq \left(\text{Re}(\tilde{\mathbf{h}}_i^T\mathbf{w}) - \sqrt{\Gamma_i\left(N_0 + \frac{N_C}{\rho_i}\right)}\right)\tan\theta, i \in \mathcal{K}, \tag{10.12b}$$

$$||\tilde{\mathbf{h}}_i^T\mathbf{w}||^2 \geq \frac{E_i}{1-\rho_i}, i \in \mathcal{K}, \tag{10.12c}$$

$$0 < \rho_i < 1, i \in \mathcal{K}. \tag{10.12d}$$

Problem 3 is a virtual multicast channel with common messages to all users [22]. This problem is nontrivial to solve because of the non-convex constraint (10.12c). The rest of this section is devoted to solving Problem 3.

10.4.2 Upper Bounding SOCP Algorithm

In this section an upper bound solution to Problem 3 is derived by approximating the problem using convex SOCP. Towards this end, we begin by reformulating (10.12b) for $i \in \mathcal{K}$ using SOCP constraints. If we define

$$v_i = |\text{Im}(\tilde{\mathbf{h}}_i^T\mathbf{w})|, \tag{10.13}$$

$$y_i^R = \text{Re}(\tilde{\mathbf{h}}_i^T\mathbf{w}) = \sum_{k=1}^{K} \text{Re}(\tilde{h}_{i,k})w_k^R - \text{Im}(\tilde{h}_{i,k})w_k^I, \tag{10.14}$$

$$y_i^I = \text{Im}(\tilde{\mathbf{h}}_i^T\mathbf{w}) = \sum_{k=1}^{K} \text{Im}(\tilde{h}_{i,k})w_k^R + \text{Re}(\tilde{h}_{i,k})w_k^I, \tag{10.15}$$

then it is true that the absolute term of (10.12b) can be equivalently represented by two linear constraints as:

$$y_i^I \leq v_i, \ -y_i^I \leq v_i, i \in \mathcal{K}. \tag{10.16}$$

This is true because, on the one hand, constraint (10.16) forces $v_i \geq |y_i^I|, i \in \mathcal{K}$, and, on the other hand, (10.12b) forces v_i to be as small as possible, which is achieved for $v_i = |y_i^I|$.

To deal with the square root, the terms in (10.12b) are rearranged and both sides of the constraint are squared yielding $(y_i^R - v_i/\tan\theta)^2 \geq \Gamma_i(N_0 + N_C/\rho_i)$, which by introducing two new variables z_i^+ and z_i^- is equivalent to

$$z_i^+ = y_i^R - v_i/\tan\theta + \sqrt{\Gamma_i N_0}, \tag{10.17}$$

$$z_i^- = y_i^R - v_i/\tan\theta - \sqrt{\Gamma_i N_0}, \tag{10.18}$$

$$z_i^+ z_i^- \geq \frac{\Gamma_i N_C}{\rho_i}. \tag{10.19}$$

From the constraint in (10.12b), it is easy to see that a solution to Problem 3 satisfies $y_i^R > v_i / \tan\theta \geq 0$, and hence $z_i^+ > 0$. Because the right-hand side (r.h.s.) of (10.19) is positive, combined with the fact that $z_i^+ > 0$, implies that a solution of Problem 3 satisfies $z_i^+, z_i^- > 0$; hence, based on the SOCP hierarchy for the geometric mean constraint [24], p. 105], (10.19) is equivalent to:

$$z_i^+ z_i^- \geq r_{1,i}^2, \quad \rho_i \geq r_{2,i}^2, \tag{10.20}$$

$$r_{1,i} r_{2,i} \geq \sqrt{\Gamma_i N_C}, \tag{10.21}$$

$$z_i^+ \geq 0, z_i^- \geq 0, r_{1,i} \geq 0, r_{2,i} \geq 0. \tag{10.22}$$

To summarize, constraint (10.12b) is equivalent to (10.14)–(10.18) and (10.20)–(10.22).

Constraint (10.12c) is not convex due to the term $||\tilde{\mathbf{h}}_i^T \mathbf{w}||^2 = (y_i^R)^2 + (y_i^I)^2$, $i \in \mathcal{K}$, nonetheless it can be convexified by eliminating the real or imaginary part. Eliminating the imaginary part is better because $y_i^R \geq |v_i| / \tan\theta + \sqrt{\Gamma_i N_0}$, yielding the constraint $(y_i^R)^2 \geq \frac{E_i}{1-\rho_i}$, which is similar to (10.19), and hence can be reformulated into SOCP, yielding the *SOCP-UB formulation* (10.23).

SOCP-UB formulation:

$$\underset{\{\mathbf{w},\rho,\mathbf{z}_i^{\pm},\mathbf{r}_1,\mathbf{r}_2,\mathbf{r}_3,\mathbf{y}^R,\mathbf{y}^I,\mathbf{v}\}}{\text{minimize}} \quad ||\mathbf{w}||^2 \tag{10.23a}$$

$$\text{s.t. constraints } (10.14) - (10.18), (10.20) - (10.22), (10.12\text{d}), \tag{10.23b}$$

$$1 - \rho_i \geq r_{3,i}^2, i \in \mathcal{K}, \tag{10.23c}$$

$$r_{3,i} y_i^R \geq \sqrt{E_i}, i \in \mathcal{K}, \tag{10.23d}$$

$$r_{3,i} \geq 0, y_i^R \geq 0 \; i \in \mathcal{K}. \tag{10.23e}$$

The solution of (10.23) provides an upper bound to the optimal solution of Problem 3, as the former is always feasible for the latter since $(y_i^R)^2 + (y_i^I)^2 \geq (y_i^R)^2$. Note that if $y_i^I = 0$, for all $i \in \mathcal{K}$, then this formulation provides an optimal solution.

10.4.3 Successive Linear Approximation Algorithm

In this section we propose an iterative approach to tackle the non-convex constraint (10.12b), i.e. $(y_i^R)^2 + (y_i^I)^2 \geq \frac{E_i}{1-\rho_i}$, using successive linear approximation [25]. To illustrate the idea, we define a quadratic function $f(v) \triangleq v^2$, which has the same form with the non-convex terms $(y_i^R)^2$ and $(y_i^I)^2$, $i \in \mathcal{K}$. The linear approximation of $f(v)$ around the point p can be expressed as

$$f(v) \approx f(p) + \left(\frac{df(v)}{dv}|_{v=p} \right) (v - p) = p^2 + 2p(v - p). \tag{10.24}$$

It is important to note that the first-order approximation $p^2 + 2(v - p)$ always approximates the quadratic function v^2 from below such that $v^2 \geq p^2 + 2(v - p)$, for all v, p.

Let $y_{i(\tau)}^R$ and $y_{i(\tau)}^I$ denote the values of variables y_i^R and y_i^I obtained using $\mathbf{w}_{(\tau)}$ at the τth iteration of the algorithm. Performing linear approximation of the non-convex constraints yields $\omega_i \geq E_i/(1 - \rho_i), i \in \mathcal{K}$ where

$$\omega_i = (y_{i(\tau)}^R)^2 + 2(y_i^R - y_{i(\tau)}^R) + (y_{i(\tau)}^I)^2 + 2(y_i^I - y_{i(\tau)}^I). \tag{10.25}$$

Hence, the problem that needs to be solved during the $\tau + 1$ iteration is given by

$$\underset{\{\mathbf{w},\rho,\mathbf{z}_i^{\pm},\mathbf{r}_1,\mathbf{r}_2,\rho_2,\mathbf{y}^R,\mathbf{y}^I,\mathbf{v},\omega\}}{\text{minimize}} \quad ||\mathbf{w}||^2 \tag{10.26a}$$

$$\text{s.t. constraints}(10.14) - (10.18), (10.20) - (10.22), (10.12d), (10.25), \tag{10.26b}$$

$$\rho_{2i} = 1 - \rho_i, i \in \mathcal{K} \tag{10.26c}$$

$$\omega_i \rho_{2i} \geq E_i, \quad u_i \geq 0, i \in \mathcal{K}. \tag{10.26d}$$

Algorithm 2 outlines the overall procedure. The algorithm starts by solving (10.23) to obtain a feasible solution as the initial point. Then, an iterative procedure is followed which involves linear approximation of the non-convex terms upon derivation of $y_{i(\tau)}^R$ and $y_{i(\tau)}^I$, followed by the solution of (10.26) to obtain a new approximate solution to the initial problem. The procedure is repeated until the relative difference of the objective function between two successive iterations is below a threshold ϵ. It is important to note that $(y_i^R)^2 + (y_i^I)^2 \geq \omega_i$ is always true, which implies that solving (10.26) always provides a feasible solution to Problem 3. Successive linear approximation is a widely used procedure in signal processing for communications that has been proven to converge to a local optimum [25]. Algorithm 2 starts from the solution of (10.23) and progressively improves so that its solution is at least as good as the one of (10.23). Hence, there is a computational complexity/quality tradeoff between the solutions of the SOCP-UB formulation and Algorithm 2 as the former has lower computational complexity, while the latter yields results closer to optimality.

Algorithm 2 Successive linear approximation

1: **Input:** N_0, N_C, θ, ϵ, $\tilde{\mathbf{h}}_i$, E_i, Γ_i, $i \in \mathcal{K}$.
2: **Output:** $\mathbf{w}^*$, ρ^*
3: **Init.:** Set $\tau = 0$. Solve (10.23) to obtain $\mathbf{w}_{(0)}$ and $\rho_{(0)}$.
4: **repeat**
5: Compute$y^R_{i(\tau)}$ and $y^I_{i(\tau)}$ based on (10.14) and (10.15).
6: Set $\tau = \tau + 1$.
7: Solve problem (10.26) to obtain $\mathbf{w}^{(\tau)}$ and $\rho^{(\tau)}$.
8: **until** $\left| \| \mathrm{w}_{(\tau)} \|_2^2 - \| \mathrm{w}_{(\tau-1)} \|_2^2 \right| / \| \mathrm{w}_{(\tau-1)} \|_2^2 \leq \epsilon)$.
9: Set $\mathbf{w}^* = \mathrm{w}_{(\tau)}$, $\rho^* \rho^{(\tau)}$.

10.4.4 Lower Bounding SOCP Formulation

In order to obtain an SOCP LB solution, we need to approximate from below the non-convex term in (10.12c). For this reason, we consider that $y_i^R \geq v_i / \tan\theta$, which implies that $(y_i^R)^2 + y_i^R v_i \tan\theta \leq (y_i^R)^2 + (y_i^I)^2$. Because $(y_i^R)^2 + y_i^R v_i \tan\theta$ can be expressed as the product of two positive linear terms, i.e. $y_i^R(y_i^R + v_i \tan\theta)$, the resulting constraint can be expressed into a convex SOCP form similar to (10.19), yielding the *SOCP-LB formulation* in (10.27).

SOCP-LB formulation:

$$\underset{\{\mathbf{w},\rho,\mathbf{z}_i^{\pm},\mathbf{r}_1,\mathbf{r}_2,\mathbf{r}_3,\mathbf{r}_4,\mathbf{u},\mathbf{y}^R,\mathbf{y}^I,\mathbf{v}\}}{\text{minimize}} \quad ||\mathbf{w}||^2 \tag{10.27a}$$

$$\text{s.t. constraints}(10.14) - (10.18), (10.20) - (10.22), (10.12\text{d}), \tag{10.27b}$$

$$1 - \rho_i \geq r_{3,i}^2, i \in \mathcal{K}, \tag{10.27c}$$

$$u_i = y_i^R + v_i \tan\theta, i \in \mathcal{K}, \tag{10.27d}$$

$$y_i^R u_i \geq r_{4,i}^2, i \in \mathcal{K}, \tag{10.27e}$$

$$r_{3,i} r_{4,i} \geq \sqrt{E_i}, i \in \mathcal{K}, \tag{10.27f}$$

$$r_{3,i} \geq 0, r_{4,i} \geq 0, u_i \geq 0, i \in \mathcal{K}. \tag{10.27g}$$

Table 10.1 System parameters

Path-loss model	Log-distance (reference distance 1 m)
Carrier center frequency	915 MHz
Path-loss exponent	2.5
BS and receiver antenna gain	8 dBi and 3 dBi
BS–receiver distance	$\mathcal{U}(2,7)$ m
BS–receiver direction	$\mathcal{U}(-\pi,\pi)$
Multipath fading distribution	Rician fading (see [4])
Rician factor	5 dB
Antenna array model	Uniform linear with distance$\lambda/2$ between antenna elements ([27])
Channel noise variance	N_0 =−70 dBm
RF-baseband conversion noise variance	N_C =−50 dBm
Modulation scheme	QPSK

10.5 Simulation Results

In this section we examine the performance of the developed algorithms and investigate the benefits of using CI. The considered setting involves $K = 4$ SWIPT-enabled receivers randomly located around a BS with $N = 4$ antennas, unless otherwise stated. We consider all EH and SINR thresholds to be the same for all receivers, i.e. $\Gamma_i = \Gamma$, $E_i = E$, $i \in \mathcal{K}$. The main simulation parameters are listed in Table 10.1.

Figures 10.3, 10.4, and 10.5 depict the transmitted power difference between the upper bound (obtained via Algorithm 2) and lower bound (formulation SOCP-LB) of Problem 3, as well as between the solution of Problems 2 (via the SDP algorithm proposed in [5]) and 3 (Algorithm 2) for varying Γ, E, and K. From the figures it can be clearly seen that the required total transmitted power without consideration of the CI, resulting from the solution of Problem 2, is 4–13 dB larger than the corresponding quantity with consideration of the CI, resulting from the solution of Problem 3. Figures 10.4 and 10.5 further indicate that increasing the requirements for harvested power and the number of users, respectively, leads to higher performance gains when considering CI. Another important observation regards the performance of Algorithm 2; in all scenarios considered the transmitted power difference between the upper and lower bound of Problem 3, respectively, is less than 1 dB, which highlights the excellent performance of Algorithm 2. In terms of asymptotic optimality, Figure 10.3 shows that as the SINR threshold increases, the gap between the LB and UB solution of Problem 3 tends to zero.

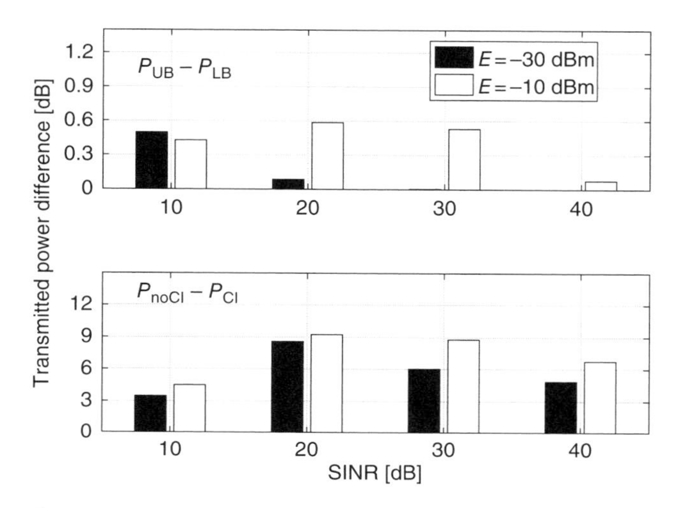

Figure 10.3 Transmitted power difference between the upper and lower bounds of Problem 3 (top), as well as between the solution of Problems 2 and 3 (bottom) for varying Γ when $E = \{-30, -10\}$ dBm and $K = 4$.

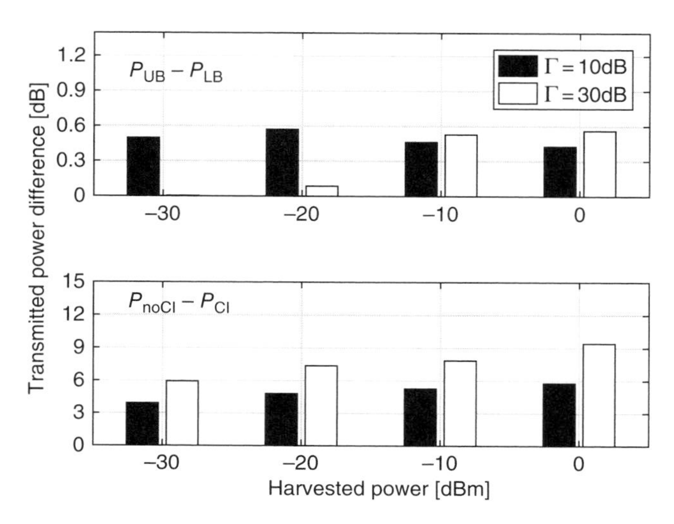

Figure 10.4 Transmitted power difference between the upper and lower bounds of Problem 3 (top), as well as between the solution of Problems 2 and 3 (bottom) for varying E when $\Gamma = \{10, 30\}$ dB and $K = 4$.

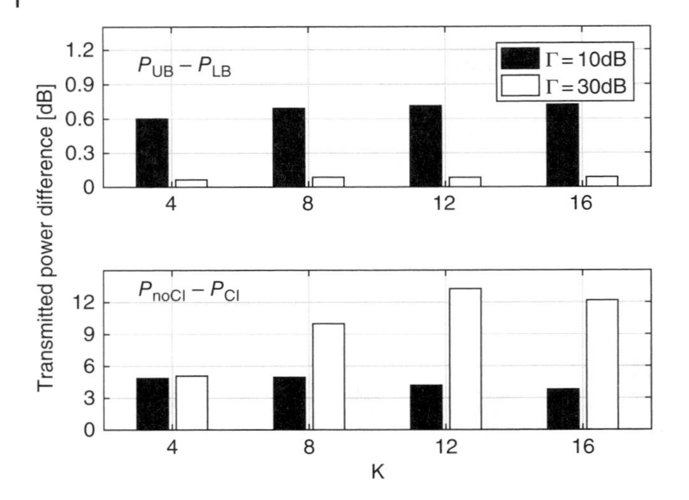

Figure 10.5 Transmitted power difference between the upper and lower bounds of Problem 3 (top), as well as between the solution of Problems 2 and 3 (bottom) for varying K when $\Gamma = \{10, 30\}$ dB, $E = -20$ dBm, and $N = K$.

10.6 Conclusions

This chapter presents a new perspective of harnessing interference in SWIPT systems to improve performance in terms of both information decoding and energy harvesting. In this context, we introduced constructive interference in MISO downlink systems by means of symbol-level precoding, which requires beamforming design on a symbol-by-symbol basis. Because the formulated problem is non-convex, lower and upper bounds have been derived by approximating the considered problem with convex SOCP formulations which can be solved to optimality using standard solvers [26]. Elaborating further on these formulations led to the development of a successive linear approximation algorithm. Simulation results demonstrated the effectiveness of the developed algorithm in yielding results close to optimality, and highlighted the advantage of precoding design using constructive interference on a symbol basis over conventional precoding design, yielding approximately an order of magnitude less transmitted power.

Bibliography

1 I. Krikidis, S. Timotheou, S. Nikolaou, G. Zheng, D.W.K. Ng, and R. Schober (2014) Simultaneous wireless information and power transfer in modern communication systems. *IEEE Commun. Mag.*, **52** (11): 104–110.

2 R. Zhang and C.K. Ho (2013) MIMO broadcasting for simultaneous wireless information and power transfer. *IEEE Trans. Wireless Commun.*, **12** (5): 1989–2001.

3 S. Timotheou, I. Krikidis, S. Karachontzitis and K. Berberidis (2015) Spatial domain simultaneous information and power transfer for MIMO channels. *IEEE Trans. Wireless Commun.*, **14** (8): 4115–4128.

4 Q. Shi, L. Liu, W. Xu, and R. Zhang (2014) Joint transmit beamforming and receive power splitting for MISO SWIPT systems. *IEEE Trans. Wireless Commun.*, **13** (6): 3269–3280.

5 S. Timotheou, I. Krikidis, G. Zheng, and B. Ottersten (2014) Beamforming for MISO interference channels with QoS and RF energy transfer. *IEEE Trans. Wireless Commun.*, **13** (5): 2646–2658.

6 Q. Shi, W. Xu, T.-H. Chang, Y. Wang, and E. Song (2014) Joint beamforming and power splitting for MISO interference channel with SWIPT: An SOCP relaxation and decentralized algorithm. *IEEE Trans. Signal Process.*, **62** (23): 6149–6208.

7 M. Mohammadi, B.K. Chalise, H.A. Suraweera, C. Zhong, G. Zheng, and I. Krikidis (2016) Throughput analysis and optimization of wireless-powered multiple antenna full-duplex relay systems. *IEEE Trans. Commun.*, **64** (4): 1769–1785.

8 B. Clerckx (2018) Wireless information and power transfer: Nonlinearity, waveform design and rate–energy tradeoff. *IEEE Trans. Signal Process.*, **66** (4): 847–862.

9 M. Bengtsson and B. Ottersten (2001) Optimal and suboptimal transmit beamforming. In *Handbook of Antennas in Wireless Communications*, L. Godara (ed.). Boca Raton, CRC Press, ch. 18.

10 D. Nguyen, L. Tran, P. Pirinen, and M. Latva-aho (2013) Precoding for full duplex multiuser MIMO systems: Spectral and energy efficiency maximization. *IEEE Trans. Signal Process.*, **61** (16): 4038–4050.

11 Q. Shi, M. Razaviyayn, Z.Q. Luo, and C. He (2011) An iteratively weighted MMSE approach to distributed sum-utility maximization for a MIMO interfering broadcast channel. *IEEE Trans. Signal Process.*, **59** (9): 4331–4340.

12 W. Yu, T. Kwon, and C. Shin (2013) Multicell coordination via joint scheduling, beamforming, and power spectrum adaptation. *IEEE Trans. Wireless Commun.*, **12** (7): 3300–3313.

13 M. Payaro, A. Pascual, and M. Lagunas (2007) Robust power allocation designs for multiuser and multiantenna downlink communication systems through convex optimization. *IEEE J. Selected Areas Commun.*, **25** (7): 1390–1401.

14 G. Zheng, K.K. Wong, and T.S. Ng (2008) Robust linear MIMO in the downlink: A worst-case optimization with ellipsoidal uncertainty regions. *EURASIP J Adv. Signal Process.*, **2008** (1): 609028:1–15.

15 C. Masouros, and E. Alsusa (2009) Dynamic linear precoding for the exploitation of known interference in MIMO broadcast systems. *IEEE Trans. Wireless Commun.*, **8** (3): 1396–1404.

16 C. Masouros (2011) Correlation rotation linear precoding for MIMO broadcast communications. *IEEE Trans. Signal Process.*, **59** (1): 252–262.

17 C. Masouros, T. Ratnarajah, M. Sellathurai, C. Papadias, and A. Shukla (2013) Known interference in wireless communications: a limiting factor or a potential source of green signal power? *IEEE Commun. Mag.*, **51** (10): 162–171.

18 G. Zheng, I. Krikidis, C. Masouros, S. Timotheou, D.A. Toumpakaris, and Z. Ding (2014) Rethinking the role of interference in wireless networks. *IEEE Commun. Mag.*, **52** (11): 152–158.

19 M. Alodeh, S. Chatzinotas, and B. Ottersten (2015) Constructive multiuser interference in symbol level precoding for the MISO downlink channel. *IEEE Trans. Signal Process.*, **63** (9): 2239–2252.

20 C. Masouros and G. Zheng (2015) Exploiting known interference as green signal power for downlink beamforming optimization. *IEEE Trans. Signal Process.*, **63** (14): 3668–3680.

21 S. Timotheou, G. Zheng, C. Masouros, and I. Krikidis (2016) Exploiting constructive interference for simultaneous wireless information and power transfer in multiuser downlink systems. In *IEEE J. Selected Areas Commun.*, **34** (5): 1772–1784.

22 N.D. Sidiropoulos, T.N. Davidson, and Z.Q. Luo (2006) Transmit beamforming for physical-layer multicasting. *IEEE Trans. Signal Process.*, **54** (6): 2239–2251.

23 F. Alizadeh and D. Goldfarb (2003) Second-order cone programming. *Math. Program., Ser. B*, **95** (1): 3–51.

24 A. Ben-Tal and A. Nemirovski (2001) *Lectures on modern convex optimization: analysis, algorithms and engineering applications.* MPS-SIAM Series on Optimization, SIAM, Philadelphia.

25 L. Tran, M. Hanif, and M. Juntti (2014) A conic quadratic programming approach to physical layer multicasting for large-scale antenna arrays. *IEEE Signal Process. Lett.*, **21** (1): 114–117.

26 S. Boyd and L. Vandenberghe (2004) *Convex Optimization.* Cambridge University Press.

27 E. Karipidis, N.D. Sidiropoulos, and Z.Q. Luo (2007) Far-field multicast beamforming for uniform linear antenna arrays. *IEEE Trans. Signal Process.*, **55** (10): 4916–4927.

11

Physical Layer Security in SWIPT Systems with Nonlinear Energy Harvesting Circuits

Yuqing Su[1], *Derrick Wing Kwan Ng*[1*], *and Robert Schober*[2]

[1] *School of Electrical Engineering and Telecommunications, The University of New South Wales, Australia*
[2] *Institute for Digital Communications, Friedrich-Alexander-University Erlangen-Nuremberg (FAU), Germany*

11.1 Introduction

In recent decades the rapid development of wireless communication technologies has triggered a massive growth in the number of wireless communication devices and sensors for emerging applications such as video-conferencing, e-health, entertainment, smart city, and energy management, etc. Moreover, the popularity of wireless communication systems has fueled the roll-out of the Internet of Things (IoT) [1]. Thereby, wireless communication modules are unobtrusively and invisibly integrated into physical objects such as clothing, walls, and vehicles, which are connected to advanced computing systems (e.g., cloud computing) to provide intelligent services in daily life. It is expected that by 2020 the number of devices interconnected via the Internet may reach up to 50 billion. However, this tremendous increase in the number of devices has also led to a huge demand for energy at both service operators and end-users. As a result, providing self-sustainable and high speed communication with guaranteed quality of service (QoS) has become a major goal for the next generation communication systems [2, 3].

Recently, various green technologies and resource allocation methods have been proposed in the literature for maximizing the energy efficiency (bit-per-joule) of wireless communication systems [4]–[6]. In [4], the maximization of the energy efficiency of a multicell system was studied. In [5], the authors proposed an energy-efficient resource allocation algorithm for the maximization of the system efficiency of a massive multiple-input multiple-output (MIMO) system. In [6], a novel energy-spectrum trading mechanism was proposed. Specifically, a small-cell transmitter consumes

*Corresponding author: Derrick Wing Kwan Ng; w.k.ng@unsw.edu.au

Wireless Information and Power Transfer: Theory and Practice, First Edition.
Edited by Derrick Wing Kwan Ng, Trung Q. Duong, Caijun Zhong, and Robert Schober.

additional power to serve the users of an energy-limited macro-cell transmitter, while the macro-cell transmitter shares its dedicated bandwidth with the small-cell transmitter. Although the optimization algorithms/methods proposed in [4]–[6] can improve the efficiency in utilizing the limited system resources, the availability of a continuous power supply is required such that a sufficient amount of energy can be exploited for system operation. In practice, transceivers may not be connected to the power grid, such that the energy requirement is difficult to satisfy. In such situations, harvesting energy from external energy sources is an appealing solution. Conventionally, solar, wind, and geothermal heat are suitable natural energy sources to enable sustainability of energy-limited devices and to reduce the operating costs of the service providers [7]–[9]. However, the availability of these natural energy sources is often limited by location, weather, or climate and may be problematic in indoor environments. In other words, these energy harvesting technologies can support only limited mobility of wireless communication devices. In addition, it is difficult to guarantee stable communication services when communication devices are powered by uncontrollable natural energy sources. Thus, a new form of controllable energy source for portable wireless devices is needed to extend the lifetime of communication networks.

In practice, ambient background electromagnetic radiation in radio frequency (RF) is also a potential source of energy for energy scavenging [10]–[14]. Nowadays, RF-based energy harvesting circuits are able to harvest microwatts to milliwatts of power over the range of several meters for a transmit power of 1 W and a carrier frequency of less than 1 GHz [15]. Thus, RF energy is a viable energy source for devices with low-power consumption, e.g., wireless sensors [12, 13], and is considered as a practical solution to realize the IoT. In particular, a large portion of the IoT wireless devices, some of which may be inaccessible for battery replacement, could be powered wirelessly by dedicated power stations via RF-based wireless energy transfer (WET) technology to power their information transmissions. Moreover, RF-based energy harvesting enables simultaneous wireless information and power transfer (SWIPT), since RF signals can serve as a carrier for both information and energy. Hence, SWIPT has attracted much attention recently [10]–[23]. For instance, the fundamental tradeoff between WET and wireless information transfer (WIT) was studied in [10] and [11] for frequency flat fading and frequency selective fading channels, respectively. The authors in [17] investigated the rate–energy tradeoff regions for point-to-point systems. However, the efficiency of WET is still low. In particular, the path loss of communication channels severely attenuate wireless signals, leading to a small amount of harvested energy at the receiver. For instance, for the short distance of 10 m in free space, the attenuation of a wireless signal can be up to 50 dB for a carrier frequency of 915 MHz. Hence, multiple-antenna technologies have been introduced to

SWIPT systems to improve the efficiency of WET. Specifically, by exploiting the spatial degrees of freedom offered by multiple antennas, wireless energy can be focused more accurately on the desired receivers, which can improve the efficiency of WET substantially.

On the other hand, in SWIPT systems, RF-based energy harvesting receivers are generally located closer to the transmitter than an information receiver for more efficient energy harvesting. Thus, such scenarios raise communication security issues due to the broadcast nature of wireless channels and the relatively high transmit powers needed for SWIPT. Traditionally, cryptographic encryption algorithms are employed at the application layer to guarantee communication security. However, these algorithms were designed based on the assumption of perfect secret key management and distribution, which may not be possible in future wireless IoT networks with a massive number of wireless sensor nodes. As an alternative solution, information-theoretic physical layer (PHY) security offers a complementary technology to cryptographic encryption [24]–[30]. PHY security is used to exploit the physical characteristics of the wireless channels, e.g., interference, fading, and noise, to ensure perfect secrecy of communication. It has been shown in [24] that in a wiretap channel, a source, and a destination can exchange perfectly secure information with a nonzero rate if the desired receiver enjoys better channel conditions than the passive eavesdropper(s). This motivated the investigation of beamforming designs for multiple-antenna SWIPT systems with the consideration of PHY security in [31]–[34]. In [31] and [32], taking into account potential eavesdropping by energy harvesting receivers, the authors designed the beamformers for minimization of the total transmit power for the scenarios of perfect channel state information (CSI) and imperfect CSI, respectively. In [33], a multi-objective framework was proposed to accommodate the desire of efficient WET and WIT while taking into account the concerns of communication security. In [34], beamforming design was investigated for the maximization of secrecy rate in SWIPT systems. The key to ensure secure SWIPT in [31]–[34] is to inject optimized artificial noise signals via multiple antennas to degrade the channel quality of the potential eavesdroppers. However, in [31]–[34] an over-simplified linear energy harvesting model was adopted for characterization of the RF energy-to-direct current (DC) power conversion. Indeed, various experiments for practical energy harvesting circuits have shown that their input-output characteristic is highly nonlinear [35, 36]. The discrepancy between the properties of practical nonlinear harvesting circuits and the linear harvesting model conventionally assumed in the literature may cause a severe mismatch for resource allocation, leading to insecure SWIPT systems. Hence, in this chapter we study novel beamforming designs enabling secure SWIPT based on a practical nonlinear energy harvesting model.

The remainder of this chapter is organized as follows. In Section 11.2 we introduce the adopted channel model and a practical nonlinear RF-based energy harvesting model. Section 11.3 studies the beamforming design for secure communication in SWIPT systems taking into account the impact of imperfect CSI. Simulation results are provided in Section 11.4, and we conclude this chapter with a brief summary in Section 11.5.

Notation. We use boldface capital and lower case letters to denote matrices and vectors, respectively. $\mathbf{A}^H$, $\text{Tr}(\mathbf{A})$, and $\text{Rank}(\mathbf{A})$ represent the Hermitian transpose, the trace, and the rank of matrix $\mathbf{A}$, respectively. $\mathbf{A} \succeq \mathbf{0}$ indicates that $\mathbf{A}$ is a positive semi-definite matrix. $\mathbf{I}_N$ is the $N \times N$ identity matrix. $\mathbb{C}^{N\times M}$ and $\mathbb{R}^{N\times M}$ denote the set of all $N \times M$ matrices with complex and real entries, respectively. $\mathbb{H}^N$ denotes the set of all $N \times N$ Hermitian matrices. $|\cdot|$ and $\|\cdot\|$ denote the absolute value of a complex scalar and the l_2-norm of a vector, respectively. The circularly symmetric complex Gaussian (CSCG) distribution is denoted by $\mathcal{CN}(\mu, \sigma^2)$ with mean μ and variance σ^2. $\sim$ stands for "distributed as". $[x]^+ = \max\{0, x\}$.

11.2 Channel Model

We consider a time division duplex (TDD) SWIPT system where the communication channels are slowly time-varying fading and frequency flat. There are a base station (BS), an information receiver (IR), and J energy harvesting receivers (ERs), cf. Figure 11.1. We assume that the BS is equipped with $N_\mathrm{T} \geq 1$ transmit antennas serving both the IR and ERs, which are single-antenna devices. In practice, due to the broadcast nature of the RF channels, the signal intended for the IR is overheard by the ERs. In addition, ERs are usually located close to the BS for a more efficient RF-based energy harvesting, which also facilitates eavesdropping. As a result, SWIPT systems are more vulnerable to potential eavesdropping compared to traditional communication systems, e.g., [27]–[30]. To ensure communication security in SWIPT systems, the ERs are treated as potential eavesdroppers for resource allocation algorithm design. The signals received at the IR and ER $j \in \{1, \dots, J\}$ are modeled as

$$y = \mathbf{h}^H(\mathbf{w}s + \mathbf{v}) + n \quad \text{and} \tag{11.1}$$

$$y_{\mathrm{ER}_j} = \mathbf{g}_j^H(\mathbf{w}s + \mathbf{v}) + n_{\mathrm{ER}_j}, \quad \forall j \in \{1, \dots, J\}, \tag{11.2}$$

respectively, where $s \in \mathbb{C}$ and $\mathbf{w} \in \mathbb{C}^{N_\mathrm{T}\times 1}$ are the information symbol and the corresponding information beamforming vector, respectively. Without loss of generality, we assume that $\mathcal{E}\{|s|^2\} = 1$. $\mathbf{v} \in \mathbb{C}^{N_\mathrm{T}\times 1}$ is a pseudo-random artificial noise (AN) vector generated by the BS. In fact, the AN facilitates both wireless

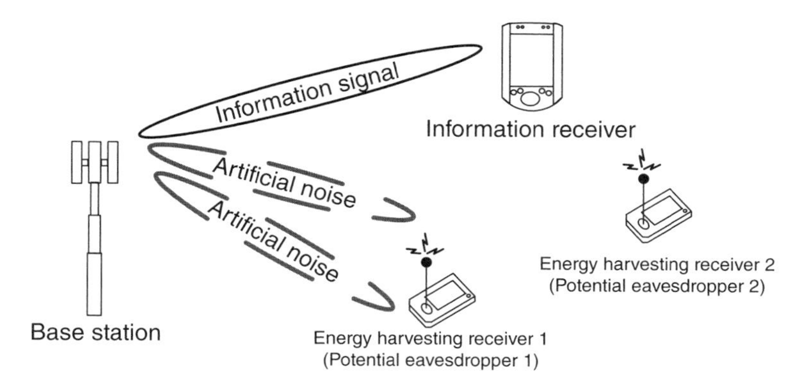

Figure 11.1 A SWIPT downlink model with a BS, an IR, and $J = 2$ ERs. The ERs harvest energy from the received RF signals to extend their lifetimes.

energy transfer and communication security. In particular, $\mathbf{v}$ is modeled as a random vector with circularly symmetric complex Gaussian distribution

$$\mathbf{v} \sim \mathcal{CN}(\mathbf{0}, \mathbf{V}), \tag{11.3}$$

where $\mathbf{V} \in \mathbb{H}^{N_\mathrm{T}}, \mathbf{V} \succeq \mathbf{0}$, denotes the covariance matrix of the AN. The channel vector between the BS and the IR is denoted by $\mathbf{h} \in \mathbb{C}^{N_\mathrm{T}\times 1}$ and the channel vector between the BS and ER j is denoted by $\mathbf{g}_j \in \mathbb{C}^{N_\mathrm{T}\times 1}$. $n \sim \mathcal{CN}(0, \sigma_\mathrm{s}^2)$ and $n_{\mathrm{ER}_j} \sim \mathcal{CN}(0, \sigma_\mathrm{s}^2)$ are the additive white Gaussian noises (AWGNs) at the IR and ER j, respectively, where σ_s^2 denotes the noise power at the antenna of the receiver.

Remark 11.1 In general, both pseudo-random signals and constant amplitude signals are potential candidates to facilitate WET. However, pseudo-random AN signals can be shaped more easily to satisfy certain requirements on the spectrum mask of the transmit signal and provide communication security. Thus, pseudo-random AN is adopted in this chapter.

11.2.1 Energy Harvesting Model

Figure 11.2 depicts a block diagram of an ER in SWIPT systems. In general, a passive bandpass filter and a rectifying circuit are employed in an ER to convert the received RF power to direct current (DC) power. The total received RF power at ER j is given by

$$P_{\mathrm{ER}_j} = \mathrm{Tr}((\mathbf{w}\mathbf{w}^H + \mathbf{V})\mathbf{G}_j), \tag{11.4}$$

where $\mathbf{G}_j = \mathbf{g}_j\mathbf{g}_j^H$. In the SWIPT literature, for simplicity, the total harvested power at ER j is typically modeled as follows:

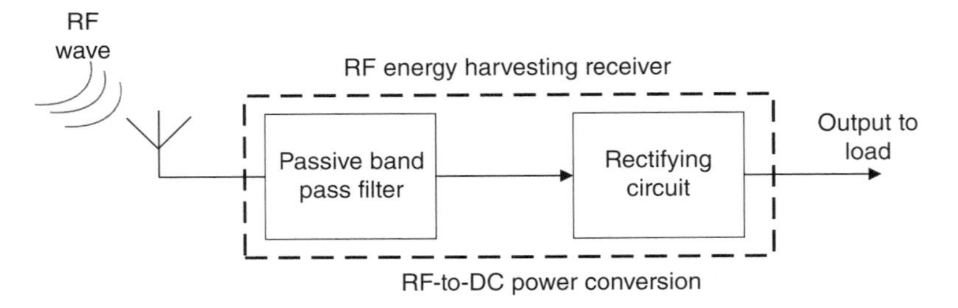

Figure 11.2 Block diagram of an ER for converting the received RF energy to electrical energy.

$$\Phi_{\mathrm{ER}_j}^{\mathrm{Linear}} = \eta_j P_{\mathrm{ER}_j}, \tag{11.5}$$

where $0 \leq \eta_j \leq 1$ denotes the energy conversion efficiency of ER j. From (11.5), it can be seen that with existing models the total harvested power at the ER is linearly and directly proportional to the received RF power. However, practical RF-based energy harvesting circuits consist of resistors, capacitors, and diodes. Experimental results have shown that these circuits [35]–[37] introduce various nonlinearities into the end-to-end WET. In order to design a resource allocation algorithm for practical secure SWIPT systems, we adopt the nonlinear parametric energy harvesting model proposed in [38]–[40]. Consequently, the total harvested power at ER j, Φ_{ER_j}, is modeled as:

$$\Phi_{\mathrm{ER}_j} = \frac{[\Psi_{\mathrm{ER}_j} - M_j\Omega_j]}{1 - \Omega_j}, \quad \Omega_j = \frac{1}{1 + \exp(a_j b_j)}, \tag{11.6}$$

$$\text{where} \quad \Psi_{\mathrm{ER}_j} = \frac{M_j}{1 + \exp(-a_j(P_{\mathrm{ER}_j} - b_j))} \tag{11.7}$$

is a sigmoid function which has the received RF power, P_{ER_j}, as input. Constant M_j denotes the maximal harvested power at ER j when the energy harvesting circuit is driven to saturation due to an exceedingly large input RF power. Constants a_j and b_j capture the joint effects of resistance, capacitance, and circuit sensitivity. In particular, a_j reflects the nonlinear charging rate (e.g., the steepness of the curve) with respect to the input power and b_j determines the minimum turn-on voltage of the energy harvesting circuit.

In practice, parameters a_j, b_j, and M_j of the proposed model in (11.6) can be obtained by applying a standard curve fitting algorithm to measurement results of a given energy harvesting hardware circuit. In Figure 11.3 we show two examples for the curve fitting for the nonlinear energy harvesting model in (11.6). For the upper and lower figure, the parameters are $\{M = 10.73$ mW, $b = 0.2308$, $a = 5.365\}$ and $\{M = 0.1071$ mW, $b = 0.6614$, $a = 0.8963\}$, for input powers in the mW and 10^{-4} W range, respectively. As can be observed, the

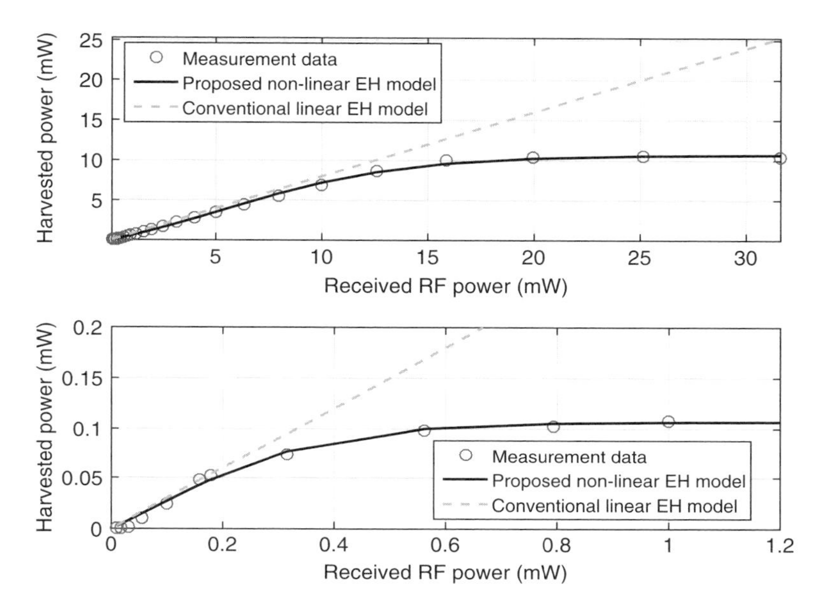

Figure 11.3 A comparison between the harvested power for the proposed model in (11.6) and measurement data obtained for two different practical energy harvesting circuits. The measurement data of the upper and lower figure are obtained from [35] and [36], respectively, showing the different dynamic ranges in harvested energy of practical energy harvesting circuits. The parameters a_j, b_j, and M_j in (11.6) are calculated with a standard curve fitting tool. The energy conversion efficiency of the ER for the linear energy harvesting model is set to $\eta_j = 0.8$ and $\eta_j = 0.3$ in (11.5) for the upper and lower figure, respectively.

parametric nonlinear model closely matches the experimental results provided in [35] and [36] for the wireless power harvested by a practical energy harvesting circuit. Figure 11.3 also illustrates the inability of the linear model in (11.5) to capture the nonlinear characteristics of practical energy harvesting circuits, especially in the high received RF power regime.

11.2.2 Channel State Information Model

In practice, the legitimate IR performs handshaking with the BS, which facilitates downlink packet transmission. In particular, the CSI of the BS-to-IR link can be estimated by measuring the uplink pilots embedded in the handshaking signals exploiting channel reciprocity. However, since the communication channels are slowly time-varying, the CSI of the IR obtained at the BS may become outdated during transmission. On the other hand, the ERs have only limited interaction with the BS to minimize their energy consumption. Hence, the CSI of the ERs is usually outdated. To design a robust secure SWIPT system, we take into account the imperfection of the CSI for beamforming design.

To this end, we adopt a deterministic model [41]–[44]. In particular, we model the CSI of the links as:

$$\mathbf{h} = \hat{\mathbf{h}} + \Delta\mathbf{h}, \tag{11.8}$$

$$\Xi \triangleq \{\Delta\mathbf{h} \in \mathbb{C}^{N_\mathrm{T}\times 1} : \|\Delta\mathbf{h}\|^2 \leq \theta^2\}, \tag{11.9}$$

$$\mathbf{g}_j = \hat{\mathbf{g}}_j + \Delta\mathbf{g}_j, \ \forall j, \tag{11.10}$$

$$\Upsilon_j \triangleq \{\Delta\mathbf{g}_j \in \mathbb{C}^{N_\mathrm{T}\times 1} : \|\Delta\mathbf{g}_j\|^2 \leq \rho_j^2\}, \tag{11.11}$$

respectively, where $\hat{\mathbf{h}}$ and $\hat{\mathbf{g}}_j$ are the estimates of channel vectors of $\mathbf{h}$ and $\mathbf{g}_j$, respectively. The channel estimation errors of $\mathbf{h}$ and $\mathbf{g}_j$ are denoted by $\Delta\mathbf{h}$ and $\Delta\mathbf{g}_j$, respectively. Ξ and Υ_j are the channel uncertainty sets which contain all possible channel estimation errors with respect to IR and ER j, respectively. Constants θ and ρ_j denote the maximum values of the norms of the CSI estimation error vectors $\Delta\mathbf{h}$ and $\Delta\mathbf{g}_j$, respectively.

11.2.3 Secrecy Rate

Assuming perfect CSI is available at the IR for signal detection, the achievable rate (bit/s/Hz) between the BS and the IR is given by

$$R = \log_2\left(1 + \frac{\mathbf{w}^H\mathbf{H}\mathbf{w}}{\sigma_\mathrm{s}^2}\right), \tag{11.12}$$

where $\mathbf{H} = \mathbf{h}\mathbf{h}^H$. Note that since the pseudo-AN signal, $\mathbf{v}$, is a deterministic sequence which is known by the IR, its impact on the achievable rate, i.e. $\mathrm{Tr}(\mathbf{HV})$, has been canceled with interference cancellation techniques before decoding the desired signal.

On the other hand, the capacity between the BS and ER j for decoding the signal intended for the IR can be expressed as

$$R_{\mathrm{ER}_j} = \log_2\left(1 + \frac{\mathbf{w}^H\mathbf{G}_j\mathbf{w}}{\mathrm{Tr}(\mathbf{G}_j\mathbf{V}) + \sigma_\mathrm{s}^2}\right), \tag{11.13}$$

where $\mathrm{Tr}(\mathbf{G}_j\mathbf{V})$ denotes the interference power received at ER j. Hence, the achievable secrecy rate of the IR is given by [26]

$$R_\mathrm{sec} = [R - \max_{\forall j}\ \{R_{\mathrm{ER}_j}\}]^+. \tag{11.14}$$

11.3 Optimization Problem and Solution

In the considered SWIPT system, we aim to maximize the system secrecy rate while taking into account the impact of imperfect CSI and guaranteeing the QoS in WET. To this end, we formulate the resource allocation algorithm design as the following optimization problem:

Problem 4: Robust resource allocation for secure SWIPT

$$\underset{\mathbf{V}\in\mathbb{H}^{N_\mathrm{T}},\mathbf{w}}{\text{maximize}}\ [\min_{\Delta\mathbf{h}\in\Xi} R - \max_{\Delta\mathbf{g}_j\in\Upsilon_j,\forall j}\{R_{\mathrm{ER}_j}\}]^+ \tag{11.15}$$

$$\text{subject to}\ \mathrm{C1}:\ \|\mathbf{w}\|_2^2 + \mathrm{Tr}(\mathbf{V}) \leq P_{\max},$$

$$\mathrm{C2}:\ \min_{\Delta\mathbf{g}_j\in\Upsilon_j} \Psi_{\mathrm{ER}_j} \geq,\quad \mathrm{C3}:\ \mathbf{V}\succeq\mathbf{0}.$$

Constant $P_{\max}$ in constraint C1 denotes the maximum transmit power budget of the transmitter. Constraint C2 ensures that the minimum required harvested power at ER j is no less than a constant, $P_{\mathrm{req}_j}^{\min}$, despite the imperfection of the CSI. Constraint C3 and $\mathbf{V}\in\mathbb{H}^{N_\mathrm{T}}$ constrain matrix $\mathbf{V}$ to be a positive semi-definite Hermitian matrix. It can be observed that the objective function in (11.15) is a non-convex function. In order to obtain a tractable solution for the considered problem, we recast it as a rank-constrained semi-definite programming (SDP) problem. To facilitate the reformulation, we define $\mathbf{W}=\mathbf{w}\mathbf{w}^H$ and rewrite (11.15) as:

Problem 5: Equivalent rank-constrained optimization problem

$$\underset{\substack{\mathbf{W},\mathbf{V}\in\mathbb{H}^{N_\mathrm{T}},\\ \tau,\beta_j,\mu}}{\text{maximize}}\ \left[\log_2\left(1+\frac{\mu}{\sigma_\mathrm{s}^2}\right)-\tau\right]^+ \tag{11.16}$$

$$\text{subject to}\ \mathrm{C1}:\ \mathrm{Tr}(\mathbf{W})+\mathrm{Tr}(\mathbf{V})\leq P_{\max},$$

$$\mathrm{C2}:\ \frac{M_j}{1+\exp(-a_j(\beta_j-b_j))}\geq P_{\mathrm{req}_j}^{\min}(1-\Omega_j)+M_j\Omega_j,\forall j,$$

$$\mathrm{C3}:\ \mathbf{V}\succeq\mathbf{0},\ \mathrm{C4}:\ \mathbf{W}\succeq\mathbf{0},$$

$$\mathrm{C5}:\ \beta_j\leq\min_{\Delta\mathbf{g}_j\in\Upsilon_j}\mathrm{Tr}((\mathbf{W}+\mathbf{V})\mathbf{G}_j),\ \forall j,$$

$$\mathrm{C6}:\ \max_{\Delta\mathbf{g}_j\in\Upsilon_j}\log_2\left(1+\frac{\mathrm{Tr}(\mathbf{W}\mathbf{G}_j)}{\mathrm{Tr}(\mathbf{G}_j\mathbf{V})+\sigma_\mathrm{s}^2}\right)\leq\tau,\forall j,$$

$$\mathrm{C7}:\ \mu\leq\min_{\Delta\mathbf{h}\in\Xi}\mathrm{Tr}(\mathbf{W}\mathbf{H}),\quad \mathrm{C8}:\ \mathrm{Rank}(\mathbf{W})\leq 1,$$

where μ is an auxiliary optimization variable denoting the maximum received signal strength of the information signal at the IR and $\beta_j,\forall j\in\{1,\dots,J\}$, are auxiliary optimization variables denoting the received power at ER j. Here, "equivalent" means that both the optimization problems have the same optimal solution. Constraints C4, C8, and $\mathbf{W}\in\mathbb{H}^{N_\mathrm{T}}$ are imposed to guarantee that $\mathbf{W}=\mathbf{w}\mathbf{w}^H$ holds after optimization. We note that the optimization problem is still non-convex due to constraint C6. In addition, there are infinitely many constraints in C5–C7 due to the continuity of channel uncertainty sets Ξ and Υ_j.

Now, we handle the non-convexity due to constraint C6. First, we perform a one-dimensional search with respect to τ. In particular, for a given value of τ, we optimize the resource allocation for the maximization of the secrecy rate. We repeat the procedure for all possible values of τ and record the corresponding achieved secrecy rate. Finally, we select the value of τ which provides the maximum system secrecy rate from all the trials. For a fixed τ, the optimization problem in (11.16) can be transformed into the following equivalent form:

Problem 6: Equivalent rank-constrained optimization problem

$$\begin{aligned} \underset{\mathbf{W},\mathbf{V}\in\mathbb{H}^{N_{\mathrm{T}}},\beta_j,\mu}{\text{maximize}} \quad & \mu \\ \text{subject to} \quad & \text{C1–C4}, \\ & \text{C5}: \quad \beta_j \le \min_{\Delta\mathbf{g}_j\in\Upsilon_j} \mathrm{Tr}((\mathbf{W}+\mathbf{V})\mathbf{G}_j), \ \forall j, \\ & \text{C6}: \quad \max_{\Delta\mathbf{g}_j\in\Upsilon_j} \mathrm{Tr}((\mathbf{W}-\mathbf{V}\psi)\mathbf{G}_j) \le \psi\sigma_{\mathrm{s}}^2, \forall j, \\ & \text{C7}: \quad \min_{\Delta\mathbf{h}\in\Xi} \mathrm{Tr}(\mathbf{H}\mathbf{W}) \ge \mu, \\ & \text{C8}: \quad \mathrm{Rank}(\mathbf{W}) \le 1, \end{aligned} \tag{11.17}$$

where $\psi = 2^{\tau} - 1$. Then, we handle constraints C5, C6, and C7 by transforming them into linear matrix inequalities (LMIs) using the following lemma:

Lemma 11.1 ***(S-Procedure [45])*** Let a function $f_m(\mathbf{x}), m \in \{1,2\}, \mathbf{x} \in \mathbb{C}^{N\times 1}$, be defined as

$$f_m(\mathbf{x}) = \mathbf{x}^H\mathbf{A}_m\mathbf{x} + 2\mathrm{Re}\{\mathbf{b}_m^H\mathbf{x}\} + c_m, \tag{11.18}$$

where $\mathbf{A}_m \in \mathbb{H}^N$, $\mathbf{b}_m \in \mathbb{C}^{N\times 1}$, and $c_m \in \mathbb{R}$. Then, the implication $f_1(\mathbf{x}) \le 0 \Rightarrow f_2(\mathbf{x}) \le 0$ holds if and only if there exists a $\delta \ge 0$ such that

$$\delta \begin{bmatrix} \mathbf{A}_1 & \mathbf{b}_1 \\ \mathbf{b}_1^H & c_1 \end{bmatrix} - \begin{bmatrix} \mathbf{A}_2 & \mathbf{b}_2 \\ \mathbf{b}_2^H & c_2 \end{bmatrix} \succeq \mathbf{0}, \tag{11.19}$$

provided that there exists a point $\hat{\mathbf{x}}$ such that $f_m(\hat{\mathbf{x}}) < 0$.

Applying Lemma 11.1, the original constraint C5 holds if and only if there exists $\delta_j \geq 0, \forall j \in \{1, \ldots, J\}$, such that the following LMI constraint holds:

$$\overline{\text{C5}} : \mathbf{S}_{\overline{\text{C}}_{5_j}}(\mathbf{W}, \mathbf{V}, \beta_j, \delta_j) = \begin{bmatrix} \delta_j \mathbf{I}_{N_\text{T}} & \mathbf{0} \\ \mathbf{0} & -\delta_j \rho_j^2 - \beta_j \end{bmatrix} + \mathbf{U}_{\hat{\mathbf{g}}_j}^H (\mathbf{W} + \mathbf{V}) \mathbf{U}_{\hat{\mathbf{g}}_j} \succeq \mathbf{0}, \tag{11.20}$$

where $\mathbf{U}_{\hat{\mathbf{g}}_j} = [\mathbf{I}_{N_\text{T}} \quad \hat{\mathbf{g}}_j]$. Similarly, constraints C6 and C7 can be equivalently written as

$$\overline{\text{C6}}: \mathbf{S}_{\overline{\text{C}}_{6_j}}(\mathbf{W}, \mathbf{V}, \varphi_j) = \begin{bmatrix} \varphi_j \mathbf{I}_{N_\text{T}} & \mathbf{0} \\ \mathbf{0} & \psi \sigma_\text{s}^2 - \rho_j^2 \varphi_j \end{bmatrix} - \mathbf{U}_{\hat{\mathbf{g}}_j}^H (\mathbf{W} - \mathbf{V}\psi) \mathbf{U}_{\hat{\mathbf{g}}_j} \succeq \mathbf{0} \quad \text{and}$$

$$\overline{\text{C7}}: \mathbf{S}_{\overline{\text{C}}_7}(\mathbf{W}, \mu, \nu) = \begin{bmatrix} \nu \mathbf{I}_{N_\text{T}} & \mathbf{0} \\ \mathbf{0} & -\nu \theta^2 - \mu \end{bmatrix} + \mathbf{U}_{\hat{\mathbf{h}}}^H \mathbf{W} \mathbf{U}_{\hat{\mathbf{h}}} \succeq \mathbf{0}, \tag{11.21}$$

respectively, where $\varphi_j \geq 0$ and $\nu \geq 0$ are the extra optimization variables introduced via applying Lemma 11.1, and $\mathbf{U}_{\hat{\mathbf{h}}} = [\mathbf{I}_{N_\text{T}} \quad \hat{\mathbf{h}}]$. We note that now constraints $\overline{\text{C5}}$, $\overline{\text{C6}}$, and $\overline{\text{C7}}$ involve only a finite number of LMI constraints, which facilitates the design of a computationally efficient resource allocation algorithm.

After substituting (11.20) and (11.21) into (11.17), the only non-convexity in (11.17) is due to the rank constraint in C8. To circumvent this difficulty, we adopt semi-definite programming relaxation (SDR) to obtain a tractable solution. Specifically, we remove the non-convex constraint C8 from (11.17), which yields

Problem 7: SDR optimization problem

$$\begin{aligned} \underset{\substack{\mathbf{W}, \mathbf{V} \in \mathbb{H}^{N_T}, \mu, \beta_j, \\ \delta_j, \nu, \varphi_j \geq 0}}{\text{maximize}} \quad & \mu \\ \text{subject to} \quad & \text{C1} - \text{C4}, \overline{\text{C5}} - \overline{\text{C7}}, \\ & \cancel{\text{C8}: \text{Rank}(\mathbf{W}) \leq 1.} \end{aligned} \tag{11.22}$$

In fact, Problem 7 is a standard convex optimization problem that can be solved by standard numerical convex program solvers such as Sedumi or SDPT3 [46]. Now, we study the tightness of the adopted rank constraint relaxation in (11.22).

Theorem 11.1 For $P_{\max} > 0$ and if the considered problem is feasible, the condition of $\text{Rank}(\mathbf{W}) \leq 1$ is always satisfied, despite the fact that the rank constraint is removed.

Proof: Please refer to the Appendix.

In other words, the optimal resource allocation can be obtained by beamforming. Specifically, the optimal $\mathbf{w}$ can be obtained by performing eigenvalue decomposition of $\mathbf{W}$ and select the principal eigenvector as the beamformer.

11.4 Results

In this section, simulation results are presented to illustrate the performance of the proposed resource allocation algorithm. In the simulation, the IR and the J ERs are located at 100 m and 10 m from the BS, respectively. For the nonlinear energy harvesting circuit, we set $M_j = 10.7$ mW, which corresponds to the maximum harvested power per ER. In addition, we adopt $a_j = 0.2308$ and $b_j = 5.365$. Furthermore, we denote the normalized maximum channel estimation errors of ER j and IR as $\sigma^2_{\text{est}_{g_j}} \geq \frac{\rho_j^2}{\|\mathbf{g}_j\|^2}, \forall j$, and $\sigma^2_{\text{est}_h} \geq \frac{\theta^2}{\|\mathbf{h}\|^2}$, respectively. In the following simulation results, we set $\sigma^2_{\text{est}_{g_j}} = \sigma^2_{\text{est}_h} = \sigma^2_{\text{est}}$ for simplicity. Other important simulation parameters are summarized in Table 11.1.

In Figure 11.4 we study the average system secrecy rate (bit/s/Hz) versus the average minimum harvested power per ER for different numbers of transmit antennas. The normalized maximum channel estimation errors and the maximum transmit power are $\sigma^2_{\text{est}} = 1\%$ and $P_{\max} = 36$ dBm, respectively. There is $J = 1$ ER located at 10 m from the BS. As can be observed, there is a nontrivial tradeoff between the achievable system secrecy rate and the average minimum harvested power per ER. In other words, improving the system secrecy rate and improving the harvested energy per ER are two conflicting system design objectives. In addition, for the proposed optimal resource allocation, the system achievable secrecy rate and the minimum harvested power per ER are enlarged significantly by increasing N_T. This is due to the fact that the extra degrees of freedom offered by additional transmit antennas help the transmitter in focusing the energy of the information and energy signals on the desired receivers, thus improve the beamforming efficiency. However, there is a diminishing return in the system achievable secrecy rate when N_T is large due to the channel hardening effect [47] in the channel of the IR. Also, there is only a marginal gain in terms of the minimum harvested power at the ER when the numbers of transmit antennas is large, e.g., $N_\text{T} > 8$. In fact, a large number of antennas leads to an exceedingly large received power at the ER, causing saturation in the nonlinear energy harvesting circuit. Hence, the corresponding

Table 11.1 Simulation parameters

System bandwidth	200 kHz
Carrier center frequency	915 MHz
Transceiver antenna gain	18 dBi
Noise powerσ_s^2	−120 dBm
BS-to-ER fading distribution	Rician with Rician factor 6 dB
BS-to-IR fading distribution	Rayleigh

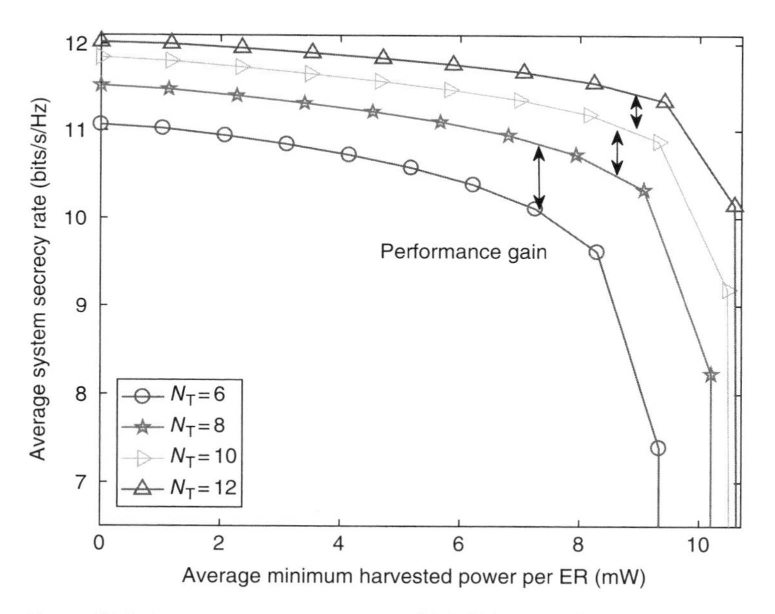

Figure 11.4 Average system secrecy rate (bit/s/Hz) versus the average minimum harvested power per ER (mW). The double-sided arrows represent the performance gains due to additional transmit antennas.

amount of harvested power approaches the maximal possible harvested power of the circuits, i.e. M_j. Furthermore, we verified by simulation that Rank($\mathbf{W}$) $= 1$ can be obtained for all considered channel realizations, which confirms the correctness of Theorem 11.1.

Figure 11.5 illustrates the average system secrecy rate versus the normalized maximum channel estimation errors, σ^2_{est}, for different numbers of transmit antennas, N_{T}. The maximum transmit power and the minimum required harvested power per ER are $P_{\max} = 36$ dBm and $P^{\min}_{\text{req}_j} = 5$ dBm, $\forall j$, respectively.

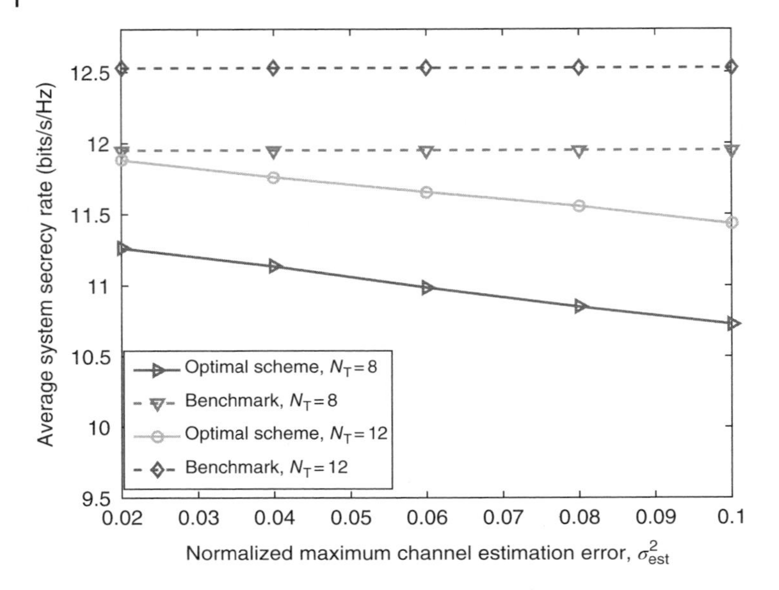

Figure 11.5 Average system secrecy rate (bit/s/Hz) versus the normalized maximum channel estimation error, σ^2_{est}, for different numbers of transmit antennas, N_T.

There are $J = 2$ ERs located at 10 m from the BS. For comparison, we also consider the performance of a benchmark scheme and a baseline scheme. For the benchmark scheme, we assume that the perfect CSI is available for resource allocation. For the baseline scheme, the resource allocation is performed subject to the same constraint set as (11.22) except that the linear energy harvesting model in (11.5) is adopted. Besides, the imperfect CSI as provided if it was perfect CSI. Then, the performance of the baseline scheme was evaluated in the presence of the imperfect CSI. However, for the adopted simulation settings, the baseline scheme could not satisfy the QoS requirements in constraint C2 for all considered channel realizations. Therefore, performance results for the baseline scheme are not shown in Figure 11.5. This finding underlines the importance of taking imperfect CSI and the nonlinearity of the energy harvesting circuit into account for resource allocation design. On the other hand, it can be observed from Figure 11.5 that the average system secrecy rate decreases only slowly with increasing σ^2_{est} due to the proposed robust optimization. In fact, the CSI quality degrades with increasing σ^2_{est}. In particular, for a larger value of σ^2_{est} it is more difficult for the transmitter to steer the information signal towards the IR accurately to improve the secrecy rate. Besides, less accurate CSI would also increase the possibility of information leakage to the ERs. Hence, a higher artificial noise transmit power is required to neutralize the higher potential for information leakage, leading to a smaller power allocated to the information signal. Also, the achievable secrecy rate of both the proposed optimal scheme and the benchmark scheme improves with increasing number of transmit antennas equipped at the BS. In fact, the extra spatial degrees of freedom offered by the increasing number of antennas provide a greater flexibility in utilizing the limited system resources. Furthermore, the proposed optimal scheme is able to fulfill the minimum required harvested power per ER in all considered scenarios, despite the imperfect CSI knowledge.

11.5 Conclusions

In this chapter we presented a beamforming design enabling secure communication in SWIPT systems. The algorithm design based on a practical nonlinear energy harvesting model was formulated as a non-convex optimization problem taking into account the impact of imperfect CSI. We aimed to maximize the system achievable secrecy rate while guaranteeing a required minimum harvested power at the ERs. The optimization problem was solved optimally by a one-dimensional search and SDR. Numerical results unveiled the nontrivial tradeoff between the system secrecy rate and the minimum harvested power per ER. In addition, our results illustrated the potential performance gain in secrecy rate brought by the proposed optimization and its robustness against CSI imperfectness.

Appendix-Proof of Theorem 11.1

We follow a similar approach as in [39] to prove Theorem 11.1. We note that the relaxed problem in (11.22) is jointly convex with respect to the optimization variables. In addition, it can be verified that the problem satisfies the Slater's constraint qualification and thus has a zero duality gap. Therefore, to reveal the structure of $\mathbf{W}$, we consider the Lagrangian of (11.17), which is given by

$$\begin{aligned} L =& \mathrm{Tr}(\mathbf{S}_{\overline{\mathrm{C}_7}}(\mathbf{W},\mu,\nu)\mathbf{D}_{\overline{\mathrm{C}_7}}) + \alpha(\mathrm{Tr}(\mathbf{W}_{\mathrm{E}}) + \mathrm{Tr}(\mathbf{W}) - P_{\max}) - \mathrm{Tr}(\mathbf{WY}) \\ & - \sum_{j=1}^{J} \mathrm{Tr}(\mathbf{S}_{\overline{\mathrm{C}_{6_j}}}(\mathbf{W},\mathbf{V},\varphi_j)\mathbf{D}_{\overline{\mathrm{C}_{6_j}}}) + \sum_{j=1}^{J} \mathrm{Tr}(\mathbf{S}_{\overline{\mathrm{C}_{5_j}}}(\mathbf{W},\mathbf{V},\beta_j,\delta_j)\mathbf{D}_{\overline{\mathrm{C}_{5_j}}}) + \Delta, \end{aligned} \tag{11.23}$$

where $\alpha \geq \mathbf{0}, \mathbf{Y} \succeq \mathbf{0}, \mathbf{D}_{\overline{\mathrm{C}_{5_j}}} \succeq \mathbf{0}, \mathbf{D}_{\overline{\mathrm{C}_{6_j}}} \succeq \mathbf{0}, \forall j \in \{1,\dots,J\}$, and $\mathbf{D}_{\overline{\mathrm{C}_7}} \succeq \mathbf{0}$ are the dual variables for constraints C1, C4, $\overline{\mathrm{C5}}$, $\overline{\mathrm{C6}}$, and $\overline{\mathrm{C7}}$, respectively. In addition, Δ is a collection of variables and constants that are not relevant to the proof. For notational convenience, we denote the optimal primal and dual variables of the SDP relaxed version of (11.17) by the corresponding variables with an asterisk superscript in the following. Now, we focus on those Karush–Kuhn–Tucker (KKT) conditions [45], which are needed for the proof:

$$\mathbf{Y}^*, \mathbf{D}^*_{\overline{\mathrm{C}}_{5_j}}, \mathbf{D}^*_{\overline{\mathrm{C}}_{6_j}}, \mathbf{D}_{\overline{\mathrm{C}}_7} \succeq \mathbf{0}, \quad \alpha^* \geq 0, \tag{11.24a}$$

$$\mathbf{Y}^*\mathbf{W}^* = \mathbf{0}, \tag{11.24b}$$

$$\mathbf{Y}^* = \alpha^*\mathbf{I}_{N_\mathrm{T}} - \boldsymbol{\Xi}, \tag{11.24c}$$

$$\boldsymbol{\Xi} = \mathbf{U}_{\hat{\mathbf{h}}}\mathbf{D}^*_{\overline{\mathrm{C}}_7}\mathbf{U}^H_{\hat{\mathbf{h}}} - \sum_{j=1}^{J}\mathbf{U}_{\hat{\mathbf{g}}_j}(\mathbf{D}^*_{\overline{\mathrm{C}}_{6_j}} - \mathbf{D}^*_{\overline{\mathrm{C}}_{5_j}})\mathbf{U}^H_{\hat{\mathbf{g}}_j}. \tag{11.24d}$$

From (11.24b) we know that the columns of $\mathbf{W}^*$ lie in the null space of $\mathbf{Y}^*$. In order to reveal the rank of $\mathbf{W}^*$, we investigate the structure of $\mathbf{Y}^*$. First, it can be shown that $\alpha^* > 0$ since constraint C1 is active at the optimal solution. Then, we focus on $\boldsymbol{\Xi}$. Suppose Ξ is a negative definite matrix, then from (11.24c), $\mathbf{Y}^*$ has to be a full-rank and positive definite matrix. By (11.24b), $\mathbf{W}^*$ is forced to be the zero matrix and Rank($\mathbf{W}^*$) $= \mathbf{0}$. Thus, in the following, we focus on the case $\boldsymbol{\Xi} \succeq \mathbf{0}$. Since matrix $\mathbf{Y}^* = \alpha^*\mathbf{I}_{N_\mathrm{T}} - \boldsymbol{\Xi}$ is positive semi-definite, the following inequality holds:

$$\alpha^* \geq \lambda^{\max}_{\boldsymbol{\Xi}} \geq 0, \tag{11.25}$$

where $\lambda^{\max}_{\boldsymbol{\Xi}}$ is the maximum eigenvalue of matrix Ξ. From (11.24c), if $\alpha^* > \lambda^{\max}_{\boldsymbol{\Xi}}$, matrix $\mathbf{Y}^*$ has to be a positive definite matrix with full rank. However, this will again yield the solution $\mathbf{W}^* = \mathbf{0}$ and Rank($\mathbf{W}^*$) $= 0$. On the other hand, if $\alpha^* = \lambda^{\max}_{\boldsymbol{\Xi}}$ holds at the optimal solution, then in order to have a bounded optimal dual solution, it follows that the null space of $\mathbf{Y}^*$ is spanned by vector $\mathbf{u}_{\boldsymbol{\Xi},\max} \in \mathbb{C}^{N_\mathrm{T}\times 1}$, which is the unit-norm eigenvector of $\boldsymbol{\Xi}$ associated with eigenvalue $\lambda^{\max}_{\boldsymbol{\Xi}}$. As a result, the optimal beamforming matrix $\mathbf{W}^*$ has to be a rank-one matrix and is given by

$$\mathbf{W}^* = \zeta\mathbf{u}_{\boldsymbol{\Xi},\max}\mathbf{u}^H_{\boldsymbol{\Xi},\max}. \tag{11.26}$$

where ζ is a parameter such that the power consumption satisfies constraint C1.

Acknowledgements. This work was supported in part by the AvH Professorship Program of the Alexander von Humboldt Foundation, the Australian Research Councils Discovery Early Career Researcher Award funding scheme (DE170100137), and the Linkage Project (LP160100708).

Bibliography

1 M. Zorzi, A. Gluhak, S. Lange, and A. Bassi (2010) From today's INTRAnet of Things to a future INTERnet of Things: a Wireless- and mobility-related view. *IEEE Wireless Commun.*, **17**: 44–51.

2 V.W.S. Wong, R. Schober, D.W.K. Ng, and L.-C. Wang (2017) *Key Technologies for 5G Wireless Systems.* Cambridge University Press.

3 Q. Wu, G.Y. Li, W. Chen, D.W.K. Ng, and R. Schober (2017) An overview of sustainable green 5G networks. *IEEE Wireless Commun.*, **24** (4): 72–80.

4 D.W.K. Ng, E.S. Lo, and R. Schober (2012) Energy-efficient resource allocation in multi-cell OFDMA systems with limited backhaul capacity. *IEEE Trans. Wireless Commun.*, **11**: 3618–3631.

5 D.W.K. Ng, E.S. Lo, and R. Schober (2012) Energy-efficient resources allocation in OFDMA systems with large numbers of base station antennas. *IEEE Trans. Wireless Commun.*, **11** (9): 3292–3304.

6 Q. Wu, G.Y. Li, W. Chen, and D.W.K. Ng (20160 Energy-efficient small cell with spectrum-power trading. *IEEE J. Selected Areas Commun.*, **34** (12): 3394–3408.

7 J. Yang, O. Ozel, and S. Ulukus (2012) Broadcasting with an energy harvesting rechargeable transmitter. *IEEE Trans. Wireless Commun.*, **11**: 571–583.

8 D.W.K. Ng, E.S. Lo, and R. Schober (2013) Energy-efficient resource allocation in OFDMA systems with hybrid energy harvesting base station. *IEEE Trans. Wireless Commun.*, **12**: 3412–3427.

9 I. Ahmed, A. Ikhlef, D.W.K. Ng, and R. Schober (2013) Power allocation for an energy harvesting transmitter with hybrid energy sources. *IEEE Trans. Wireless Commun.*, **12**: 6255–6267.

10 L. Varshney (2008) Transporting information and energy simultaneously. In *Proceedings of the IEEE International Symposium on Information Theory*, pp. 1612 –1616.

11 P. Grover and A. Sahai (2010) Shannon meets Tesla: wireless information and power transfer. In *Proceedings of the IEEE International Symposium on Information Theory*, pp. 2363–2367.

12 I. Krikidis, S. Timotheou, S. Nikolaou, G. Zheng, D.W.K. Ng, and R. Schober (2014) Simultaneous wireless information and power transfer in modern communication systems. *IEEE Commun. Mag.*, **52** (11): 104–110.

13 Z. Ding, C. Zhong, D.W.K. Ng, M. Peng, H.A. Suraweera, R. Schober, and H.V. Poor (2015) Application of smart antenna technologies in simultaneous wireless information and power transfer. *IEEE Commun. Mag.*, **53** (4): 86–93.

14 X. Chen, Z. Zhang, H.-H. Chen, and H. Zhang (2015) Enhancing wireless information and power transfer by exploiting multi-antenna techniques. *IEEE Commun. Mag.*, **4**: 133–141.

15 Powercast Corporation (2011) RF energy harvesting and wireless power for low-power applications. [Online]. Available: http://www.mouser.com/pdfdocs/Powercast-Overview-2011-01-25.pdf.

16 R. Zhang and C.K. Ho (2011) MIMO broadcasting for simultaneous wireless information and power transfer. In *Proceedings of the IEEE Global Telecommunications Conference*, pp. 1–5.

17 L. Liu, R. Zhang, and K.-C. Chua (2013) Wireless information transfer with opportunistic energy harvesting. *IEEE Trans. Wireless Commun.*, **12** (1): 288–300.

18 X. Zhou, R. Zhang, and C.K. Ho (2013) Wireless information and power transfer: Architecture design and rate–energy tradeoff. *IEEE Trans. Commun.*, **61**: 4754–4767.

19 Q. Wu, M. Tao, D.W.K. Ng, W. Chen, and R. Schober (2016) Energy-efficient resource allocation for wireless powered communication networks. *IEEE Trans. Wireless Commun.*, **15** (3): 2312–2327.

20 Q. Wu, W. Chen, and J. Li (2015) Wireless powered communications with initial energy: QoS guaranteed energy-efficient resource allocation. *IEEE Wireless Commun. Lett.*, **19**.

21 D.W.K. Ng, E.S. Lo, and R. Schober (2013) Wireless information and power transfer: Energy efficiency optimization in OFDMA systems. *IEEE Trans. Wireless Commun.*, **12**: 6352–6370.

22 D.W.K. Ng and R. Schober (20130 Spectral efficient optimization in OFDM systems with wireless information and power transfer. In *21st European Signal Processing Conference (EUSIPCO)*, pp. 1–5.

23 D.W.K. Ng, E.S. Lo, and R. Schober (2013) Energy-efficient power allocation in OFDM systems with wireless information and power transfer. In *Proceedings of the IEEE International Communications Conference*, pp. 4125–4130.

24 A.D. Wyner (1975) *The Wire-Tap Channel*, Technical Report.

25 J. Zhu, R. Schober, and V. Bhargava (2014) Secure transmission in multicell massive MIMO systems. *IEEE Trans. Wireless Commun.*, **13**: 4766–4781.

26 S. Goel and R. Negi (2008) Guaranteeing secrecy using artificial noise. *IEEE Trans. Wireless Commun.*, **7**: 2180–2189.

27 H.M. Wang, C. Wang, D. Ng, M. Lee, and J. Xiao (2016) Artificial noise assisted secure transmission for distributed antenna systems. *IEEE Trans. Signal Process.*, **64** (15): 4050–4064.

28 J. Chen, X. Chen, W.H. Gerstacker, and D.W.K. Ng (2016) Resource allocation for a massive MIMO relay aided secure communication. *IEEE Trans. Inf. Forensics and Security*, **11** (8): 1700–1711.

29 D.W.K. Ng, E.S. Lo, and R. Schober (2012) Efficient resource allocation for secure OFDMA systems. *IEEE Trans. Veh. Technol.*, **61**: 2572–2585.

30 H.M. Wang, C. Wang, and D.W.K. Ng (2015) Artificial noise assisted secure transmission under training and feedback. *IEEE Trans. Signal Process.*, **63** (23): 6285–6298.

31 D.W.K. Ng and R. Schober (2013) Resource allocation for secure communication in systems with wireless information and power transfer. In *Proceedings of the IEEE Global Telecommunications Conference.*

32 D.W.K. Ng, E.S. Lo, and R. Schober (2014) Robust beamforming for secure communication in systems with wireless information and power transfer. *IEEE Trans. Wireless Commun.*, **13**: 4599–4615.

33 R.T. Marler and J.S. Arora (2004) Survey of multi-objective optimization methods for engineering. *Structural and Multidisciplinary Optimization*, **26**: 369–395.

34 L. Liu, R. Zhang, and K.-C. Chua (2014) Secrecy wireless information and power transfer with MISO beamforming. *IEEE Trans. Signal Process.*, **62**: 1850–1863.

35 D. Wang and R. Negra (2013) Design of a dual-band rectifier for wireless power transmission. In *2013 IEEE Wireless Power Transfer (WPT)*, pp. 127–130.

36 T. Le, K. Mayaram, and T. Fiez (2008) Efficient far-field radio frequency energy harvesting for passively powered sensor networks. *IEEE J. Solid-State Circuits*, **43**: 1287–1302.

37 C. Valenta and G. Durgin (2014) Harvesting wireless power: Survey of energy-harvester conversion efficiency in far-field, wireless power transfer systems. *IEEE Microwave Mag.*, **15**: 108–120.

38 E. Boshkovska, D. Ng, N. Zlatanov, and R. Schober (20150 Practical non-linear energy harvesting model and resource allocation for SWIPT systems. *IEEE Commun. Lett.*, **19**: 2082–2085.

39 E. Boshkovska, D.W.K. Ng, N. Zlatanov, A. Koelpin, and R. Schober (2017) Robust resource allocation for MIMO wireless powered communication networks based on a non-linear EH model. *IEEE Trans. Commun.*, **65** (5): 1984–1999.

40 E. Boshkovska, D.W.K. Ng, L. Dai, and R. Schober (2017) Power-efficient and secure WPCNs with hardware impairments and non-Linear EH circuit. *IEEE Trans. Commun.*, submitted.

41 T.A. Le, Q.T. Vien, H.X. Nguyen, D.W.K. Ng, and R. Schober (2017) Robust chance-constrained optimization for power-efficient and secure SWIPT systems. *IEEE Trans. Green Commun. Network.*, **1** (3): 333–346.

42 D.W.K. Ng and R. Schober (2015) Secure and green SWIPT in distributed antenna networks with limited backhaul capacity. *IEEE Trans. Wireless Commun.*, **14** (9): 5082–5097.

43 Y. Sun, D.W.K. Ng, J. Zhu, and R. Schober (2016) Multi-objective optimization for robust power efficient and secure full-duplex wireless communication systems. *IEEE Trans. Wireless Commun.*, **15** (8): 5511–5526.

44 J. Wang and D. Palomar (2009) Worst-case robust MIMO transmission with imperfect channel knowledge. *IEEE Trans. Signal Process.*, **57**: 3086–3100.

45 S. Boyd and L. Vandenberghe (2004) *Convex Optimization.* Cambridge University Press.

46 M. Grant and S. Boyd (2013) CVX: Matlab software for disciplined convex programming, version 2.0 Beta. [Online]. https://cvxr.com/cvx.

47 D. Tse and P. Viswanath (2005) *Fundamentals of Wireless Communication.* Cambridge University Press.

12

Wireless-Powered Cooperative Networks with Energy Accumulation

*Yifan Gu**, *He Chen, and Yonghui Li*

School of Electrical and Information Engineering, The University of Sydney, Australia

12.1 Introduction

Background. The energy harvesting technique has been regarded as a sustainable solution to prolonging the lifetime of energy constrained wireless networks, see [1] and references therein. Conventional studies of energy harvesting focus on collecting energy from renewable sources such as wind and solar. Very recently, radio frequency (RF) signals have been used as a new viable source for energy harvesting. In this sense, wireless signals can now be used to deliver information as well as energy. The feasibility of wireless energy transfer in practical applications has been demonstrated in [2]. Additionally, [3] has realized the successful communication between two energy harvesting nodes solely exploiting the ambient RF signals. Apart from ambient signals, dedicated power transmitters can also be deployed to implement wireless energy transfer in some applications, e.g., passive radio frequency identification (RFID) networks [4].

On the other hand, the cooperative diversity technique has drawn significant interest from the wireless communication community over the past decade. The key idea of this technique is that single-antenna nodes can transmit signals cooperatively and create a virtual multiple-input multiple-output (MIMO) array. With this technology, spatial diversity can be achieved without the need to equip multiple antennas at each node [5, 6]. The advantages of this technique, such as increasing system capacity, coverage, and energy efficiency, have been demonstrated by numerous papers in open literature. The RF energy transfer and harvesting technique has brought new opportunities to conventional cooperative networks due to its various advantages [7]. For instance, the relay

*Corresponding author: Yifan Gu; yigu6254@uni.sydney.edu.au

Wireless Information and Power Transfer: Theory and Practice, First Edition.
Edited by Derrick Wing Kwan Ng, Trung Q. Duong, Caijun Zhong, and Robert Schober.

node can now harvest energy from ambient RF signals and then use the harvested energy to assist the information transmission between the source and destination. In this sense, the relay is more willing to cooperate with the source since it does not need to consume its own energy. In this chapter, we refer to a cooperative communication network with wireless-powered (energy harvesting) relay(s) as a wireless-powered cooperative network (WPCN).

Literature Survey. Nasir et al. [8] first investigated a classical three-node WPCN consisting of one source–destination pair and one energy harvesting relay. Two practical relaying protocols, namely time switching (TS) relaying and power splitting (PS) relaying, were proposed. The outage probability and ergodic capacity of the system were then derived. It was shown that the TS-based relaying protocol outperforms the PS-based relaying protocol when the transmission rate is high or the signal-to-noise ratio (SNR) is relatively low. Otherwise, PS relaying is superior to TS relaying. Chen et al. [9] considered an alternative scenario where both source and relay harvest energy from a hybrid access point (AP). Based on the considered system model, a harvest-then-cooperate (HTC) protocol was developed, in which the source and relay harvest energy at the same time and work cooperatively for information transmission operation. Compared to a non-cooperative scheme, the HTC protocol can yield considerable performance gain by introducing cooperation between the source and relay. Gu et al. [10] then extended the works in [9] and proposed an adaptive transmission (AT) protocol that the information transmission between source and AP can be performed either directly or cooperatively through the relay. The decision depends on the instantaneous channel state information (CSI) between the source and the AP.

Different from all the above works where the wireless-powered devices exhaust the harvested energy to perform information transmission/forwarding, some researchers investigated the energy accumulation (EA) behaviors of the wireless-powered devices in the sense that they can accumulate the harvested energy and perform information transmission/forwarding in an appropriate time slot. The motivation behind this can be explained as follows. When the channel of the wireless-powered node to the energy source suffers from deep fading, it can only harvest a small amount of energy and thus it may not be able to perform effective information transmission/forwarding even if it exhausts all the harvested energy. On the other hand, when the transmitting channel of the wireless-powered node is in good condition, it should use part of its harvested energy for information transmission/forwarding and save the rest of the energy for future use. Specifically, in [11], a classical three-node WPCN, including one source, one wireless-powered relay, and one destination, was investigated. Considering that the relay is equipped with a rechargeable battery and can perform EA, the authors proposed a simple greedy switching policy that the relay will only forward the information when the stored energy can guarantee an outage-free transmission for the second hop channel. Otherwise, it will harvest energy from the source and store the harvested energy. The authors studied the performance of a wireless-powered two-way relay network with EA behavior at the relay in [12], wherein a power splitting-based energy accumulation (PS-EA) scheme was then developed. It was concluded in [12] that the proposed PS-EA scheme outperforms the conventional time switching-based energy accumulation (TS-EA) scheme. Krikidis [13], Liu [14], and Gu et al. [15] studied the relay selection problem for multiple wireless powered relays with energy accumulation capability between a source node and a destination node. It was shown that the energy storage significantly affects the performance of the system and the proposed protocols outperform the existing ones without energy accumulation.

In this chapter we investigate the advantages of cooperation and energy accumulation in WPCNs. Specifically, we extend the work of [10] and evaluate the performance of the AT protocol when the wireless-powered relay can accumulate the harvested energy. We then compare the results with the non-energy accumulation case as well as the non-cooperative harvest-then-transmit (HTT) protocol proposed in [16].

12.2 System Model

As shown in Figure 12.1, we consider a WPCN with one hybrid AP, one source and one relay to assist information transmission. Only the hybrid AP is connected to external energy supplies while the source and relay have no embedded

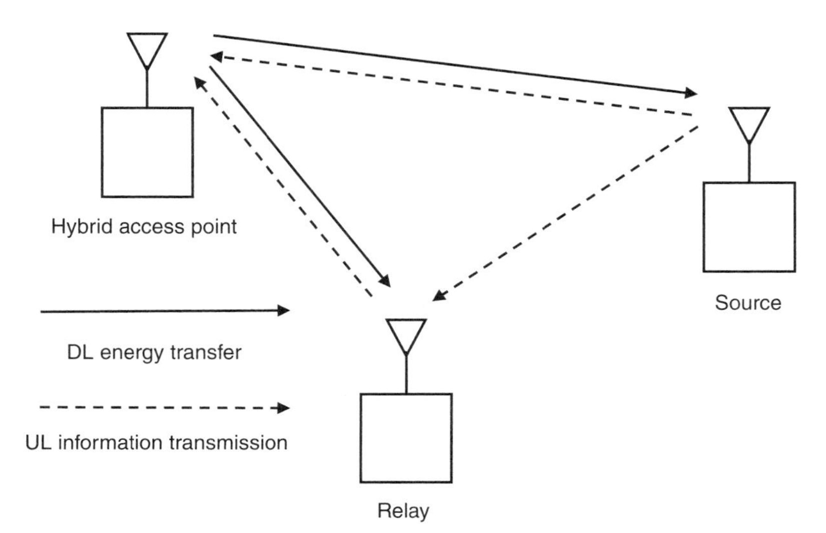

Figure 12.1 The three-node wireless-powered cooperative network.

energy supplies. However, they could store the harvested energy from the RF signals radiated by the AP. Thus, the source and relay need first to harvest energy from the AP in the downlink (DL), then transmit information cooperatively in the uplink (UL). Let the subscripts A, S, and R denote the hybrid AP, source, and relay, respectively. We consider all the channels between node X and node Y to be Nakagami-m fading channels with a fading severity parameter[1] m_{XY} and a second parameter Ω_{XY} where $X, Y \in \{A, S, R\}$. Moreover, we use h_{XY} to denote the channel gain from X to Y. With reference to [17], the probability density function (PDF) and the cumulative density function (CDF) of the channel gain can be expressed as

$$f_{h_{XY}}(h) = \frac{{\beta_{XY}}^{m_{XY}}}{\Gamma(m_{XY})} h^{m_{XY}-1} \exp(-\beta_{XY} h), \tag{12.1}$$

$$F_{h_{XY}}(h) = \frac{\gamma(m_{XY}, \beta_{XY} h)}{\Gamma(m_{XY})}, \tag{12.2}$$

where $\Gamma(z) = \int_0^\infty \exp(-t) t^{z-1} dt$ is the Gamma function, $\beta_{XY} = m_{XY}/\Omega_{XY}$, and $\gamma(\alpha, x) = \int_0^x \exp(-t) t^{\alpha-1} dt$ is the lower incomplete Gamma function. In addition, we assume that all channels in both DL and UL experience independent slow and frequency flat fading where the channel gains remain constant during each transmission block but may change from one block to the other. In the AT protocol, the source will adaptively transmit information either directly to the AP or cooperatively via the relay node based on the channel information $h_{AS}h_{SA}$. More specifically, according to the value of $h_{AS}h_{SA}$ and a fixed transmission rate, the source can judge whether an outage will occur when it transmits information directly to the AP. If an outage happens, cooperative transmission with the assist of a relay is invoked, otherwise the source will choose direct transmission. The reason behind this is that cooperation may create outage when no outage happens in the direct transmission between the source and AP, since the information transmission time for the source is halved.

Figure 12.2 shows the time diagram of the AT protocol, in which Figures 12.2(a) and 12.2(b) indicate the direct transmission and cooperative transmission, respectively. The allocated time $T\tau$ for energy harvesting is the same in both cases where $0 < \tau < 1$. In the direct transmission, the source transmits information to the AP in the remaining time $T(1-\tau)$ and the relay accumulates the harvested energy in this time block. In the cooperative transmission, the remaining time is further divided into two time slots with equal length $T(1-\tau)/2$. In the first slot the source transmits information to the relay and AP, and in the second slot the relay forwards the received information to the AP. Amplify-and-forward relaying is assumed to be implemented at the

1 For the purpose of exploration, we consider that all the fading severity parameters are integers in this chapter.

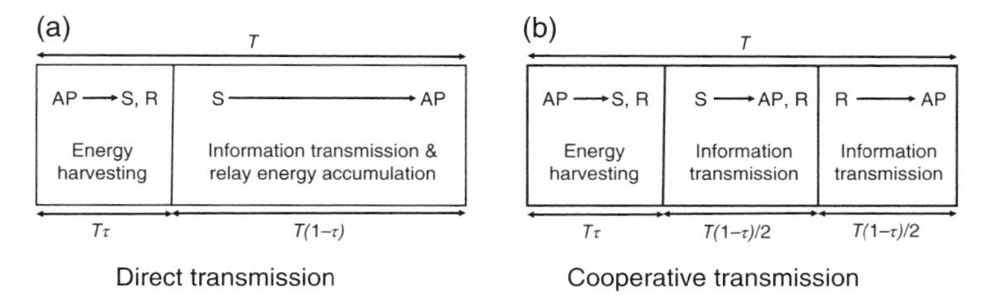

Figure 12.2 Time diagram of AT protocol with energy accumulation.

relay due to its simplicity and the selection combining (SC) technique [18] is employed for information processing at the AP for cooperative transmission. We first characterize the selection criterion between direct transmission and cooperative transmission in the AT protocol by evaluating the condition of a zero outage probability in the direct transmission. The harvested energy at the source and relay during each transmission block can be expressed as [8]

$$E_S = \eta\tau P_A h_{AS}, \tag{12.3}$$

$$E_R = \eta\tau P_A h_{AR}, \tag{12.4}$$

where $0 < \eta < 1$ is the energy harvesting efficiency. Assuming that the source exhausts the harvested energy during each transmission block, the transmitted power of the source during the UL information transmission can be expressed as

$$P_{S,1} = E_S/[(1-\tau)] = \eta\tau P_A h_{AS}/(1-\tau). \tag{12.5}$$

The received signal-to-noise ratio (SNR) at the AP can be expressed as

$$\gamma_{A,1} = P_{S,1} h_{SA}/N_0 = \mu_1 h_{AS} h_{SA}, \tag{12.6}$$

where N_0 is the power of the noise and $\mu_1 = \eta(P_A/N_0)\tau/(1-\tau)$. The mutual information between source and AP for the direct transmission is given by $I_{SA,1} = \log_2(1+\gamma_{A,1})$. Let σ denote the transmission rate and the condition for a zero outage probability can be expressed $I_{SA,1} > \sigma$, which can be expressed as $h_{AS}h_{SA} > \frac{\upsilon_1}{\mu_1}$ after some manipulation and $\upsilon_1 = 2^\sigma - 1$. With the above analysis, we can now conclude that if the acquired channel information between the source and AP $h_{AS}h_{SA} > \frac{\upsilon_1}{\mu_1}$, the source will transmit information directly to the AP. When $h_{AS}h_{SA} \leq \frac{\upsilon_1}{\mu_1}$, the source will transmit information cooperatively to the AP via the relay node.

In the following we will express the received SNR at the AP for the cooperative transmission case of the AT protocol in order to further evaluate the outage probability. The received SNR for the direct transmission is ignored since the outage probability is always zero in that case. Let E_A denote the accumulated energy at the relay node available for information transmission and assume the relay exhausts the stored energy E_A when it is invoked. Note that for the special case that no energy accumulation is considered, the relay exhausts the harvested energy during each transmission block and $E_A = \eta\tau P_A h_{AR}$. The transmitted power at the source and relay can be expressed as

$$P_{S,2} = E_S/[(1-\tau)/2] = 2\eta\tau P_A h_{AS}/(1-\tau), \tag{12.7}$$

$$P_{R,2} = E_A/[(1-\tau)/2] = 2E_A/(1-\tau). \tag{12.8}$$

The received SNR from the S-AP link can be expressed as

$$\gamma_{SA2} = \mu_2 h_{AS} h_{SA}, \tag{12.9}$$

where $\mu_2 = 2\eta(P_A/N_0)\tau/(1-\tau)$. In the last time slot, the relay will amplify and forward the received signal to the AP using the power $P_{R,2}$ given in (12.8) with an amplification factor $\beta = 1/\sqrt{P_{S,2}h_{SR} + N_0}$ [19]. After some algebraic manipulations, we can express the received SNR from S-R-A link as

$$\gamma_{SRA,2} = \frac{\mu_2 h_{AS} h_{SR} \frac{E_A}{\eta\tau P_A} h_{RA}}{h_{AS} h_{SR} + \frac{E_A}{\eta\tau P_A} h_{RA} + 1/\mu_2}, \tag{12.10}$$

With the SC technique, the received SNR at the AP for the cooperative transmission is given by

$$\gamma_{A,2} = \max(\gamma_{SA,2}, \gamma_{SRA,2}). \tag{12.11}$$

12.3 Energy Accumulation of Relay Battery

In this section we model the energy accumulation behavior of the relay battery by modeling the available energy at the relay E_A as a discrete Markov chain (MC). We consider $L+1$ discrete levels of E_A with increments ε, i.e. $E_A \in \{0, \varepsilon, 2\varepsilon, \cdots, L\varepsilon\}$. The MC approach can evaluate a stationary distribution of E_A at each discrete value $\{0, \varepsilon, 2\varepsilon, \cdots, L\varepsilon\}$. We define the states S_i as $E_A = i\varepsilon$ with integer $0 \le i \le L$. The transition probability $T_{i,j}$ is defined as the transition from S_i to S_j. For simplicity, we use the notation that $\varepsilon_i = i\varepsilon$, $i \in \{0, 1, \cdots, L\}$. With the adopted discrete-level battery model, the amount of harvested energy can only be one of the discrete energy levels. Thus, the discretized amount of harvested energy at the relay during the energy harvesting phase is defined as

$$E_D \Delta = \varepsilon_j, \quad \text{where} \quad j = \arg \max_{i\in\{0,1,\cdots,\ L\}} \{\varepsilon_i : \varepsilon_i \le E_R\}. \tag{12.12}$$

12.3.1 Transition Matrix of the MC

In the following we evaluate the transition matrix for two operation modes of the considered system, either direct transmission or cooperative transmission.

12.3.1.1 Direct Transmission

When the system operates in direct transmission, the relay will only harvest energy from the AP's signal and transits from S_i to S_j, $j \in \{i, i+1, \cdots, L\}$ due to the fact that the battery is not discharged. Specifically, $j = i$ represents the case where the harvested energy is discretized to zero and the battery remains the same, and $j = L$ denotes the case where the battery is fully charged by the AP. From the definition of discretization given in (12.12), the transition probabilities are given by

$$T_{i,j} = \begin{cases} \Pr\{E_D = \varepsilon_{j-i}\}, & \text{if } i \le j < L \\ \Pr\{E_D > \varepsilon_{L-i}\}, & \text{if } j = L \\ 0, & \text{otherwise} \end{cases}$$
$$= \begin{cases} \Pr\{\varepsilon_{j-i} \le E_R < \varepsilon_{j-i+1}\}, & \text{if } i \le j < L \\ \Pr\{E_R > \varepsilon_{L-i}\}, & \text{if } j = L \\ 0, & \text{otherwise} \end{cases}$$
$$= \begin{cases} F_{h_{AR}}\left(\dfrac{\varepsilon_{j-i+1}}{\eta\tau P_A}\right) - F_{h_{AR}}\left(\dfrac{\varepsilon_{j-i}}{\eta\tau P_A}\right), & \text{if } i \le j < L \\ 1 - F_{h_{AR}}\left(\dfrac{\varepsilon_{L-i}}{\eta\tau P_A}\right), & \text{if } j = L \\ 0, & \text{otherwise} \end{cases} . \tag{12.13}$$

12.3.1.2 Cooperative Transmission

When the relay works in the cooperative mode, it will first exhaust all the stored energy to forward information from the source and then harvest energy from the AP's signal prior to the next transmission phase of the relay. The transition probabilities for this case thus only depend on the end state j since the relay has exhausted its stored energy and it definitely harvests ε_j amount of energy during the transition for any initial state i. The transition probabilities are given by

$$T_{i,j} = \begin{cases} \Pr\{E_D = \varepsilon_j\}, & \text{if } j < L \\ \Pr\{E_D > \varepsilon_L\}, & \text{if } j = L \end{cases}$$
$$= \begin{cases} \Pr\{\varepsilon_j \le E_R < \varepsilon_{j+1}\}, & \text{if } j < L \\ \Pr\{E_R > \varepsilon_L\}, & \text{if } j = L \end{cases}$$
$$= \begin{cases} F_{h_{AR}}\left(\frac{\varepsilon_{j+1}}{\eta\tau P_A}\right) - F_{h_{AR}}\left(\frac{\varepsilon_j}{\eta\tau P_A}\right), & \text{if } j < L \\ 1 - F_{h_{AR}}\left(\frac{\varepsilon_L}{\eta\tau P_A}\right), & \text{if } j = L \end{cases} . \tag{12.14}$$

12.3.2 Stationary Distribution of the Relay Battery

Let $\mathbf{Z_D} = (T_{i,j})$ and $\mathbf{Z_C} = (T_{i,j})$ denote the $(L+1) \times (L+1)$ state transition matrix of the MC derived from direct transmission in Section 12.3.1.1 and cooperative transmission in Section 12.3.1.2, respectively. The final transition matrix of the system can now be expressed as $\mathbf{Z} = (1-p_1)\mathbf{Z_D} + p_1\mathbf{Z_C}$, where p_1 denotes the probability that cooperative transmission is adopted, which is calculated in (12.27). By using similar methods to [13], we can easily verify that the MC transition matrix $\mathbf{Z}$ derived from the above MC model is irreducible and row stochastic. Thus, there must exist a unique stationary distribution π that satisfies the following equation

$$\boldsymbol{\pi} = (\pi_0, \pi_1, \cdots, \pi_L)^T = (\mathbf{Z})^T \boldsymbol{\pi}, \tag{12.15}$$

where π_i, $i \in \{0, 1, \cdots, L\}$, is the ith component of π representing the stationary distribution of the ith energy level at relay. The battery stationary distribution of relay can be solved from (12.15) and expressed as

$$\boldsymbol{\pi} = ((\mathbf{Z})^T - \mathbf{I} + \mathbf{B})^{-1}\mathbf{b}, \tag{12.16}$$

where $\mathbf{B}_{i,j} = 1, \forall i, j$ and $\mathbf{b} = (1, 1, \cdots, 1)^T$.

12.4 Throughput Analysis

In this section we analytically derive the outage probability and throughput of the proposed AT scheme for both the case where the relay can accumulate the harvested energy and the case where no energy accumulation is considered.

Lemma 12.1 Considering energy accumulation, an approximated expression of the outage probability of the AT protocol over Nakagami-m fading channels can be expressed as

$$\begin{aligned} P_{out,1} \approx 1 - S_{m_{AS},m_{SA}}\left(\frac{\beta_{AS}\beta_{SA}\upsilon_1}{\mu_1}\right) - \sum_{i=0}^{L}\left[1 - F_{h_{RA}}\left(\frac{\upsilon_2\eta\tau P_A}{\mu_2\varepsilon_i}\right)\right]\pi_i \\ \times\left[S_{m_{SR},m_{AS}}\left(\frac{\beta_{SR}\beta_{AS}\upsilon_2}{\mu_2}\right) - \Theta\left(\beta_{SA}\frac{\upsilon_1}{\mu_1}, \beta_{SR}\frac{\upsilon_2}{\mu_2}\right)\right] \end{aligned} \tag{12.17}$$

where

$$S_{m_1,m_2}(x) = \frac{2}{\Gamma(m_2)}\sum_{i=0}^{m_1-1}\frac{x^{\frac{m_2+i}{2}}}{i!}K_{m_2-i}(2\sqrt{x}), \tag{12.18}$$

$$\begin{aligned} \Theta(x_1, x_2) = \frac{2}{\Gamma(m_{AS})}\sum_{i=0}^{m_{SA}-1}\sum_{j=0}^{m_{SR}-1}\left[\frac{(x_1)^i}{i!}\frac{(x_2)^j}{j!}(x_1+x_2)^{\frac{m_{AS}-i-j}{2}}\right. \\ \left.\times(\beta_{AS})^{\frac{m_{AS}+i+j}{2}}K_{m_{AS}-i-j}(2\sqrt{(x_1+x_2)\beta_{AS}})\right], \end{aligned} \tag{12.19}$$

$\upsilon_2 = 2^{2\sigma} - 1$ and $K_\upsilon(z)$ is the modified Bessel function of third kind with order υ [20], Eq. (8.407)].

Proof: See Appendix A of this chapter.

Lemma 12.2 Without energy accumulation, an approximated expression of the outage probability of the AT protocol over Nakagami-m fading channels can be expressed as

$$\begin{aligned} P_{out,2} \approx 1 - S_{m_{AS},m_{SA}}\left(\frac{\beta_{AS}\beta_{SA}\upsilon_1}{\mu_1}\right) - S_{m_{AR},m_{RA}}\left(\frac{\beta_{AR}\beta_{RA}\upsilon_2}{\mu_2}\right) \\ \times \left[S_{m_{SR},m_{AS}}\left(\frac{\beta_{SR}\beta_{AS}\upsilon_2}{\mu_2}\right) - T\left(\beta_{SA}\frac{\upsilon_1}{\mu_1}, \beta_{SR}\frac{\upsilon_2}{\mu_2}\right)\right]. \end{aligned} \quad (12.20)$$

Proof: In this case, the relay exhausts the harvested energy during each transmission and $E_A = \eta\tau P_A h_{AR}$. Compared with the analysis for the energy accumulation case, it is clear that only the second probability term in (12.25) changes to $\Pr\left(h_{AR}h_{RA} \geq \frac{\upsilon_2}{\mu_2}\right)$. From (12.27), it can be easily calculated that

$$\Pr\left(h_{AR}h_{RA} \geq \frac{\upsilon_2}{\mu_2}\right) = S_{m_{AR},m_{RA}}\left(\frac{\beta_{AR}\beta_{RA}\upsilon_2}{\mu_2}\right). \quad (12.21)$$

Substituting (12.21) into (12.25), we have the desired form given in (12.20) and this completes the proof. Based on the calculated outage probability $P_{out} \in \{P_{out,1}, P_{out,2}\}$ for the AT protocol, the approximated average throughput is given by

$$\Psi \approx R(1 - P_{out})(1 - \tau). \quad (12.22)$$

Remark 12.1 The expression of outage probability is given in (12.24) and it is inversely proportional to the received SNRs $\gamma_{SA,2}$, $\gamma_{SRA,2}$, and μ_1. Additionally, the received SNRs given in (12.9) and (12.10) are proportional to μ_1 and μ_2. Furthermore, μ_1 and μ_2 are both proportional to the allocated time τ for energy harvesting. Thus it can be concluded that the outage probability is proportional to the allocated time τ for energy harvesting. This observation is understandable since the more time allocated for energy harvesting, the more energy is harvested. Thus, the source transmits a higher power which results in a low outage probability. However, the expression (12.22) shows that the average throughput still depends on the information transmission time $1 - \tau$, which is inversely proportional to τ. Hence, there must be an optimal value for the allocated time τ between 0 and 1 such that the average throughput is maximized.

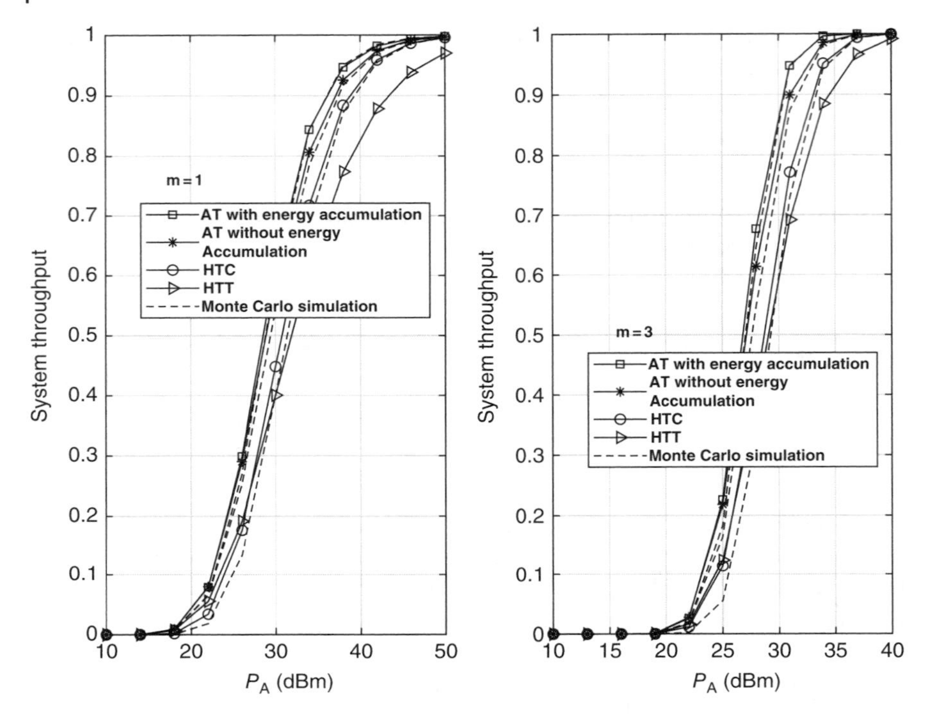

Figure 12.3 Throughput versus transmitted power for $m = 1$ and $m = 3$, where $d_{AR} = 8$, $\tau = 0.5$, and $\sigma = 2$.

12.5 Numerical Results

In this section we present some numerical results to illustrate and validate the above theoretical analysis. We investigate the channel model that $m_{XY} = m$ for simplicity but their variance Ω_{XY} may differ from each other. In order to capture the effect of path loss, we use the model that $\Omega_{XY} = 10^{-3}(d_{XY})^{-\alpha}$, where d_{XY} denotes the distance between X and Y and $\alpha \in [2, 5]$ is the path-loss exponent [22]. We consider a linear topology that the relay is located on a straight line between the source and the hybrid AP. In all following simulations, we set the distance between AP and S to $d_{AS} = 10$ m, the path-loss exponent is $\alpha = 2$, the noise power is $N_0 = -80$ dBm, and the energy harvesting efficiency is $\eta = 0.5$. In addition, the battery capacity of the relay is chosen to be proportional to the average harvested energy as $C = 2\eta\tau P_A \times 10^{-3}$ and the level L is set to 500.

First, we compare the analytical throughput derived in the above sections with the Mont Carlo simulation results. From Figure 12.3 we can see that the derived approximated expression of throughput becomes very tight at medium

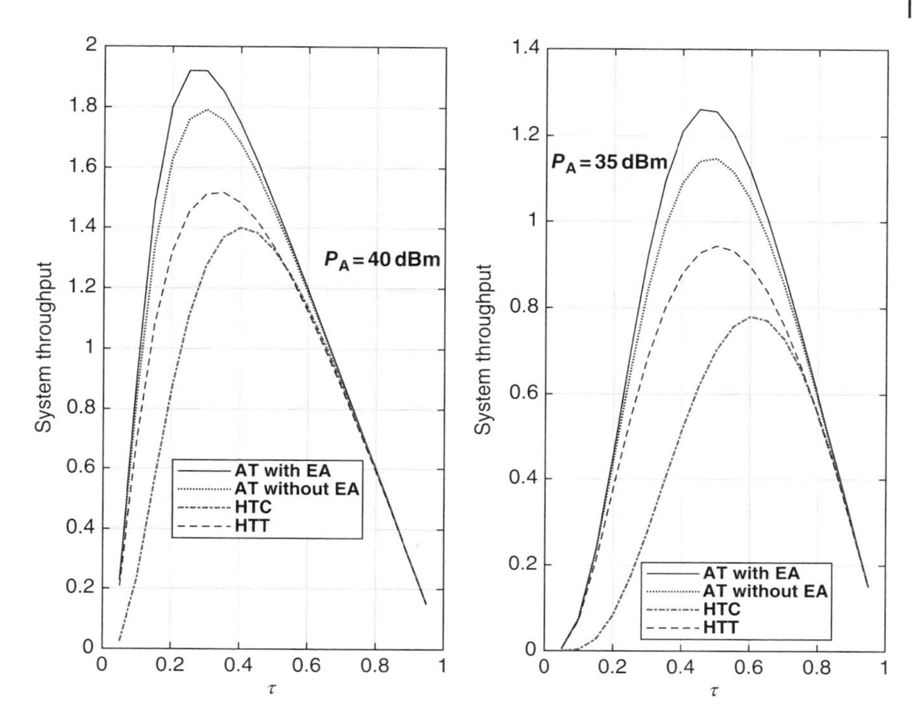

Figure 12.4 Throughput versus τ for different transmit power P_A, where $\sigma = 3$, $d_{AR} = 8$, and $m = 2$.

and high transmit power regime for all the three protocols. This numerical result validates our theoretical throughput analysis. Since the theoretical analysis agrees with the simulations when the transmit power is high enough, we will only plot the analytical results in the remaining figures.

In Figure 12.4 we plot the throughput curves versus time with different transmitted power $P_A = 40$ dB and 35 dB. It validates our analyses in Remark 14.1 that the AT protocol has an optimal τ to achieve its maximum throughput. Additionally, Figure 12.4 shows that the optimal value of τ will shift to the left as the transmitted power increases. This makes sense since the source can harvest the same amount of energy with a shorter time when the transmitted power is higher.

Figure 12.5 illustrates the effect of relay location on the throughput in which the throughput curves are plotted versus d_{AR} with optimal values of τ. Note that the optimal value of τ can be easily found by a one-dimension exhaustive search. Since HTT protocol does not assist with the external relay, the throughput is plotted as a straight line for comparison. It can be concluded that the AT

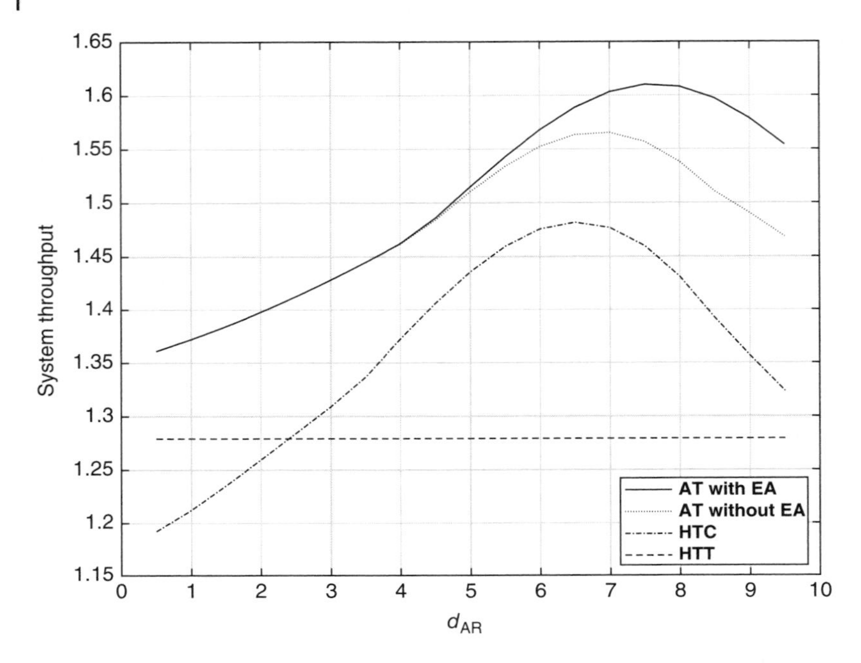

Figure 12.5 Throughput versus the distance between AP and R d_{AR}, where $P_A = 40$ dBm, $m = 2$, and $\sigma = 2$.

protocol outperforms the HTC and HTT protocols in all distance cases. Moreover, our results show that the energy accumulation of relay will only improve the performance of AT protocol when the relay is close to the source and it will become more significant as the relay approaches the source. This is understandable since the energy accumulation can only improve the performance of the R-AP link as the transmitted power at the relay is higher with energy accumulation. However, if the channel condition of the S-R link is poor, the S-R-AP link will still have an outage even if the power provided by the relay is high.

12.6 Conclusion

In this chapter we analyzed the performance of a typical three-node wireless-powered cooperative network (WPCN) implementing an adaptive transmission protocol. In addition, we also studied the energy accumulation process of the wireless-powered relay by modeling the changing/discharging behavior of the battery as a finite-state Markov chain. Our results demonstrated the superiority of cooperation and energy accumulation over non-cooperation and non-energy accumulation in WPCNs.

12.7 Appendix

Let A and A^c denote the event that direct transmission and cooperative transmission are chosen in the AT protocol, respectively. Let O denote the event that an outage happens. Since outage will only happen when cooperative transmission is invoked, from the Bayes' formula [21, Eq. (1.7)] the outage probability for the AT protocol can be expressed as

$$P_{out,1} = \Pr(O, A) + \Pr(O, A^c) = \Pr(O, A^c). \tag{12.23}$$

Now we consider the cooperative transmission of the AT protocol. The mutual information between source and AP is given by $I_{SA,2} = \frac{1}{2}\log_2(1 + \gamma_{A,2})$. The outage probability can be further evaluated as

$$P_{out,1} = \Pr(I_{SA,2} < R, A^c) = \Pr\left(\gamma_{SA,2} < \upsilon_2, \gamma_{SRA,2} < \upsilon_2, h_{AS}h_{SA} \leq \frac{\upsilon_1}{\mu_1}\right). \tag{12.24}$$

Since the received SNR given in (12.11) is not analytically tractable, we propose an approximation $\gamma_{SRA2} \approx \mu_2 \min\left(h_{AS}h_{SR}, \frac{E_A}{\eta\tau P_A}h_{RA}\right)$ in order to calculate the outage probability. With the fact that $2^\sigma - 1 < \frac{2^{2\sigma}-1}{2}$ for all $\sigma > 0$, it can be easily shown that $\frac{\upsilon_1}{\mu_1} < \frac{\upsilon_2}{\mu_2}$ and the outage probability can be evaluated as

$$\begin{aligned}
P_{out,1} &\approx \Pr\left(h_{AS}h_{SA} < \frac{\upsilon_2}{\mu_2}, \min\left(h_{AS}h_{SR}, \frac{E_A}{\eta\tau P_A}h_{RA}\right) < \frac{\upsilon_2}{\mu_2}, h_{AS}h_{SA} \leq \frac{\upsilon_1}{\mu_1}\right)\\
&= \Pr(\min\left(h_{AS}h_{SR}, \frac{E_A}{\eta\tau P_A}h_{RA}\right) < \frac{\upsilon_2}{\mu_2}, h_{AS}h_{SA} \leq \frac{\upsilon_1}{\mu_1})\\
&= \Pr(h_{AS}h_{SA} \leq \frac{\upsilon_1}{\mu_1}) - \Pr(\frac{E_A}{\eta\tau P_A}h_{RA} \geq \frac{\upsilon_2}{\mu_2})\\
&\times \Pr\left(h_{AS}h_{SA} \leq \frac{\upsilon_1}{\mu_1}, h_{AS}h_{SR} \geq \frac{\upsilon_2}{\mu_2}\right),
\end{aligned} \tag{12.25}$$

where the last equality follows the fact that $\Pr(A, B) = \Pr(A) - \Pr(A, \overline{B})$ [21, Eq. (1.7)]. Now we calculate the three probabilities in the final step of (12.25). Let p_1 denote the first probability, which is also the probability that cooperative transmission happens in the AT protocol and

$$\begin{aligned}
p_1 &= \Pr\left(h_{AS}h_{SA} \leq \frac{\upsilon_1}{\mu_1}\right) = \int_0^\infty \Pr\left(h_{AS} \leq \frac{\upsilon_1}{\mu_1 h}\right) p_{h_{SA}}(h)dh\\
&= \int_0^\infty F_{h_{AS}}\left(\frac{\upsilon_1}{\mu_1 h}\right) p_{h_{SA}}(h)dh = \int_0^\infty \frac{\gamma(m_{AS}, \frac{\beta_{AS}\upsilon_1}{\mu_1 h})}{\Gamma(m_{AS})} p_{h_{SA}}(h)dh.
\end{aligned} \tag{12.26}$$

We now expand the incomplete Gamma function with integer m as $\frac{\gamma(m,x)}{\Gamma(m)} = 1 - \exp(-x)\sum_{i=0}^{m-1}\frac{x^i}{i!}$ [20, Eq.(8.352.6)]. The above integral can be further evaluated as

$$
\begin{aligned}
p_1 &= \int_0^\infty \left(1 - \exp(-\frac{\beta_{AS}\upsilon_1}{\mu_1 h})\sum_{i=0}^{m_{AS}-1}\frac{\left(\frac{\beta_{AS}\upsilon_1}{\mu_1}\right)^i}{i!h^i}\right) p_{h_{SA}}(h)dh \\
&= 1 - \frac{\beta_{SA}{}^{m_{SA}}}{\Gamma(m_{SA})}\sum_{i=0}^{m_{AS}-1}\frac{\left(\frac{\beta_{AS}\upsilon_1}{\mu_1}\right)^i}{i!}\int_0^\infty h^{m_{SA}-i-1}\exp\left(-\frac{\beta_{AS}\upsilon_1}{\mu_1 h} - \beta_{SA}h\right)dh \\
&= 1 - S_{m_{AS},m_{SA}}\left(\frac{\beta_{AS}\beta_{SA}\upsilon_1}{\mu_1}\right),
\end{aligned}
\tag{12.27}
$$

where the last integral is solved by [20, Eq.(3.471-9)].

Next, we evaluate the second probability term in (12.25). The discrete distribution of E_A is given in (12.16). Thus the second probability can be calculated from the total probability theorem

$$
\begin{aligned}
\Pr\left(\frac{E_A}{\eta\tau P_A}h_{RA} \geq \frac{\upsilon_2}{\mu_2}\right) &= \sum_{i=0}^{L}\Pr\left(h_{RA} \geq \frac{\upsilon_2\eta\tau P_A}{\mu_2\varepsilon_i}\right)\Pr(E_A = \varepsilon_i) \\
&= \sum_{i=0}^{L}\left[1 - F_{h_{RA}}\left(\frac{\upsilon_2\eta\tau P_A}{\mu_2\varepsilon_i}\right)\right]\pi_i.
\end{aligned}
\tag{12.28}
$$

For the last term, we have

$$
\begin{aligned}
&\Pr\left(h_{AS}h_{SA} \leq \frac{\upsilon_1}{\mu_1}, h_{AS}h_{SR} \geq \frac{\upsilon_2}{\mu_2}\right) = \int_0^\infty \Pr\left(h_{SA} \leq \frac{\upsilon_1}{\mu_1 h}\right)\Pr\left(h_{SR} \geq \frac{\upsilon_2}{\mu_2 h}\right)p_{h_{AS}}(h)dh \\
&= \underbrace{\int_0^\infty \Pr\left(h_{SR} \geq \frac{\upsilon_2}{\mu_2 h}\right)p_{h_{AS}}(h)dh}_{I_1} - \underbrace{\int_0^\infty \Pr\left(h_{SA} > \frac{\upsilon_1}{\mu_1 h}\right)\Pr\left(h_{SR} \geq \frac{\upsilon_2}{\mu_2 h}\right)p_{h_{AS}}(h)dh.}_{I_2}
\end{aligned}
\tag{12.29}
$$

Similarly, from p_1 given in (12.27), I_1 can be calculated as $I_1 = S_{m_{SR},m_{AS}}\left(\frac{\beta_{SR}\beta_{AS}\upsilon_2}{\mu_2}\right)$. I_2 can be evaluated as

$$
\begin{aligned}
I_2 &= \frac{\beta_{AS}{}^{m_{AS}}}{\Gamma(m_{AS})}\sum_{i=0}^{m_{SA}-1}\sum_{j=0}^{m_{SR}-1}\frac{\left(\frac{\beta_{SA}\upsilon_1}{\mu_1}\right)^i}{i!}\frac{\left(\frac{\beta_{SR}\upsilon_2}{\mu_2}\right)^j}{j!}\times \\
&\int_0^\infty h^{m_{AS}-i-j-1}\exp\left(-\frac{\beta_{SA}\frac{\upsilon_1}{\mu_1} + \beta_{SR}\frac{\upsilon_2}{\mu_2}}{h} - \beta_{AS}h\right)dh = \Theta\left(\beta_{SA}\frac{\upsilon_1}{\mu_1}, \beta_{SR}\frac{\upsilon_2}{\mu_2}\right),
\end{aligned}
\tag{12.30}
$$

again the last integral is solved by [20], Eq.(3.471-9)]. Substitute the three calculated probabilities in (12.25) and we have the desired form given in Lemma 12.1.

Bibliography

1 O. Ozel, K. Tutuncuoglu, J. Yang, S. Ulukus, and A. Yener (2011) Transmission with energy harvesting nodes in fading wireless channels: Optimal policies. *IEEE J. Selected Areas Commun.*, **29** (8): 1732–1743.

2 N. Shinohara (2011) Power without wires.*IEEE Microwave Mag.*, **12** (7): S64–S73.

3 V. Liu, A. Parks, V. Talla, S. Gollakota, D. Wetherall, and J. Smith (2013) Ambient backscatter: Wireless communication out of thin air. *ACM SIGCOMM Computer Commun. Rev.*, **43** (4): 39–50.

4 J. Smith (2013) *Wirelessly powered sensor networks and computational RFID.* Springer, New York.

5 J. Laneman, D. Tse, and G. Wornell (2004) Cooperative diversity in wireless networks: Efficient protocols and outage behavior. *IEEE Trans. Inf. Theory*, **50** (12): 3062–3080.

6 M. Dohler and Y. Li (2010) *Cooperative Communications: Hardware, Channel and PHY*. Wiley-Blackwell, Chichester.

7 He Chen, C. Zhai, Y. Li, and B. Vucetic (2016) Cooperative strategies for wireless-powered communications: An overview. *IEEE Wireless Commun.*, accepted. [Online]. Available: http://adsabs.harvard.edu/abs/2016arXiv161003527C.

8 A. Nasir, X. Zhou, S. Durrani, and R. Kennedy (2013) Relaying protocols for wireless energy harvesting and information processing. *IEEE Trans. Wireless Commun.*, **12** (7): 3622–3636.

9 He Chen, Yonghui Li, J. Luiz Rebelatto, B. Uchoa-Filho, and B. Vucetic (2015) Harvest-then-cooperate: Wireless-powered cooperative communications. *IEEE Trans. Signal Process.*, **63** (7): 1700–1711.

10 Y. Gu, H. Chen, Y. Li, and B. Vucetic (2015) An adaptive transmission protocol for wireless-powered cooperative communications. In *2015 IEEE International Conference on Communications (ICC)*, pp. 4223–4228.

11 I. Krikidis, S. Timotheou, and S. Sasaki (2012) RF energy transfer for cooperative networks: Data relaying or energy harvesting? *IEEE Commun. Lett.*, **16**: 1772–1775.

12 Y. Gu, H. Chen, Y. Li, and B. Vucetic (2016) Wireless-powered two-way relaying with power splitting-based energy accumulation. In *2016 IEEE Global Communications Conference (GLOBECOM)*, pp. 1–6.

13 I. Krikidis (2015) Relay selection in wireless powered cooperative networks with energy storage. *IEEE J. Selected Areas Commun.*, **33**: 2596–2610.

14 K.H. Liu (2016) Performance analysis of relay selection for cooperative relays based on wireless power transfer with finite energy storage. *IEEE Trans Veh. Technol.*, **65**: 5110–5121.

15 Y. Gu, H. Chen, Y. Li, Y.C. Liang, and B. Vucetic (2018) Distributed multi-relay selection in accumulate-then-forward energy harvesting relay networks. *IEEE Trans. Green Commun. Network.*, **2** (2): 74–86.

16 H. Ju and R. Zhang (2014) Throughput maximization in wireless powered communication networks. *IEEE Trans. Wireless Commun.*, **13**: 418–428.

17 M. Simon and M. Alouini (2005) *Digital communication over fading channels.* Wiley-Interscience, Hoboken, NJ.

18 J. Laneman, D. Tse, and G. Wornell (2004) Cooperative diversity in wireless networks: Efficient protocols and outage behavior. *IEEE Trans. Inf. Theory*, **50** (12): 3062–3080.

19 S.S. Ikki and M.H. Ahmed (2009) Performance analysis of decode-and-forward cooperative diversity using differential EGC over Nakagami-m fading channels. *VTC Spring 2009 – IEEE 69th Vehicular Technology Conference*, Barcelona, pp. 1–6.

20 A. Jeffrey and D. Zwillinger (2000) *Table of Integrals, Series, and Products.* Elsevier, Burlington.

21 S. Ross (2014) *Introduction to probability models.* Elsevier, Amsterdam.

22 H. Chen, J. Liu, L. Zheng, C. Zhai, and Y. Zhou (2010) Approximate SEP analysis for DF cooperative networks with opportunistic relaying. *IEEE Signal Process. Lett.*, **17**: 779–782.

13

Spectral and Energy-Efficient Wireless-Powered IoT Networks

Qingqing Wu[1*], *Wen Chen*[2], *and Guangchi Zhang*[3]

[1] *Department of Electrical and Computer Engineering, National University of Singapore, Singapore*
[2] *Department of Electronic Engineering, Shanghai Jiao Tong University, China*
[3] *School of Information Engineering, Guangdong University of Technology, China*

13.1 Introduction

The number of connected devices globally will skyrocket to 30 billion by 2025, giving rise to the well-known Internet of Things (IoT) [1]. With such a huge number of IoT devices, the lifetime of networks becomes a critical issue and conventional battery-based solutions may no longer be sustainable due to the high cost of battery replacement as well as environmental concerns. As a result, wireless power transfer, which enables energy harvesting from ambient radio frequency (RF) signals, is envisioned as a promising solution for powering massive IoT devices [2–5]. However, due to the significant signal attenuation in wireless communication channels, the harvested RF energy at the devices is generally limited [6, 7]. Therefore, how to efficiently utilize the scarce harvested energy becomes particularly crucial for realizing sustainable and scalable IoT networks. To this end, a "harvest and then transmit" protocol is proposed in [8, 9] for wireless-powered communication networks (WPCNs), where devices first harvest energy in the downlink (DL) for wireless energy transfer (WET) and then transmit information signals in the uplink (UL) for wireless information transmission (WIT). To improve the spectral efficiency (SE), this work is further extended to a full-duplex communication network in [10], where devices can harvest energy and transmit information at the same time. However, it may not be feasible to implement the full-duplex functionality in IoT devices due to the resulting high complexity, energy consumption, and cost. As a result, the most recent narrowband IoT (NB-IoT) standard requires NB-IoT devices to support only the half-duplex protocol for simplicity [1].

*Corresponding author: Qingqing Wu; wuqq1010@gmail.com

Wireless Information and Power Transfer: Theory and Practice, First Edition.
Edited by Derrick Wing Kwan Ng, Trung Q. Duong, Caijun Zhong, and Robert Schober.

Meanwhile, non-orthogonal multiple access (NOMA) has been proposed to improve the SE as well as user fairness by allowing multiple users simultaneously to access the same spectrum. With successive interference cancellation (SIC) performed at the receiver, NOMA has been demonstrated to be superior to orthogonal multiple access (OMA) in terms of the ergodic sum rate [11]. As such, NOMA has recently been pursued for UL WIT in WPCNs [12, 13], where the decoding order of the users is exploited to enhance the throughput fairness among users. However, the conclusions drawn in [11] are only applicable for the DL scenario and may not hold for UL IoT networks with energy-constrained devices. Furthermore, [12] and [13] focus only on improving the system/individual user throughput without considering the total system energy consumption. In fact, due to rapidly rising energy costs and the large carbon footprint of existing systems, energy consumption is gradually accepted as an important design criterion for future communication systems. As such, the energy consumption of different wireless networks has been extensively studied in prior works. e.g., for energy harvesting and wireless power transfer [9, 14–22], heterogeneous networks [23–25], relaying networks [26–28], device-to-device (D2D) communications [29], orthogonal frequency division multiplexing access (OFDMA) [30–39], multiple-input multiple-output (MIMO) [40–45], and secure and cognitive networks [46–48]. However, a theoretical total energy consumption comparison between NOMA and TDMA is missing and important since the efficiency of WET is generally low in practice, which is unlike conventional wireless networks. Also, the circuit energy consumption of users is completely ignored in [8, 12, 13]. However, the circuit power consumption is often comparable to the transmit power and thus is important for short-range IoT applications, such as wearable devices. As multiple users access the same spectrum simultaneously in NOMA, the circuit energy consumption of each user inevitably increases, which may contradict a fundamental design requirement of future IoT networks, i.e. ultra low power consumption [4]. For example, in NOMA-based WPCN (N-WPCN) with a fixed total available harvested energy, if devices consume more energy for operating their electronic circuits than in time-division multiple access (TDMA)-based WPCN (T-WPCN), then less energy will be left for signal transmission [49]. As a result, a natural question arises: Does NOMA improve the SE and/or reduce the total energy consumption of such wireless powered IoT networks in practice compared to TDMA?

Driven by the above question, we investigate the following in this chapter: (i) By taking into account the circuit energy consumption, we first derive the optimal time allocation for the SE maximization problem for T-WPCN, based on which the corresponding problem for N-WPCN can be cast as the single user case for T-WPCN, (ii) we prove that N-WPCN in general requires a longer DL WET time duration than T-WPCN, which implies that N-WPCN is more energy demanding, and (iii) we prove that N-WPCN in general achieves a lower

SE than T-WPCN. Given (ii) and (iii), NOMA may not be a good candidate for realizing spectral and energy-efficient wireless-powered IoT networks if the circuit energy consumption is not negligible.

The remainder of this chapter is organized as follows. In Section 13.2 we describe the adopted system model and problem formulation. In Section 13.3 we analyze the performance of T-WPCN and N-WPCN. Section 13.4 presents the numerical results. Finally, we conclude the chapter in Section 13.5.

13.2 System Model and Problem Formulation

13.2.1 System Model

We consider a WPCN that consists of one power beacon (PB), $K > 1$ wireless-powered IoT devices, and one information access point (AP), as shown in Figure 13.1. The total available transmission time is denoted by $T_{\max}$. The "harvest and then transmit" protocol [8] is adopted where the devices first harvest energy from the signal sent by the PB and then transmit information to the AP. We note that the "doubly near-far phenomenon" [8] can be avoided by using separated PB and AP, as in our model [9, 50]. For the ease of practical implementation, the PB and all users are assumed to operate in the time division manner over the same frequency band. To compare the upper bound performance of T-WPCN and N-WPCN, we assume that perfect channel state information (CSI) is available for resource allocation. The DL channel gain between the PB and device $k \in \{1, 2, \ldots, K\}$, and the UL channel gain between device k and the AP are denoted by h_k and g_k, respectively.

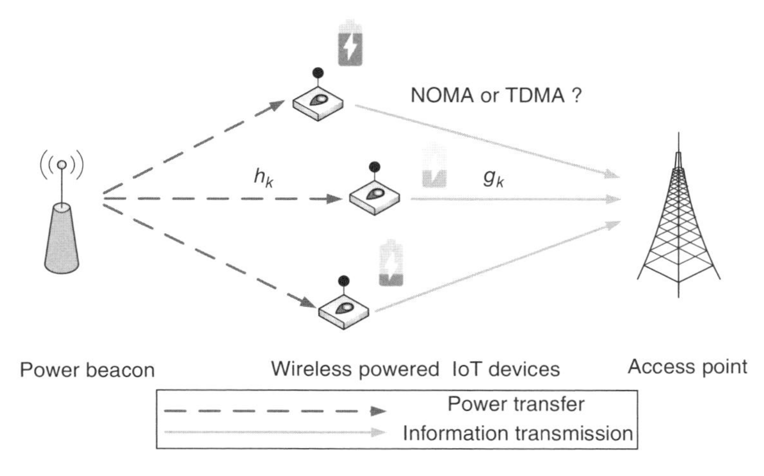

Figure 13.1 System model of a wireless-powered IoT network [49].

During DL WET, the PB broadcasts the energy signal with a constant transmit power $P_{\rm E}$ for time τ_0. The energy harvested from the noise and the received UL WIT signals from other devices are assumed to be negligible, since both the noise power and the device transmit power are much smaller than the transmit power of the PB in practice [8]. Thus, the amount of harvested energy at device k can be expressed as

$$E_k^h = \eta_k P_{\rm E} h_k \tau_0, \tag{13.1}$$

where $\eta_k \in (0, 1]$ is the constant energy conversion efficiency of device k. During UL WIT, device k transmits its information signal to the AP with transmit power p_k. In addition to the transmit power, each device also consumes a constant circuit power accounting for the power needed to operate its transmit filter, mixer, frequency synthesizers, etc., denoted by $p_{{\rm c},k} \geq 0$ [3, 9]. For the multiple access scheme in UL WIT, we consider two schemes, i.e. TDMA and NOMA. For T-WPCN, device k exclusively accesses the spectrum for a duration of τ_k, while for N-WPCN, all the devices access the spectrum simultaneously for a duration of $\overline{\tau}_1$. Then, the energy consumed by device k during UL WIT for T-WPCN and N-WPCN can be expressed as $(p_k + p_{{\rm c},k})\tau_k$ and $(p_k + p_{{\rm c},k})\overline{\tau}_1$, respectively. Denote $\gamma_k = \frac{g_k}{\sigma^2}$ as the normalized UL channel gain of device k, where σ^2 is the additive white Gaussian noise power at the AP. For convenience, we assume that the normalized UL channel power gains are sorted in ascending order, i.e. $0 < \gamma_1 \leq \gamma_2 \cdots \leq \gamma_K$.

13.2.2 T-WPCN and Problem Formulation

For T-WPCN, the achievable throughput of device k in bits/Hz can be expressed as

$$r_k = \tau_k \log_2(1 + p_k \gamma_k). \tag{13.2}$$

Then, the system throughput of T-WPCN is given by

$$R_{\rm TDMA} = \sum_{k=1}^{K} r_k = \sum_{k=1}^{K} \tau_k \log_2(1 + p_k \gamma_k). \tag{13.3}$$

Accordingly, the SE maximization problem is formulated as

$$\underset{\tau_0, \{\tau_k\}, \{p_k\}}{\text{maximze}} \quad \sum_{k=1}^{K} \tau_k \log_2(1 + p_k \gamma_k) \tag{13.4a}$$

$$\text{s.t.} \quad (p_k + p_{{\rm c},k})\tau_k \leq \eta_k P_{\rm E} h_k \tau_0, \forall\, k, \tag{13.4b}$$

$$\tau_0 + \sum_{k=1}^{K} \tau_k \leq T_{\max}, \tag{13.4c}$$

$$\tau_0 \geq 0, \tau_k \geq 0, p_k \geq 0, \forall\, k. \tag{13.4d}$$

In problem (13.4), (13.4b) is the energy causality constraint which ensures that the energy consumed for WIT does not exceed the total energy harvested during WET. (13.4c) and (13.4d) are the total time constraint and the non-negativity constraint on the optimization variables, respectively.

13.2.3 N-WPCN and Problem Formulation

For N-WPCN, since all the K devices share the same spectrum, SIC is employed at the AP to eliminate multi-user interference [11]. Specifically, for detecting the message of the kth device, the AP first decodes the message of the ith device, $\forall\, i < k$, and then removes this message from the received signal, in the order of $i = 1, 2, \ldots, k-1$. The message of the ith user, $\forall\, i > k$, is treated as noise. Hence, the achievable throughput of device k in bits/Hz in N-WPCN can be expressed as

$$r_k = \overline{\tau}_1 \log_2 \left(1 + \frac{p_k \gamma_k}{\sum_{i=k+1}^{K} p_i \gamma_i + 1} \right). \tag{13.5}$$

Then, the system throughput of T-WPCN is given by

$$R_{\text{NOMA}} = \sum_{k=1}^{K} r_k = \overline{\tau}_1 \log_2 \left(1 + \sum_{k=1}^{K} p_k \gamma_k \right). \tag{13.6}$$

Accordingly, the SE maximization problem is formulated as

$$\underset{\tau_0, \overline{\tau}_1, \{p_k\}}{\text{maximize}} \quad \overline{\tau}_1 \log_2 \left(1 + \sum_{k=1}^{K} p_k \gamma_k \right) \tag{13.7a}$$

$$\text{s.t.} \qquad (p_k + p_{c,k}) \overline{\tau}_1 \leq \eta_k P_{\text{E}} h_k \tau_0, \forall\, k, \tag{13.7b}$$

$$\tau_0 + \overline{\tau}_1 \leq T_{\max}, \tag{13.7c}$$

$$\tau_0 \geq 0, \overline{\tau}_1 \geq 0, p_k \geq 0, \forall\, k. \tag{13.7d}$$

Similar to problem (13.4), (13.7b), (13.7c), and (13.7d) represent the energy causality constraint, total time constraint, and non-negativity constraints, respectively.

13.3 T-WPCN or N-WPCN?

In this section we first derive the optimal solutions to problems (13.4) and (13.7), respectively. Then, we theoretically analyze and compare the system energy consumed and the SE achieved by both T-WPCN and N-WPCN.

13.3.1 Optimal Solution for T-WPCN

It can be shown that each device will deplete all of its energy at the optimal solution, i.e. constraint (13.4b) holds with equality, since otherwise p_k can always be increased to improve the objective value such that (13.4b) is active. Thus, problem (13.4) is simplified to

$$\underset{\tau_0,\{\tau_k\}}{\text{maximize}} \quad \sum_{k=1}^{K} \tau_k \log_2\left(1 - p_{c,k}\gamma_k + \frac{\eta_k P_E h_k \gamma_k}{\tau_k}\tau_0\right) \tag{13.8a}$$

$$\text{s.t.} \quad \tau_0 + \sum_{k=1}^{K} \tau_k \le T_{\max}, \tag{13.8b}$$

$$\tau_0 \ge 0, \tau_k \ge 0, \forall\, k. \tag{13.8c}$$

It is easy to verify that problem (13.8) is a convex optimization problem and also satisfies Slater's condition [49]. Thus, the optimal solution can be obtained efficiently by applying the Lagrange dual method. To this end, we need the Lagrangian function of problem (13.8), which can be written as

$$\mathcal{L}(\tau_0, \{\tau_k\}) = \sum_{k=1}^{K} \tau_k \log_2\left(1 - p_{c,k}\gamma_k + \frac{\eta_k P_E h_k \gamma_k}{\tau_k}\tau_0\right) + \lambda\left(T_{\max} - \tau_0 - \sum_{k=1}^{K}\tau_k\right), \tag{13.9}$$

where $\lambda \ge 0$ is the Lagrange multiplier associated with (13.8b). (13.8c) is naturally satisfied since the PB is activated in the DL and each user is scheduled in the UL. Taking the partial derivative of $\mathcal{L}$ with respect to τ_0 and τ_k, respectively, yields

$$\frac{\partial \mathcal{L}}{\partial \tau_0} = \sum_{k=1}^{K} \frac{\eta_k P_E h_k \gamma_k \log_2(e)}{1 - p_{c,k}\gamma_k + x_k} - \lambda, \tag{13.10}$$

$$\frac{\partial \mathcal{L}}{\partial \tau_k} = \log_2(1 - p_{c,k}\gamma_k + x_k) - \frac{x_k \log_2(e)}{1 - p_{c,k}\gamma_k + x_k} - \lambda, \tag{13.11}$$

where $x_k = \frac{\eta_k P_E h_k \gamma_k}{\tau_k}\tau_0, \forall k$. Since $\tau_0 > 0$ and $\tau_k > 0, \forall k$, always hold at the optimal solution, we have $\frac{\partial \mathcal{L}}{\partial \tau_0} = 0$ and $\frac{\partial \mathcal{L}}{\partial \tau_k} = 0, \forall k$. As a result, the optimal values of x_k, $\forall k$, can be obtained by solving the following set of equations

$$\mathcal{G}_k(x_k^*) \triangleq \log_2(1 - p_{c,k}\gamma_k + x_k^*) - \frac{x_k^* \log_2(e)}{1 - p_{c,k}\gamma_k + x_k^*} - \sum_{k=1}^{K} \frac{\eta_k P_E h_k \gamma_k \log_2(e)}{1 - p_{c,k}\gamma_k + x_k^*} = 0, \forall\, k. \tag{13.12}$$

Note that the first two terms of $\mathcal{G}_k(x_k^*)$ monotonically increase with x_k^* while the last term is the same for all users. Thus, x_k^* can be efficiently obtained by the bisection method. It can be shown that (13.8b) is active at the optimal solution, i.e. $\tau_0 + \sum_{k=1}^K \tau_k = \tau_0 + \sum_{k=1}^K \frac{P_E h_k \eta_k \gamma_k}{x_k^*}\tau_0 = T_{\max}$. With x_k^*, $\forall$ k, from (13.12), the optimal time allocation for T-WPCN is given by

$$\tau_0^* = \frac{T_{\max}}{1 + \sum_{k=1}^K \frac{\eta_k P_E h_k \gamma_k}{x_k^*}}, \tag{13.13}$$

$$\tau_k^* = \frac{\eta_k P_E h_k \gamma_k}{x_k^*}\tau_0^*, \forall\, k. \tag{13.14}$$

13.3.2 Optimal Solution for N-WPCN

Similarly, problem (13.7) can be simplified to the following problem:

$$\underset{\tau_0,\overline{\tau}_1}{\text{maximize}} \quad \overline{\tau}_1 \log_2\left(1 - \sum_{k=1}^K p_{c,k}\gamma_k + \frac{\sum_{k=1}^K \eta_k P_E h_k \gamma_k}{\overline{\tau}_1}\tau_0\right) \tag{13.15a}$$

$$\text{s.t.} \quad \tau_0 + \overline{\tau}_1 \le T_{\max}, \tag{13.15b}$$

$$\tau_0 \ge 0, \overline{\tau}_1 \ge 0. \tag{13.15c}$$

It is interesting to observe that problem (13.15) has the same structure as problem (13.8) when $K = 1$ with only minor changes in constant terms. As such, the proposed solution for T-WPCN can be immediately extended to N-WPCN. Specifically, the optimal time allocation for N-WPCN is given by

$$\tau_0^\star = \frac{T_{\max}}{1 + \frac{\sum_{k=1}^K \eta_k P_E h_k \gamma_k}{x^\star}}, \quad \overline{\tau}_1^\star = \frac{\sum_{k=1}^K \eta_k P_E h_k \gamma_k}{x^\star}\tau_0^\star, \tag{13.16}$$

where $x^\star$ is the unique root satisfying

$$\mathcal{G}(x^\star) \triangleq \log_2\left(1 - \sum_{k=1}^K p_{c,k}\gamma_k + x^\star\right) - \frac{x^\star \log_2(e)}{1 - \sum_{k=1}^K p_{c,k}\gamma_k + x^\star}$$
$$- \frac{\sum_{k=1}^K \eta_k P_E h_k \gamma_k \log_2(e)}{1 - \sum_{k=1}^K p_{c,k}\gamma_k + x^\star} = 0. \tag{13.17}$$

The solutions proposed in the above sections serve as the theoretical foundation for the comparison between T-WPCN and N-WPCN.

13.3.3 TDMA versus NOMA

For notational simplicity, we first denote by E^*_{TDMA} and $E^\bigstar_{\text{NOMA}}$ the total energy consumption of T-WPCN and N-WPCN at the optimal solutions to problems (13.8) and (13.15), respectively. The corresponding SEs are denoted by R^*_{TDMA} and $R^\bigstar_{\text{NOMA}}$, respectively.

Theorem 13.1 At the optimal solution, (i) the DL WET time of N-WPCN in (13.16) is greater than or equal to that of T-WPCN in (13.13), i.e. $\tau_0^\bigstar \geq \tau_0^*$ and (ii) the energy consumption of N-WPCN is larger than or equal to that of T-WPCN, i.e.

$$E^\bigstar_{\text{NOMA}} \geq E^*_{\text{TDMA}}, \tag{13.18}$$

where "=" holds when $p_{c,k} = 0, \forall k$.

Proof: Since $\sum_{k=1}^{K} p_{c,k}\gamma_k \geq p_{c,k}\gamma_k$, it is easy to show that $x^\bigstar \geq x_k^*$, $\forall k$, from (13.17) and (13.12), where "=" holds when $p_{c,k} = 0, \forall k$. Then, it follows from (13.16) and (13.13) that $\tau_0^\bigstar \geq \tau_0^*$. Furthermore, since each device depletes all of its harvested energy, then the total energy consumption of N-WPCN and T-WPCN satisfies $E^\bigstar_{\text{NOMA}} = P_{\text{E}}\tau_0^\bigstar \geq E^*_{\text{TDMA}} = P_{\text{E}}\tau_0^*$.

Theorem 13.1 implies that N-WPCN is more energy demanding than T-WPCN in terms of the total energy consumption. This is fundamentally due to simultaneous transmissions of multiple devices during UL WIT, which thereby leads to a higher circuit energy consumption. Furthermore, since $\tau_0^\bigstar \geq \tau_0^*$, more energy is also wasted during DL WET for N-WPCN than for T-WPCN. Next, we compare the SE of the two networks.

Theorem 13.2 The maximum SE of T-WPCN is greater than or equal to that of N-WPCN, i.e.

$$R^*_{\text{TDMA}} \geq R^\bigstar_{\text{NOMA}}, \tag{13.19}$$

where "=" holds when $p_{c,k} = 0, \forall\, k$.

Proof: Assume that $\{\tau_0^\bigstar, \overline{\tau}_1^\bigstar\}$ achieves the maximum SE of problem (13.15), $R^\bigstar_{\text{NOMA}}$. Then, we can construct a new solution $\{\tilde{\tau}_0, \{\tilde{\tau}_k\}\}$ satisfying $\tilde{\tau}_0 = \tau_0^\bigstar$ and $\sum_{k=1}^{K} \tilde{\tau}_k = \overline{\tau}_1^\bigstar$ such that all devices achieve the same signal-to-noise ratio (SNR) in T-WPCN, i.e.

$$\begin{aligned}\text{SNR} =& \frac{(\eta_k P_\text{E} h_k \tilde{\tau}_0 - p_{\text{c},k}\tilde{\tau}_k)\gamma_k}{\tilde{\tau}_k} = \frac{(\eta_m P_\text{E} h_m \tilde{\tau}_0 - p_{\text{c},m}\tilde{\tau}_m)\gamma_m}{\tilde{\tau}_m} \\ =& \frac{\sum_{k=1}^{K}(\eta_k P_\text{E} h_k \tilde{\tau}_0 - p_{\text{c},k}\tilde{\tau}_k)\gamma_k}{\sum_{k=1}^{K}\tilde{\tau}_k}, \forall\ m \neq k.\end{aligned} \tag{13.20}$$

It can be verified that the constructed solution always exists and is also feasible for problem (13.8). Denote the SEs achieved by the optimal solution $\{\tau_0^*, \{\tau_k^*\}\}$ and the constructed solution $\{\tilde{\tau}_0, \{\tilde{\tau}_k\}\}$ as R_{TDMA}^* and $\tilde{R}_{\text{TDMA}}$, respectively. Then, it follows that

$$\begin{aligned}R_{\text{TDMA}}^* &\geq \tilde{R}_{\text{TDMA}} \\ &= \sum_{k=1}^{K} \tilde{\tau}_k \log_2\left(1 + \frac{(\eta_k P_\text{E} h_k \tilde{\tau}_0 - p_{\text{c},k}\tilde{\tau}_k)\gamma_k}{\tilde{\tau}_k}\right) \\ &= \sum_{k=1}^{K} \tilde{\tau}_k \log_2\left(1 + \frac{\sum_{m=1}^{K}(\eta_m P_\text{E} h_m \tilde{\tau}_0 - p_{\text{c},m}\tilde{\tau}_m)\gamma_m}{\sum_{m=1}^{K}\tilde{\tau}_m}\right) \\ &\overset{(a)}{\geq} \overline{\tau}_1^\star \log_2\left(1 + \frac{\sum_{m=1}^{K}(\eta_m P_\text{E} h_m \tau_0^\star - p_{\text{c},m}\overline{\tau}_1^\star)\gamma_m}{\overline{\tau}_1^\star}\right) \\ &= R_{\text{NOMA}}^\star,\end{aligned} \tag{13.21}$$

where inequality "(a)" holds due to $\sum_{k=1}^{K} \tilde{\tau}_k = \overline{\tau}_1^\star$ and $0 < \tilde{\tau}_k < \overline{\tau}_1^\star$, $\forall\ k$, and the equality holds when $p_{\text{c},k} = 0$, $\forall\ k$. Thus, if $\exists\ k, p_{\text{c},k} > 0$, it follows that $R_{\text{TDMA}}^* > R_{\text{NOMA}}^\star$. Next, we prove that when $p_{\text{c},k} = 0$, $\forall\ k$, the constructed solution is the optimal solution to problem (13.8), i.e. $\tau_0^* = \tilde{\tau}_0$ and $\tau_k^* = \tilde{\tau}_k$. The SE of T-WPCN is given by

$$\begin{aligned}R_{\text{TDMA}} &= \sum_{k=1}^{K} \tau_k \log_2\left(1 + \frac{\eta_k P_\text{E} h_k \gamma_k}{\tau_k}\tau_0\right) \\ &\overset{(b)}{\leq} \sum_{k=1}^{K} \tau_k \log_2\left(1 + \frac{\sum_{m=1}^{K}\eta_m P_\text{E} h_m \gamma_m}{\sum_{m=1}^{K}\tau_m}\tau_0\right) \\ &= (1-\tau_0)\log_2\left(1 + \frac{\sum_{m=1}^{K}\eta_m P_\text{E} h_m \gamma_m}{1-\tau_0}\tau_0\right) \\ &\overset{(c)}{\leq} (1-\tau_0^\star)\log_2\left(1 + \frac{\sum_{m=1}^{K}\eta_m P_\text{E} h_m \gamma_m}{1-\tau_0^\star}\tau_0^\star\right) \\ &= R_{\text{NOMA}}^\star,\end{aligned} \tag{13.22}$$

where "(b)" holds due to the concavity of the logarithm function and "=" holds when $\frac{\eta_k P_E h_k \gamma_k}{\tau_k}\tau_0 = \frac{\eta_m P_E h_m \gamma_m}{\tau_m}\tau_0$, $\forall\, k$, which is exactly the same as (13.20) for $p_{c,k} = 0, \forall\, k$. Thus, we have $\tau_k^* = \tilde{\tau}_k$. Equality in "$(c)$" is due to the optimality of $\overline{\tau}_0^\star$ for N-WPCN. Thus, it follows that $\tau_0^* = \overline{\tau}_0^\star = \tilde{\tau}_0$.

Theorem 13.2 answers the question raised in the introduction regarding to the SE comparison of T-WPCN and N-WPCN. Specifically, TDMA in general achieves a higher SE than NOMA for wireless-powered IoT devices. This seems contradictory to the conclusions of previous works, e.g., [11], which have shown that NOMA always outperforms OMA schemes such as TDMA. Such a conclusion, however, was based on the conventional transmit power limited scenario where more transmit power is always beneficial for improving the SE by leveraging SIC. To show this, suppose that the transmit power of device k is limited by p_k and the energy causality constraints in (13.4) are removed. By setting $\tau_0 = 0$ in (13.4c), we have

$$\begin{aligned}
R_{\mathrm{TDMA}} &= \sum_{k=1}^{K} \tau_k \log_2(1 + p_k\gamma_k) \\
&\overset{(d)}{\leq} \sum_{k=1}^{K} \tau_k \log_2\left(1 + \sum_{m=1}^{K} p_m\gamma_m\right) \\
&= T_{\max}\log_2\left(1 + \sum_{k=1}^{K} p_k\gamma_k\right) = R_{\mathrm{NOMA}},
\end{aligned} \tag{13.23}$$

where strict inequality "(d)" holds if $p_k > 0, \forall\, k$. Accordingly, $E_{\mathrm{TDMA}} = \sum_{k=1}^{K} \tau_k p_k \leq \sum_{k=1}^{K} T_{\max} p_k = T_{\max} \sum_{k=1}^{K} p_k = E_{\mathrm{NOMA}}$. This suggests that the potential SE gain achieved by NOMA depends on the considered scenario. When each user has a maximum transmit power limitation p_k, which we refer to as the transmit power limited scenario, all users would transmit at p_k for the entire duration $T_{\max}$. The resulting SE gain of NOMA is at the expense of a higher energy consumption, as shown above. On the other hand, if the total available energy of each device is constrained, which we refer to as the energy limited scenario, NOMA provides no SE gain over TDMA, as shown in Theorem 13.2, which is consistent with the observations in [12, 13]. More importantly, when the circuit power consumption is taken into account for practical IoT devices, NOMA achieves a strictly lower SE than TDMA. Recall that the key principle of NOMA for enhancing the SE is to allow devices to access the same spectrum simultaneously. This, however, inevitably leads to a higher circuit energy consumption for NOMA because of the longer transmission time compared to TDMA, which is particularly detrimental to IoT devices that are energy limited in general.

13.4 Numerical Results

This section provides simulation results to demonstrate the effectiveness of the proposed solutions and validate our theoretical findings. There are 10 IoT devices randomly and uniformly distributed inside a disc with the PB in the center. The carrier frequency is 750 MHz and the bandwidth is 180 kHz as in typical NB-IoT systems [4]. The reference distance is 1 m and the maximum service distance is 5 m [50]. The AP is located 50 m away from the PB. Both the DL and UL channel power gains are modeled as $10^{-3}\rho^2 d^{-\alpha}$ [8], where ρ^2 is an exponentially distributed random variable (i.e., Rayleigh fading is assumed) with unit mean and d is the link distance. The path-loss exponent is set as $\alpha = 2.2$. Without loss of generality, it is assumed that all IoT devices have identical parameters which are set as $\eta_k = 0.9$ and $p_{c,k} = 0.1$ mW, $\forall\, k$ [51]. Other important parameters are set as $\sigma^2 = -117$ dBm, $P_E = 40$ dBm, and $T_{max} = 0.1$ s.

13.4.1 SE versus PB Transmit Power

Figure 13.2 shows the achievable throughput and energy consumption versus the PB transmit power, respectively. For comparison, two baseline schemes adopting TDMA and NOMA are considered, where $\tau_0 = \frac{T_{max}}{2}$ is set for both of them. This corresponds to the case where only $E_k^h = \frac{\eta_k P_E h_k T_{max}}{2}$ joules of energy is available for device k, i.e. energy-constrained IoT networks. Yet, the UL WIT is still optimized for maximizing the SE. In Figure 13.2a, the throughputs of both T-WPCN and N-WPCN improve with P_E. This is intuitive since with larger P_E, the wireless-powered IoT devices are able to harvest more energy during DL WET and hence achieve a higher throughput in UL WIT. In addition, the baseline schemes suffer from a throughput loss for both TDMA and NOMA compared to the corresponding optimal scheme due to the fixed time allocation for DL WET, which implies that optimizing the DL WET duration is also important for maximizing the SE of wireless-powered IoT networks. Furthermore, as suggested by Theorem 13.2, T-WPCN outperforms N-WPCN significantly and the performance gap between them becomes larger as P_E increases. This is because larger P_E will reduce the DL WET time and thereby leave more time for UL WIT. Since all the devices in N-WPCN are scheduled simultaneously for UL WIT, the circuit energy consumption will be significantly increased compared to that of T-WPCN, which thus leads to a larger performance gap. Figure 13.2b shows that N-WPCN is in general more energy demanding compared to T-WPCN for the optimal scheme, which verifies our theoretical finding in Theorem 13.2. Since $\tau_0 = \frac{T_{max}}{2}$ is set for both baseline schemes, they have the same total energy consumption. In addition, when $P_E = 28$ dBm, the energy consumption of optimal N-WPCN is close to that of optimal T-WPCN, which implies that each device k, $\forall\, k$, basically

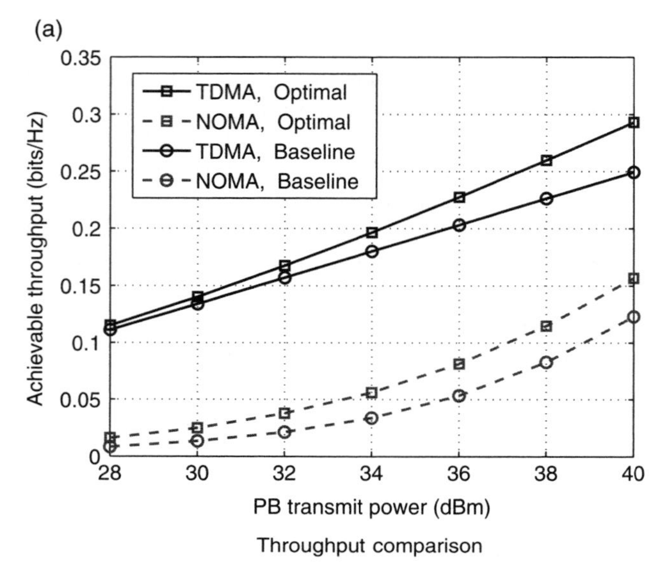

Throughput comparison

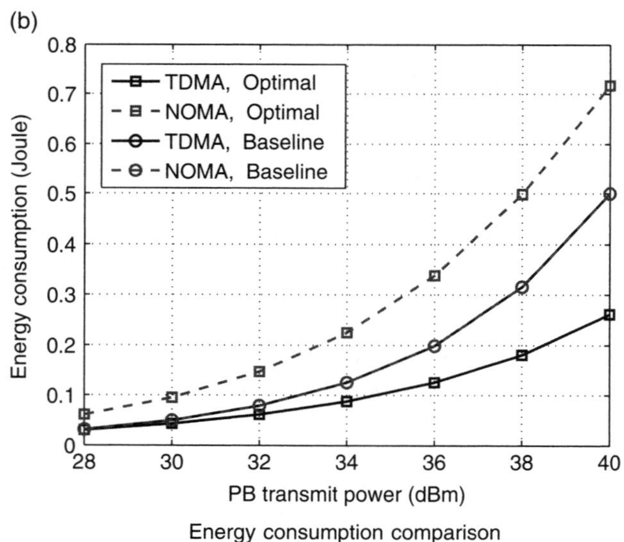

Energy consumption comparison

Figure 13.2 Throughput and energy consumption versus PB transmit power.

harvests a similar amount of energy in the DL of T-WPCN and N-WPCN. As such, the substantial SE loss in Figure 13.2a indicates that a significant portion of the harvested energy is consumed by the circuit rather than for signal transmission, due to the simultaneous transmission feature of NOMA.

13.4.2 SE versus Device Circuit Power

Figure 13.3 depicts the throughput and energy consumption versus the device circuit power consumption. Several observations are made as follows. First, for $p_{c,k} = 0$ in Figure 13.3, T-WPCN and N-WPCN achieve the same throughput and energy consumption for $K = 10$ and $K = 50$, which coincides with our findings in Theorems 1 and 2. Second, for $K = 10$ and $K = 50$, the throughput and energy consumption for T-WPCN moderately decrease and increase with $p_{c,k}$, respectively, while for N-WPCN they decrease and increase sharply with $p_{c,k}$, respectively. This suggests that the performance of N-WPCN is sensitive to $p_{c,k}$. In fact, for T-WPCN, when a device suffers from a worse DL channel condition, the corresponding harvested energy is also less. Then, the device will be allocated a short UL WIT duration such that the energy causality constraint is satisfied. However, for N-WPCN, since all devices transmit in the UL simultaneously, to meet the energy causality constraints of all the devices, i.e. $(p_k + p_{c,k})\overline{\tau}_1 \leq \eta_k P_E h_k \tau_0 = \eta_k P_E h_k (1 - \overline{\tau}_1), \forall\, k$, it follows that $\overline{\tau}_1 \leq \frac{\eta_k P_E h_k}{p_k + p_{c,k} + \eta_k P_E h_k} \leq \frac{\eta_k P_E h_k}{p_{c,k} + \eta_k P_E h_k}, \forall\, k$. As can be seen, the UL WIT duration $\overline{\tau}_1$ is always limited by the worst DL channel gain of all devices for $p_{c,k} > 0$, a phenomenon which we refer to as "worst user bottleneck problem". In addition, concurrent transmissions also lead to higher circuit energy consumption. As a result, the throughput and energy consumption of N-WPCN are significantly reduced and increased, respectively, as $p_{c,k}$ increases. Third, given the "worst user bottleneck problem", it is expected that when K increases from 10 to 50, the performance of N-WPCN decreases in Figure 13.3a and b. In contrast, for T-WPCN, since the UL WIT duration of each user can be individually allocated based on the DL and UL channel gains of each device, multi-user diversity can be exploited to improve the performances as K increases from 10 to 50.

13.5 Conclusions

In this chapter we have answered a fundamental question: Does NOMA improve SE and/or reduce the total energy consumption of the wireless-powered IoT networks? By taking into account the circuit energy consumption of the IoT devices, we have found that N-WPCN is neither spectrally efficient nor energy efficient, compared to T-WPCN. This suggests that NOMA may not be a practical solution for spectral and energy-efficient wireless IoT networks with energy-constrained devices. However, the case with user fairness

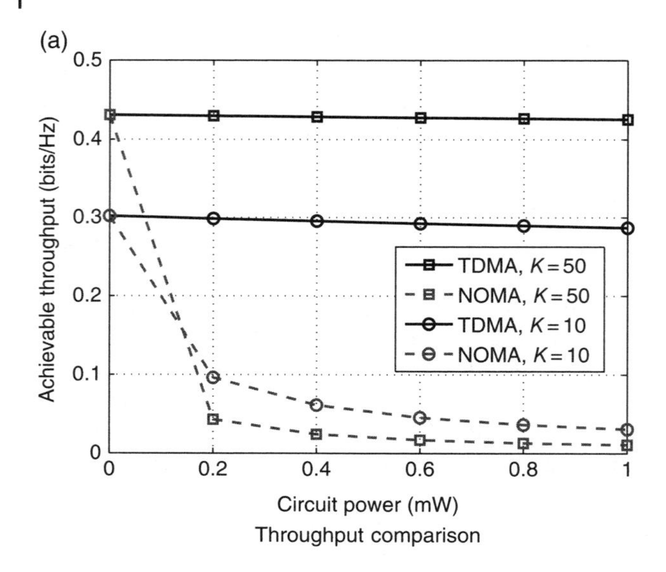

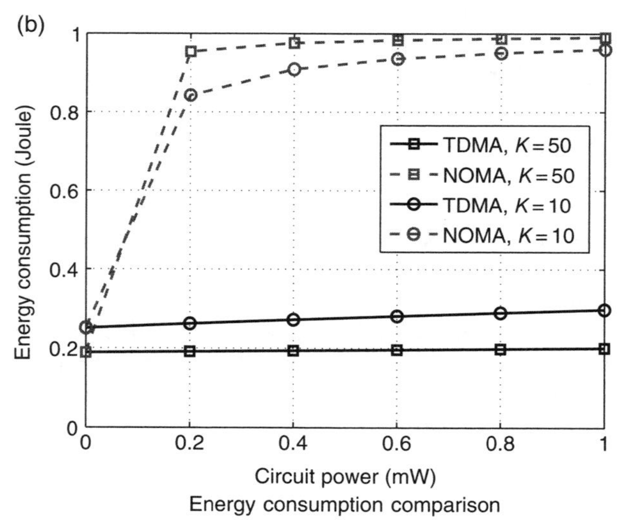

Figure 13.3 Throughput and energy consumption versus device circuit power.

consideration is an interesting topic for future work, which may lead to a different conclusion. Furthermore, for T-WPCN, the energy consumption can be further considered by optimizing the energy efficiency, measured in bits per joule, from the network-centric and user-centric perspectives, respectively. Interested readers can refer to [9] and [17] for details.

13.6 Future Work

In order to provide ubiquitous wireless energy supply to massive low-power energy-constrained devices, fixed PBs need to be deployed in an ultra-dense manner. This, however, would tremendously increase the cost, and hinder the large-scale implementation of wireless energy transfer systems. Recently, unmanned aerial vehicles (UAVs) have drawn significant research interest due to their appealing advantages such as swift and cost-effective deployment, line-of-sight (LOS) aerial-to-ground link, and controllable mobility in three-dimensional space. They are thus highly promising for numerous use cases in wireless communications [51–58], including ground BS traffic offloading, mobile relaying and edge computing, information/energy broadcasting and data collection for IoT devices, and fast network recovery after natural disasters. It has been shown in [59] that by exploiting the fully controllable mobility introduced by UAVs via proper trajectory design, the WET efficiency can be significantly improved while reducing the number of required PBs as compared to the conventional WET system with PBs deployed at fixed locations on the ground.

Bibliography

1 Novotech. *Nokia: Optimizing LTE for the internet of things.* [Online] Available: https://novotech.com/docs/default-source/default-document-library/lte-m-optimizing-lte-for-the-internet-of-things.pdf?sfvrsn=0.

2 Q. Wu, G.Y. Li, W. Chen, D.W.K. Ng, and R. Schober (2017) An overview of sustainable green 5G networks. *IEEE Wireless Commun.*, **24** (4): 72–80.

3 S. Zhang, Q. Wu, S. Xu, and G. Li (2017) Fundamental green tradeoffs: Progresses, challenges, and impacts on 5G networks. *IEEE Commun. Surveys & Tutorials*, **19** (1): 33–56.

4 Y.-P. E. Wang, X. Lin, A. Adhikary, A. Grovlen, Y. Sui, Y. Blankenship, J. Bergman, and H.S. Razaghi (2017) A primer on 3GPP narrowband internet of things. *IEEE Commun. Mag.*, **55** (3): 117–123.

5 G. Zhang, J. Xu, Q. Wu, M. Cui, X. Li, and F. Lin (2018) Wireless powered cooperative jamming for secure OFDM system. *IEEE Trans. Veh. Technol.*, **67** (2): 1331–3461.

6 R. Zhang and C.K. Ho (2013) MIMO broadcasting for simultaneous wireless information and power transfer. *IEEE Trans. Wireless Commun.*, **12** (5): 1989–2001.

7 Q. Wu, W. Chen, J. Wu, and J. Li (2015) Wireless powered communications: Industry demands and a user-centric energy-efficient approach. In *IEEE GLOBECOM*, pp. 1–6.

8 H. Ju and R. Zhang (2014) Throughput maximization in wireless powered communication networks. *IEEE Trans. Wireless Commun.*, **13** (1): 418–428.

9 Q. Wu, M. Tao, D.W.K. Ng, W. Chen, and R. Schober (2016) Energy-efficient resource allocation for wireless powered communication networks. *IEEE Trans. Wireless Commun.*, **15** (3): 2312–2327.

10 H. Ju and R. Zhang (2014) Optimal resource allocation in full-duplex wireless-powered communication network. *IEEE Trans. Commun.*, **62** (10): 3528–3540.

11 Z. Ding, Z. Yang, P. Fan, and H.V. Poor (2014) On the performance of non-orthogonal multiple access in 5G systems with randomly deployed users. *IEEE Signal Process. Lett.*, **21** (12): 1501–1505.

12 P.D. Diamantoulakis, K.N. Pappi, Z. Ding, and G.K. Karagiannidis (2016) Wireless-powered communications with non-orthogonal multiple access. *IEEE Trans. Wireless Commun.*, **15** (12): 8422–8436.

13 H. Chingoska, Z. Hadzi-Velkov, I. Nikoloska, and N. Zlatanov (2016) Resource allocation in wireless powered communication networks with non-orthogonal multiple access. *IEEE Wireless Commun. Lett.*, **5** (6): 684–687.

14 Q. Wu, M. Tao, W. Chen, and J. Wu (2014) Optimal energy-efficient transmission for fading channels with an energy harvesting transmitter. In *Proceedings of the IEEE GLOBECOM*, IEEE.

15 Q. Wu, M. Tao, D.W.K. Ng, W. Chen, and R. Schober (2015) Energy-efficient transmission for wireless powered multiuser communication networks. In *Proceedings of the IEEE ICC*, IEEE.

16 Q. Wu, W. Chen, and J. Li (2015) Wireless powered communications with initial energy: QoS guaranteed energy-efficient resource allocation. *IEEE Commun. Lett.*, **19** (12): 2278–2281.

17 Q. Wu, W. Chen, D.W.K. Ng, J. Li, and R. Schober (2016) User-centric energy efficiency maximization for wireless powered communications. *IEEE Trans. Wireless Commun.*, **15** (19): 6898–6912.

18 X. Chen, X. Wang, and X. Chen (2013) Energy-efficient optimization for wireless information and power transfer in large-scale MIMO systems employing energy beamforming. *IEEE Wireless Commun. Lett.*, **2** (6): 667–670.

19 G. Zhang, X. Li, M. Cui, G. Li, and L. Yang (2016) Signal and artificial noise beamforming for secure simultaneous wireless information and power transfer multiple-input multiple-output relaying systems. *IET Commun.*, **10** (7): 796–804.

20 Y. Dong, M.J. Hossain, and J. Cheng (2016) Joint power control and time switching for SWIPT systems with heterogeneous QoS requirements. *IEEE Commun. Lett.*, **20** (2): 328–331.

21 X. Huang and N. Ansari (2015) Energy sharing within EH-enabled wireless communication networks. *IEEE Commun. Mag.*, **22** (3): 144–149.

22 Y. Dong, M.J. Hossain, and J. Cheng (2016) Performance of wireless powered amplify and forward relaying over Nakagami-*m* fading channels with nonlinear energy harvester. *IEEE Commun. Lett.*, **20** (4): 672–675.

23 Q. Wu, G.Y. Li, W. Chen, and D.W.K. Ng (2016) Spectrum-power trading for energy-efficient small cell. In: *IEEE GLOBECOM*, IEEE.

24 Q. Wu, G.Y. Li, W. Chen, and D.W.K. Ng (2016) Energy-efficient small cell with spectrum-power trading. *IEEE J. Selected Areas Commun.*, **34** (12): 3394–3408.

25 D.W.K. Ng, E.S. Lo, and R. Schober (2012) Energy-efficient resource allocation in multi-cell OFDMA systems with limited backhaul capacity. *IEEE Trans. Wireless Commun.*, **11** (10): 3618–3631.

26 K. Cheung, S. Yang, and L. Hanzo (2013) Achieving maximum energy-efficiency in multi-relay OFDMA cellular networks: A fractional programming approach. *IEEE Trans. Commun.*, **61** (7): 2746–2757.

27 C. Sun, Y. Cen, and C. Yang (2013) Energy efficient OFDM relay systems. *IEEE Trans. Commun.*, **61** (5): 1797–1809.

28 X. Huang and N. Ansari (2016) Optimal cooperative power allocation for energy harvesting enabled relay networks. *IEEE Trans. Veh. Technol.*, **65** (4): 2424–2434.

29 Q. Wu, G.Y. Li, W. Chen, and D.W.K. Ng (2017) Energy-efficient D2D overlaying communications with spectrum-power trading. *IEEE Trans. Wireless Commun.*, **16** (7): 4404–4419.

30 Q. Wu, W. Chen, M. Tao, J. Li, H. Tang, and J. Wu (2015) Resource allocation for joint transmitter and receiver energy efficiency maximization in downlink OFDMA systems. *IEEE Trans. Commun.*, **63** (2): 416–430.

31 Q. Wu, W. Chen, J. Li, and J. Wu (2014) Low complexity energy-efficient design for OFDMA systems with an elaborate power model. In *Proceedings of the IEEE GLOBECOM*, IEEE.

32 C. Xiong, G. Li, S. Zhang, Y. Chen, and S. Xu (2012) Energy-efficient resource allocation in OFDMA networks. *IEEE Trans. Commun.*, **60** (12): 3767–3778.

33 C. Xiong, G.Y. Li, Y. Liu, Y. Chen, and S. Xu (2013) Energy-efficient design for downlink OFDMA with delay-sensitive traffic. *IEEE Trans. Wireless Commun.*, **12** (6): 3085–3095.

34 Q. Wu, M. Tao, and W. Chen (2015) Joint Tx/Rx energy-efficient scheduling in multi-radio wireless networks: A divide-and-conquer approach. In *Proceedings of the IEEE ICC*, IEEE.

35 D.W.K. Ng, E.S. Lo, and R. Schober (2012) Energy-efficient resource allocation in OFDMA systems with large numbers of base station antennas. *IEEE Trans. Wireless Commun.*, **11** (9): 3292–3304.

36 F.-S. Chu, K.-C. Chen, and G. Fettweis (2012) Green resource allocation to minimize receiving energy in ofdma cellular systems. *IEEE Commun. Lett.*, **16** (3): 372–374.

37 Q. Wu, M. Tao, and W. Chen (2005) Joint Tx/Rx energy-efficient scheduling in multi-radio wireless networks: A divide-and-conquer approach. *IEEE Trans. Wireless Commun.*, **15** (4): 2727–2740.

38 C. Isheden and G.P. Fettweis (2010) Energy-efficient multi-carrier link adaptation with sum rate-dependent circuit power. In *Proceedings of the IEEE GLOBECOM*, IEEE.

39 Q. Shi, W. Xu, D. Li, Y. Wang, X. Gu, and W. Li (2013) On the energy efficiency optimality of OFDMA for SISO-OFDM downlink system. *IEEE Commun. Lett.*, **17** (3): 541–544.

40 G. Miao (2013) Energy-efficient uplink multi-user MIMO. *IEEE Trans. Wireless Commun.*, **12** (5): 2302–2313.

41 S. Cui, A.J. Goldsmith, and A. Bahai (2004) Energy-efficiency of MIMO and cooperative MIMO techniques in sensor networks. *IEEE J. Selected Areas Commun.*, **22** (6): 1089–1098.

42 R.S. Prabhu and B. Daneshrad (2010) Energy-efficient power loading for a MIMO-SVD system and its performance in flat fading. In *Proceedings of the IEEE GLOBECOM*, IEEE.

43 W. Mei, Z. Chen, and J. Fang (2017) Energy efficiency region for Gaussian MISO channels with integrated services. *IEEE Wireless Commun. Lett.*, **6** (1): 90–93.

44 G. Zhang, Q. Li, Q. Zhang, J. Qin, and L. Yang (2015) Signal-to-interference-plus-noise ratio-based multi-relay beamforming for multi-user multiple-input multiple-output cognitive relay networks with interference from primary network. *IET Commun.*, **9** (2): 227–238.

45 Y. Dong, X. Ge, J. Hossain, J. Cheng, and V.C.M. Leung (2017) Proportional fairness-based beamforming and signal splitting for MISO-SWIPT systems. *IEEE Commun. Lett.*, **21** (5): 1135–1138.

46 W. Mei, Z. Chen, and J. Fang (2017) Artificial noise aided energy efficiency optimization in MIMOME system with SWIPT. *IEEE Commun. Lett.*, **21** (8): 1795–1798.

47 Y. Dong, A.E. Shafie, M.J. Hossain, J. Cheng, N. Al-Dhahir, and V.C.M. Leung (2018) Secure beamforming in full-duplex swipt systems with loop-back self-interference cancellation. *Proceedings of the IEEE ICC*, accepted.

48 X. Huang, T. Han, and N. Ansari (2015) On green energy powered cognitive radio networks. *IEEE Commun. Surveys & Tutorialss*, **17** (2): 827–842.

49 Q. Wu, W. Chen, D.W.K. Ng, and R. Schober (2018) Spectral and energy efficient wireless powered IoT networks: NOMA or TDMA? *IEEE Trans. Veh. Technol.*, **67** (7).

50 S. Bi, Y. Zeng, and R. Zhang (2016) Wireless powered communication networks: An overview. *IEEE Wireless Commun. Mag.*, **23** (2): 10–18.

51 G.C. Martins, A. Urso, A. Mansano, Y. Liu, and W.A. Serdijn (2017) Energy-efficient low-power circuits for wireless energy and data transfer in IoT sensor nodes. *CoRR, preprint arXiv*:**1704.08910**.

52 Q. Wu, Y. Zeng, and R. Zhang (2018) Joint trajectory and communication design for multi-UAV enabled wireless networks. *IEEE Trans. Wireless Commun.*, **17** (3): 2109–2121.

53 Q. Wu, J. Xu, and R. Zhang (2018) Capacity characterization of UAV-enabled two-user broadcast channel. *CoRR, arXiv preprint arXiv: 1801.00443.*

54 Q. Wu and R. Zhang (2018) Common throughput maximization in UAV-enabled OFDMA systems with delay consideration. *IEEE Trans. Commun.*, submitted.

55 Q. Wu, Y. Zeng, and R. Zhang(2017) Joint trajectory and communication design for UAV-enabled multiple access. In *Proceedings of the IEEE GLOBECOM*. [Online]. Available: https://arxiv.org/abs/1704.01765.

56 Q. Wu and R. Zhang (2017) Delay-constrained throughput maximization in UAV-enabled OFDM systems. In *Proceedings of the IEEE APCC*, IEEE.

57 D. Yang, Q. Wu, Y. Zeng, and R. Zhang (2017) Energy trade-off in ground-to-UAV communication via trajectory design. *IEEE Trans. Veh. Technol.* [Online]. Available: https://arxiv.org/abs/1709.02975.

58 G.Zhang, Q. Wu, M. Cui, and R. Zhang (2017) Securing UAV communications via trajectory optimization. In *Proceedings of the IEEE GLOBECOM.* [Online] Available: https://arxiv.org/abs/1710.04389.

59 J. Xu, Y. Zeng, and R. Zhang (2017) UAV-enabled wireless power transfer: Trajectory design and energy region characterization. [Online]. Available: https://arxiv.org/abs/1709.07590.

14

Wireless-Powered Mobile Edge Computing Systems

Feng Wang[1], *Jie Xu*[1*], *Xin Wang*[2], *and Shuguang Cui*[3,4]

[1] *School of Information Engineering, Guangdong University of Technology, China*
[2] *Key Laboratory for Information Science of Electromagnetic Waves (MoE), the Shanghai Institute for Advanced Communication and Data Science, Department of Communication Science and Engineering, Fudan University, China*
[3] *Department of Electrical and Computer Engineering, University of California, Davis, USA*
[4] *Shenzhen Research Institute of Big Data, China*

14.1 Introduction

The recent advancement of the Internet of Things (IoT) has motivated various new applications (such as autonomous driving, virtual reality, and tele-surgery) to provide real-time machine-to-machine and machine-to-human interactions [2]. These emerging applications critically rely on the real-time communication and computation of massive wireless devices (e.g., sensors). As extensive existing works focus on improving their communication performances [2], how to provide these devices with enhanced computational capability is a crucial but challenging task to be tackled, especially when they are of small size and low power. To resolve this issue, mobile edge computing (MEC) has emerged as a promising technique by providing cloud-like computing at the edge of mobile networks via integrating MEC servers at wireless access points (APs) and base stations (BSs) [4]. With MEC, resource-limited wireless devices can offload some or all of their computation tasks to APs, then integrated MEC servers can compute these tasks remotely on behalf of the devices. The MEC technique facilitates the real-time implementation of computational intensive tasks for massive low-power devices, and thus has attracted growing research interests from academia and industry. For example, the authors in [4] investigated the energy-efficient MEC design for mobile devices to minimize their energy consumption subject to the computation requirements. Empowered by MEC, each wireless device decides whether a task should be offloaded to the AP or

*Corresponding author: Jie Xu; jiexu@gdut.edu.cn

Wireless Information and Power Transfer: Theory and Practice, First Edition.
Edited by Derrick Wing Kwan Ng, Trung Q. Duong, Caijun Zhong, and Robert Schober.

executed locally by itself, or how many bits should be offloaded, so as to balance the energy consumption tradeoff between offloading and local computing.

On the other hand, how to provide sustainable and cost-effective energy supply to massive computation-heavy devices is another challenge faced by the IoT. Radio-frequency (RF) signal-based wireless power transfer (WPT) provides a viable solution by deploying dedicated energy transmitters to broadcast energy wirelessly [16]. Recently, emerging wireless-powered communication networks (WPCN) and simultaneous wireless information and power transfer (SWIPT) paradigms have been proposed to achieve ubiquitous wireless communications in a self-sustainable way [10]–[13], where WPT and wireless communications are combined in a joint design. In order to improve the WPT efficiency from the energy transmitter to one or more energy receivers, *transmit energy beamforming* has been recognized as a promising solution by deploying multiple antennas at energy transmitters [7]. By properly adjusting the transmit beamforming vectors, energy transmitters can concentrate the radiative energy towards the intended receivers for efficient WPT. Motivated by these approaches, it is expected that the transmit energy beamforming-enabled WPT can also play an important role in facilitating self-sustainable computing for a large number of IoT devices.

To explore the benefits of both MEC and WPT in *ubiquitous* computing, this chapter develops a joint MEC-WPT design by considering a wireless powered *multi-user* MEC system that consists of a multi-antenna AP and multiple single-antenna users. The AP employs the energy transmit beamforming to simultaneously charge the users, and each user relies on its harvested energy to execute the respective computation task. Suppose that partial offloading is allowed such that each user can arbitrarily partition the computation task into two independent parts for local computing and offloading, respectively. We assume that the WPT and wireless communication (for computation offloading) are operated simultaneously over orthogonal frequency bands. A time division multiple access (TDMA) protocol is employed to coordinate computation offloading, where different users offload their respective tasks to the AP over orthogonal time slots. Consider a block-based operation, where each user relies on its harvested wireless energy to execute the latency-sensitive computation tasks per time block via local computing or (partial) offloading to the MEC server. Under this setup, we pursue an energy-efficient wireless-powered MEC system design by jointly optimizing the transmit energy beamformer at the AP, the central processing unit (CPU) frequency, and the offloaded bits at each user, as well as the TDMA time allocation among different users. Specifically, we minimize the energy consumption at the AP over a particular block, subject to the computation latency and energy harvesting constraints per user.

Literature Survey. Transmit energy beamforming enabled WPT has been extensively studied in the literature (see, e.g., [7]–[19] and references therein).

By considering a linear energy harvesting (EH) model, various prior works have investigated the optimal design of energy beamforming under different setups with SWIPT, e.g., in two-user multiple-input multiple-output (MIMO) systems [7], secrecy communications systems [8], multiple-input single-output (MISO) interference channels [9], and multi-user MISO downlink channels [12]–[14]. Furthermore, some recent works have investigated the transmit power allocation [21] and the transmit waveform optimization [22] for WPT by taking into account the nonlinearity of the rectifier in EH [18, 19]. In addition, the benefit of energy beamforming crucially relies on the availability of the channel state information (CSI) at the transmitter. The reverse-link channel training [15] and the energy measurement and feedback methods [16, 17] were proposed in WPT systems for the energy transmitter to practically teach the CSI to users. Furthermore, [20] developed a distributed energy beamforming system for multiple energy transmitters to charge multiple energy receivers simultaneously, with the help of the energy measurement and feedback.

On the other hand, several existing works [23]–[32] investigated the energy-efficient multi-user MEC design, where each user is powered by fixed energy sources such as battery, and the objective is to minimize the energy consumption at the users via joint computing and offloading optimization at the demand side. For example, [23] provided an overview on the applications and challenges of computation offloading. The works in [24] and [25] investigated the dynamic offloading for MEC systems based on the techniques of the Markov decision process and the Lyapunov optimization, respectively. The work in [26] considered the joint computation and communication resource allocation in single-user MEC systems, and such designs were extended to multi-user MEC systems in [27]–[30]. The work in [31] pursued a multi-user MEC design based on uplink nonorthogonal multiple access (NOMA). The work in [32] investigated a joint user cooperation of communication and computation for further improving MEC system performance. Different from these prior works that studied WPT and MEC separately, this chapter pursues a joint MEC-WPT design in a wireless-powered multi-user MEC system by jointly optimizing the WPT supply at the AP, as well as the local computing and offloading demands at the users.

It is worth noting that [33] considered the wireless-powered *single-user* MEC systems with binary offloading, where the user aims to maximize the probability of successful computation, by deciding whether a task should be fully offloaded or not, subject to the computation latency constraint. The work in [34] pursued computation rate maximization for wireless-powered multi-user MEC systems, while [35] considered a special scenario with binary offloading. By contrast, this chapter considers a more general case with resource sharing among multiple users and allows for more flexible partial offloading to improve the system performance in terms of the energy efficiency (i.e., minimizing the total energy consumption at the AP including the radiated energy for WPT and the energy for computing the offloaded tasks).

Notation. Boldface letters refer to vectors (lower case) or matrices (upper case). For a square matrix $\boldsymbol{S}$, tr($\boldsymbol{S}$) denotes its trace, while $\boldsymbol{S} \succeq \boldsymbol{0}$ means that $\boldsymbol{S}$ is positive semi-definite. For an arbitrary-size matrix $\boldsymbol{M}$, rank($\boldsymbol{M}$), $\boldsymbol{M}^{\dagger}$, and $\boldsymbol{M}^{H}$ denote its rank, transpose, and conjugate transpose, respectively. $\boldsymbol{I}$ and $\boldsymbol{0}$ denote an identity matrix and an all-zero vector/matrix, respectively, with appropriate dimensions. $\mathbb{C}^{x \times y}$ denotes the space of $x \times y$ complex matrices and $\mathbb{R}$ denotes the set of real numbers. $\mathbb{E}[\cdot]$ denotes the statistical expectation. $\| \boldsymbol{x} \|$ denotes the Euclidean norm of a vector $\boldsymbol{x}$, $|z|$ denotes the magnitude of a complex number z, and $[x]^{+} \triangleq \max(x, 0)$.

14.2 System Model

As shown in Figure 14.1, we consider a wireless-powered multi-user MEC system consisting of an N-antenna AP (integrated with an MEC server) and a set $\mathcal{K} \triangleq \{1, \ldots, K\}$ of single-antenna users. In this system, the AP employs the RF-based energy transmit beamforming to charge the K users simultaneously. Each user $i \in \mathcal{K}$ utilizes the harvested energy to execute its computation task through local computing and offloading. Suppose that the downlink WPT from

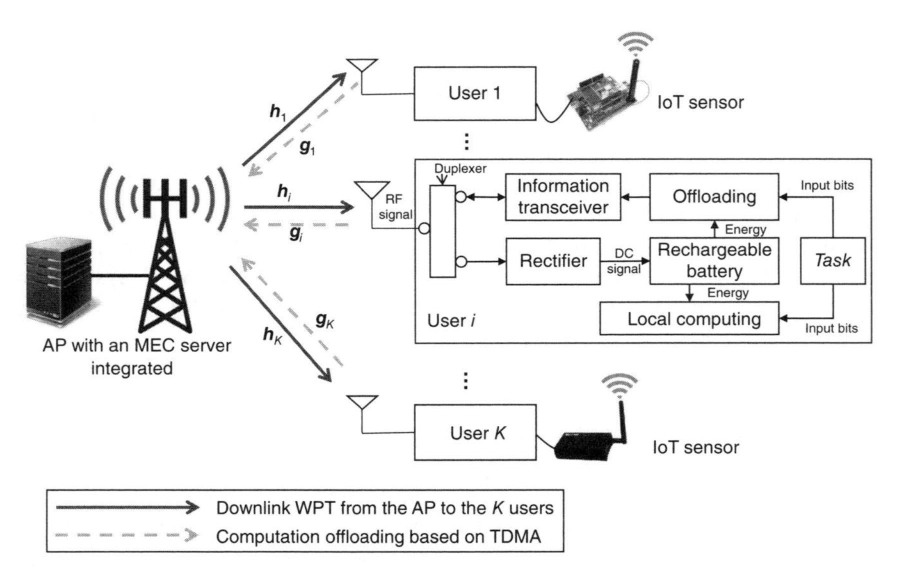

Figure 14.1 A wireless-powered multi-user MEC system with WPT in the downlink and computation offloading in the uplink [1].

the AP to the users and the uplink computation offloading are operated simultaneously over orthogonal frequency bands, and the uplink for computation offloading and the downlink for computation result downloading are operated over the same frequency band.

We consider a block-based model for both the WPT and the MEC. Let T denote the duration of each block. Assume a block-based model, and we focus on one particular block with duration T. Here, T is chosen to be no longer than the latency of the MEC application and also no longer than the channel coherence time, such that the wireless channels remain unchanged during this block. Suppose that user $i \in \mathcal{K}$ has a computation task with R_i input bits,[1] which can either be computed locally by itself or remotely computed at the MEC server via offloading. In practice, the computation task can be separated into various modules and computed in a distributed manner [4–6]. This enables partial offloading such that ℓ_i bits of the task are offloaded to the MEC server remotely, while the remaining $(R_i - \ell_i)$ bits are computed locally. For simplicity of analysis, we consider that the R_i bits at each user can be arbitrarily segmented as considered in the prior work [29]. For better capturing the AP's transmission energy for computation offloading, we assume that the AP perfectly knows the CSI from/to the K users and their computation requirements.

Energy Beamforming for WPT from AP to Users. Let $\boldsymbol{s} \in \mathbb{C}^{N\times 1}$ denote the energy-bearing signal transmitted by the AP and $\boldsymbol{Q} = \mathbb{E}[\boldsymbol{s}\boldsymbol{s}^H] \in \mathbb{C}^{N\times N}$ denote its transmit covariance matrix. In general, the AP can use multiple beams to deliver the wireless energy, i.e. $\boldsymbol{Q}$ can be of any rank. Supposing $d = \text{rank}(\boldsymbol{Q}) \leq N$, there is a total of d energy beams that can be obtained via the eigenvalue decomposition (EVD) of $\boldsymbol{Q}$ [16]. In this case, the total transmit power at the AP is given as $\text{tr}(\boldsymbol{Q})$. Let $\boldsymbol{h}_i \in \mathbb{C}^{N\times 1}$ denote the wireless channel vector from the AP to user $i \in \mathcal{K}$, and we define $\boldsymbol{H}_i \triangleq \boldsymbol{h}_i\boldsymbol{h}_i^H$, $\forall i \in \mathcal{K}$. Assuming a linear EH model, the amount of energy harvested by user i is then [8]

$$E_i = T\eta\mathbb{E}[|\boldsymbol{h}_i^H\boldsymbol{s}|^2] = T\eta\text{tr}(\boldsymbol{Q}\boldsymbol{H}_i), \quad \forall i \in \mathcal{K}, \tag{14.1}$$

where $0 < \eta \leq 1$ is the energy conversion efficiency. The harvested energy E_i is used by user $i \in \mathcal{K}$ for both computation offloading and local computing.

Computation Offloading from Users to AP. We next consider the computation offloading from the K users to the AP. In order for the MEC server to successfully compute the task on behalf of the users, each user should first offload the computation bits to the AP and then the AP sends the computation results back to the users after MEC computations. Practically, the AP with an integrated MEC server could provide sufficient CPU capability and high transmit power, while the computation results are usually of small size. Therefore, the computing time consumed at the MEC server and that consumed for

1 Each input bit can be treated as the smallest task unit, which includes the needed program codes and input parameters.

delivering the computation results are relatively small. For each user, the energy required to receive its computation result from the AP is generally negligible. For these reasons, we only consider the uplink offloading time as the total MEC latency time and ignore users' energy consumption for receiving computation results throughout this chapter.

We consider a TDMA protocol for the uplink offloading. The whole time block is divided into a total of K time slots, where user i offloads its task in the ith time slot with a duration of $t_i \in [0, T]$. Denote by $\boldsymbol{g}_i \in \mathbb{C}^{N\times 1}$ the uplink channel from user i to the AP and p_i user i's transmission power for offloading. Assume further that the AP employs the maximum ratio combining (MRC) receiver to decode the information. The achievable communication rate from user i to the AP is given by

$$r_i = B\log_2\left(1 + \frac{p_i\tilde{g}_i}{\Gamma\sigma^2}\right), \quad \forall i \in \mathcal{K}, \tag{14.2}$$

where B denotes the spectrum bandwidth for offloading, $\tilde{g}_i \triangleq \| \boldsymbol{g}_i \|^2$, σ^2 is the receiver noise power at the AP, and $\Gamma \geq 1$ is a constant accounting for the signal-to-noise ratio (SNR) gap from the channel capacity due to a practical modulation and coding scheme (MCS). For simplicity, $\Gamma = 1$ is assumed throughout this chapter. Since each user $i \in \mathcal{K}$ needs to offload a total number of ℓ_i bits over the time slot of duration t_i, without loss of generality, we have

$$\ell_i = r_i t_i, \quad \forall i \in \mathcal{K}. \tag{14.3}$$

For each user $i \in \mathcal{K}$, the power consumption consists of the transmit power p_i (for offloading) and a constant circuit power $p_{\mathrm{c},i}$ (by the digital-to-analog converter (DAC), filter, etc.). Here, the transmit power p_i can be calculated as

$$p_i = \frac{1}{\tilde{g}_i}\beta\left(\frac{\ell_i}{t_i}\right), \quad \forall i \in \mathcal{K}, \tag{14.4}$$

where $\beta(x) \triangleq \sigma^2(2^{\frac{x}{B}} - 1)$ is an increasing and convex function of $x \geq 0$; we define $\beta(\frac{\ell_i}{t_i}) = 0$ if either $\ell_i = 0$ or $t_i = 0$ holds. By combining p_i and $p_{\mathrm{c},i}$, the total energy consumption at user i to offload ℓ_i bits over the ith time slot is

$$E_{\mathrm{offl},i} \triangleq (p_i + p_{\mathrm{c},i})t_i = \frac{t_i}{\tilde{g}_i}\beta\left(\frac{\ell_i}{t_i}\right) + p_{\mathrm{c},i}t_i, \quad \forall i \in \mathcal{K}. \tag{14.5}$$

Note that the computation offloading also incurs additional energy consumption at the AP as it needs to receive the offloaded bits from the K users, execute the computation subtasks (by the integrated MEC server), and send the computation results back to the users [4]. As the AP and its integrated MEC server both are supposed to have sufficient communication and computation capacities, we assume that it consumes constant large receive/transmit power levels and a constant high CPU frequency to minimize the communication and computation time. We further assume the computation result transmission duration is proportional to the computation bits. Therefore, we simply adopt a linear model for the offloading related energy consumption at the AP:[2]

$$E_{\mathrm{MEC}} = \alpha \sum_{i=1}^{K} \ell_i, \tag{14.6}$$

where α denotes the energy consumption per offloaded bit at the AP. In practice, α is generally determined by the transceiver structure of the AP, the chip structure of the MEC server, and its operated CPU frequencies [4].

Local Computing at K Users. We now address the local computing for the remaining $(R_i - \ell_i)$ bits at each user $i \in \mathcal{K}$. Denote by C_i the number of CPU cycles required for computing one computation bit at user i. Let $f_{i,n} \in (0, f_i^{\max}]$ be the CPU frequency for the nth CPU cycle required for user i, where $f_i^{\max}$ denotes the maximum CPU frequency at user i. Then the total number of CPU cycles for local computing at each user i is $C_i(R_i - \ell_i)$ and the corresponding delay is $\sum_{n=1}^{C_i(R_i-\ell_i)} 1/f_{i,n}$. Since all the local computing should be accomplished before the end of each given time block, the computation latency for executing these $C_i(R_i - \ell_i)$ CPU cycles by user i should satisfy

$$\sum_{n=1}^{C_i(R_i-\ell_i)} \frac{1}{f_{i,n}} \le T, \quad \forall i \in \mathcal{K}. \tag{14.7}$$

Under the assumption of a low CPU voltage that normally holds for low-power devices, the consumed energy for local computing is expressed as [36]

$$E_{\mathrm{loc},i} = \sum_{n=1}^{C_i(R_i-\ell_i)} \zeta_i f_{i,n}^2, \quad \forall i \in \mathcal{K}, \tag{14.8}$$

where ζ_i is the effective capacitance coefficient that depends on the chip architecture at user i.

By combining the computation offloading energy in (14.5) and the local computation energy in (14.8), the total energy consumed by user i within the block is given as $E_{\mathrm{offl},i} + E_{\mathrm{loc},i}$. Note that user i is powered by the wireless charging from the AP. In order for each user $i \in \mathcal{K}$ to achieve self-sustainable operation, the total consumed energy $E_{\mathrm{loc},i} + E_{\mathrm{offl},i}$ cannot exceed its harvested energy E_i as in (14.1) per block. Therefore, we have

$$E_{\mathrm{loc},i} + E_{\mathrm{offl},i} \le E_i, \quad \forall i \in \mathcal{K}. \tag{14.9}$$

---

2 Note that if the MEC server adaptively adjusts its CPU frequency, then its energy consumption E_{MEC} may generally correspond to a nonlinear function of the number of offloaded bits ℓ_i values. For instance, if the MEC server adjusts the CPU frequency similarly as the users for their local computing, then E_{MEC} can be expressed as a convex quadratic function of ℓ_i values. In this case, similar approaches can be adopted to solve our WPT-MEC design problem.

14.3 Joint MEC-WPT Design

14.3.1 Problem Formulation

Targeting an energy-efficient wireless-powered MEC design, our objective is to minimize the energy consumption at the AP while ensuring the successful execution of the K users' computation tasks per time block. To this end, we jointly optimize the energy transmit covariance matrix $\boldsymbol{Q}$ at the AP, the local CPU frequencies $\{f_{i,1}, \dots, f_{i,C_i(R_i-\ell_i)}\}$, and the number ℓ_i of the offloaded bits at each user, as well as the time allocation t_i among different users. Let $\boldsymbol{t} \triangleq [t_1, \dots, t_K]^\dagger$, $\boldsymbol{\ell} \triangleq [\ell_1, \dots, \ell_K]^\dagger$, and $\boldsymbol{f} \triangleq [f_{1,1}, \dots, f_{K,C_K(R_K-\ell_K)}]^\dagger$. Mathematically, the joint MEC-WPT design problem is formulated as

$$\begin{aligned}
\textbf{Problem 1:} \quad & \underset{\boldsymbol{Q}\succeq\boldsymbol{0},\boldsymbol{t},\boldsymbol{\ell},\boldsymbol{f}}{\text{minimize}} \; T\text{tr}(\boldsymbol{Q}) + \sum_{i=1}^{K} \alpha \ell_i \\
\text{subject to} \quad & \text{C1}: \sum_{n=1}^{C_i(R_i-\ell_i)} \frac{1}{f_{i,n}} \le T \;\text{ and }\; 0 \le f_{i,n} \le f_i^{\max}, \;\; \forall n, \;\; \forall i \in \mathcal{K}, \\
& \text{C2}: \sum_{n=1}^{C_i(R_i-\ell_i)} \zeta_i f_{i,n}^2 + \frac{t_i}{\tilde{g}_i}\beta\left(\frac{\ell_i}{t_i}\right) + p_{c,i} t_i \le T\eta\text{tr}(\boldsymbol{Q}\boldsymbol{H}_i), \;\; \forall i \in \mathcal{K}, \\
& \text{C3}: 0 \le \ell_i \le R_i, \;\; \forall i \in \mathcal{K}, \\
& \text{C4}: \sum_{i=1}^{K} t_i \le T, \;\; t_i \ge 0, \;\; \forall i \in \mathcal{K}.
\end{aligned} \tag{14.10}$$

The two sets of the constraints in C1 represent the local computing latency and CPU frequency constraints at user i, respectively. The ith constraint of C2 represents the energy harvesting constraint for user $i \in \mathcal{K}$. Furthermore, C3 and C4 correspond to the constraints on the users' offloading bits and their TDMA offloading time allocation, respectively. Suppose that each user has a sufficient computing capacity, i.e. $f_i^{\max} \ge C_i R_i / T_i$, $\forall i \in \mathcal{K}$. Then Problem 1 is always feasible since one can always scale $\boldsymbol{Q}$ to satisfy the energy harvesting constraints in C2 for any $(\boldsymbol{t}, \boldsymbol{\ell}, \boldsymbol{f})$. Note that Problem 1 is nonconvex due to the nonconvexity of constraints C1 and C2.

14.3.2 Optimal Solution

To cope with the nonconvex constraints C1 and C2, we first establish the following lemma. Note that throughout this chapter we have omitted the proofs of all the lemmas and propositions. Interested readers are encouraged to refer to [1] for more details.

Lemma 14.1 Given the offloaded bits $\boldsymbol{\ell}$, the optimal solution of the CPU frequencies $\boldsymbol{f}$ to Problem 1 satisfies

$$f_{i,1} = \ldots = f_{i,C_i(R_i-\ell_i)} = C_i(R_i - \ell_i)/T, \quad \forall i \in \mathcal{K}. \tag{14.11}$$

Lemma 14.1 indicates that at each user $i \in \mathcal{K}$, the local CPU frequencies for different CPU cycles are identical in the optimal strategy. Building on (14.11), Problem 1 can be equivalently reformulated as:

$$\begin{aligned}
(\mathcal{P}1.1) : \underset{\mathbf{Q}\succeq 0, \boldsymbol{t}, \boldsymbol{\ell}}{\text{minimize}} \;\; & T\text{tr}(\mathbf{Q}) + \sum_{i=1}^{K} \alpha \ell_i \\
\text{subject to} \;\; & \text{C5} : \sum_{i=1}^{K} t_i \le T, \\
& \text{C6} : \frac{\zeta_i C_i^3 (R_i - \ell_i)^3}{T^2} + \frac{t_i}{\tilde{g}_i} \beta\left(\frac{\ell_i}{t_i}\right) + p_{c,i} t_i \le T\eta \text{tr}(\mathbf{Q}\mathbf{H}_i), \; \forall i \in \mathcal{K}, \\
& \text{C7} : 0 \le \ell_i \le R_i, \;\; \forall i \in \mathcal{K}, \\
& \text{C8} : t_i \ge 0, \;\; \forall i \in \mathcal{K}.
\end{aligned} \tag{14.12}$$

As $\beta(x)$ is a convex function of $x \ge 0$, it is obvious that its perspective function $\frac{t_i}{\tilde{g}_i}\beta\left(\frac{\ell_i}{t_i}\right)$ is a joint convex function of t_i and ℓ_i [38]. As a result, the energy harvesting constraints in C6 become convex. Furthermore, since the objective function of problem ($\mathcal{P}1.1$) is affine and the other constraints are all convex, problem ($\mathcal{P}1.1$) is convex and can thus be efficiently solved by standard convex optimization techniques such as the interior-point method [39]. To gain more insights, we next derive its semi-closed solution by leveraging the Lagrange duality method [39].

Let $\mu \ge 0$ and $\lambda_i \ge 0$ denote the dual variables associated with the time-allocation constraint in C5 and the ith energy harvesting constraint in C6, respectively. Then the partial Lagrangian of problem ($\mathcal{P}1.1$) is expressed as

$$\begin{aligned}
\mathcal{L}(\mathbf{Q}, \boldsymbol{t}, \boldsymbol{\ell}, \boldsymbol{\lambda}, \mu) = & \sum_{i=1}^{K} \lambda_i \left(\frac{\zeta_i C_i^3 (R_i - \ell_i)^3}{T^2} + \frac{t_i}{\tilde{g}_i} \beta\left(\frac{\ell_i}{t_i}\right) + p_{c,i} t_i - T\eta \text{tr}(\mathbf{Q}\mathbf{H}_i) \right) \\
& + T\text{tr}(\mathbf{Q}) + \sum_{i=1}^{K} \alpha \ell_i + \mu \left(\sum_{i=1}^{K} t_i - T \right),
\end{aligned} \tag{14.13}$$

where $\boldsymbol{\lambda} \triangleq [\lambda_1, \ldots, \lambda_K]^\dagger$. The dual function $g(\boldsymbol{\lambda}, \mu)$ is then

$$\begin{aligned}
g(\boldsymbol{\lambda}, \mu) = & \underset{\mathbf{Q}\succeq 0, \boldsymbol{t}, \boldsymbol{\ell}}{\text{minimize}} \;\; \mathcal{L}(\mathbf{Q}, \boldsymbol{t}, \boldsymbol{\ell}, \boldsymbol{\lambda}, \mu) \\
& \text{subject to} \;\; \text{C7} \;\; \text{and} \;\; \text{C8}.
\end{aligned} \tag{14.14}$$

Consequently, the dual problem of $(\mathcal{P}1.1)$ is

$$\begin{aligned}(\mathcal{D}1.1):&\underset{\lambda,\mu}{\text{maximize}}\quad g(\lambda,\mu)\\ \text{subject to } &\text{C9}: \boldsymbol{F}(\lambda)\triangleq \boldsymbol{I}-\sum_{i=1}^{K}\eta\lambda_i\boldsymbol{H}_i\succeq \boldsymbol{0},\\ &\text{C10}: \lambda_i\geq 0,\quad \forall i\in\mathcal{K},\\ &\text{C11}: \mu\geq 0.\end{aligned} \tag{14.15}$$

Note that $\boldsymbol{F}(\lambda)\succeq\boldsymbol{0}$ is needed to ensure that the dual function is bounded from below (see the proof in [1, Appendix B]). We denote by $\mathcal{X}$ the feasible set of (λ,μ), characterized by constraints C9–C11.

Since problem $(\mathcal{P}1.1)$ is convex and satisfies Slater's condition, strong duality holds between $(\mathcal{P}1.1)$ and $(\mathcal{D}1.1)$ [38]. As a result, we can solve problem $(\mathcal{P}1.1)$ by equivalently solving its dual problem $(\mathcal{D}1.1)$. In the following, we first obtain the dual function $g(\lambda,\mu)$ for any given $(\lambda,\mu)\in\mathcal{X}$, and then obtain the optimal dual variables to maximize $g(\lambda,\mu)$ using the ellipsoid method [40]. For convenience of presentation, we denote $(\boldsymbol{Q}^*,\boldsymbol{t}^*,\boldsymbol{\ell}^*)$ as the solution to problem (14.14) under given λ and μ, while $(\boldsymbol{Q}^{\text{opt}},\boldsymbol{t}^{\text{opt}},\boldsymbol{\ell}^{\text{opt}})$ denotes the primary solution to $(\mathcal{P}1.1)$ (or, equivalently, Problem 1) and $(\lambda^{\text{opt}},\mu^{\text{opt}})$ denotes the optimal dual solution to problem $(\mathcal{D}1.1)$.

Evaluating the Dual Function $g(\lambda,\mu)$. Under any given λ and μ, problem (14.14) can be decomposed into $(K+1)$ subproblems, one for optimizing $\boldsymbol{Q}$ and the other K for optimizing t_i and ℓ_i. In particular, we have

$$\begin{aligned}&\underset{\boldsymbol{Q}}{\text{minimize}}\quad \text{tr}(\boldsymbol{Q}\boldsymbol{F}(\lambda))\\ &\text{subject to}\quad \boldsymbol{Q}\succeq\boldsymbol{0}\end{aligned} \tag{14.16}$$

and

$$\begin{aligned}&\underset{t_i,\ell_i}{\text{minimize}}\quad \alpha\ell_i+\frac{\lambda_i\zeta_i C_i^3(R_i-\ell_i)^3}{T^2}+\frac{\lambda_i t_i}{\tilde{g}_i}\beta\left(\frac{\ell_i}{t_i}\right)+\lambda_i p_{c,i}t_i+\mu t_i\\ &\text{subject to}\quad 0\leq\ell_i\leq R_i,\quad t_i\geq 0,\end{aligned} \tag{14.17}$$

where the ith subproblem in (14.17) is for user i. Under the condition of $\boldsymbol{F}(\lambda)\succeq\boldsymbol{0}$, it is evident that the optimal value of (14.16) is zero and the optimal solution $\boldsymbol{Q}^*$ to (14.16) can be any positive semi-definite matrix lying in the null space of $\boldsymbol{F}(\lambda)$. Here, we simply set $\boldsymbol{Q}^*=\boldsymbol{0}$ to obtain the dual function $g(\lambda,\mu)$, which is only for the purpose of computing the optimal dual solution. Note that $\boldsymbol{Q}^*=\boldsymbol{0}$ is not a unique solution to (14.16) when $\boldsymbol{F}(\lambda)$ is rank-deficient, i.e. $\text{rank}(\boldsymbol{F}(\lambda))<N$, and it is not the primary solution to $(\mathcal{P}1.1)$ since it violates the energy harvesting constraints in C6. We will show how to retrieve the primary solution of $\boldsymbol{Q}^{\text{opt}}$ to $(\mathcal{P}1.1)$ later.

For the ith subproblem in (14.17), it is convex and satisfies Slater's condition. Based on the Karush–Kuhn–Tucker (KKT) conditions [39], one can obtain a semi-closed solution (t_i^*, ℓ_i^*) to (14.17), as stated formally below.

Lemma 14.2 For a given $\lambda_i \geq 0$ and $\mu \geq 0$, the optimal solution to (14.17) can be obtained as follows.

- If $\lambda_i = 0$, we have $\ell_i^* = 0$ and $t_i^* = 0$;
- If $\lambda_i > 0$, we have

$$\begin{cases} \ell_i^* = \left[R_i - \sqrt{\frac{T^2}{3\zeta_i C_i^3}\left(\frac{\alpha}{\lambda_i} + \frac{\sigma^2 \ln 2}{B\tilde{g}_i} 2^{\frac{r_i}{B}}\right)} \right]^+, \\ t_i^* = \ell_i^*/r_i, \end{cases} \tag{14.18}$$

where $r_i \triangleq \frac{B}{\ln 2}\left(W_0\left(\frac{\tilde{g}_i}{\sigma^2 e}\left(\frac{\mu}{\lambda_i} + p_{c,i}\right) - \frac{1}{e}\right) + 1\right)$ denotes the offloading rate of user i, $W_0(x)$ is the principal branch of the Lambert W function defined as the solution for $W_0(x)e^{W_0(x)} = x$ [37], e is the base of the natural logarithm, and $[x]^+ \triangleq \max(x, 0)$.

By combining Lemma 14.2 and $\boldsymbol{Q}^* = \boldsymbol{0}$, the dual function $g(\lambda, \mu)$ can be evaluated for any (λ, μ).

Obtaining the Optimal λ^{opt} and μ^{opt} to Maximize $g(\lambda, \mu)$. Note that the dual function $g(\lambda, \mu)$ is concave but non-differentiable in general. As a result, we use subgradient-based methods, such as the ellipsoid method [39], to obtain the optimal λ^{opt} and μ^{opt} for ($\mathcal{D}$1.1). To this end, we generate a sequence of ellipsoids of decreasing volumes, each of which is guaranteed to contain the optimal λ^{opt} and μ^{opt}, until the objective value of ($\mathcal{D}$1.1) converges. Per iteration, if λ and μ are feasible to ($\mathcal{D}$1.1), then we form a new ellipsoid using the subgradient of the objective function; otherwise, we form a new one using the subgradient of the violated constraint function. Therefore, it remains to determine the subgradients of both the objective function in ($\mathcal{D}$1.1) and the constraints in C9–C11. With $(\boldsymbol{t}^*, \boldsymbol{\ell}^*)$, one subgradient for the objective function $g(\lambda, \mu)$ in ($\mathcal{D}$1.1) is given by

$$\left[\frac{\zeta_1 C_1^3 (R_1 - \ell_1^*)^3}{T^2} + \frac{t_1^*}{\tilde{g}_1}\beta\left(\frac{\ell_1^*}{t_1^*}\right) + p_{c,1}t_1^*, \ldots, \right.$$
$$\left. \frac{\zeta_K C_K^3 (R_K - \ell_K^*)^3}{T^2} + \frac{t_K^*}{\tilde{g}_K}\beta\left(\frac{\ell_K^*}{t_K^*}\right) + p_{c,K}t_K^*, \sum_{i=1}^{K} t_i^* - T \right]^\dagger, \tag{14.19}$$

where the first K components and the last one of (14.19) correspond to the first-order derivatives of $g(\boldsymbol{\lambda}, \mu)$ with respect to $\boldsymbol{\lambda}$ and μ, respectively. For the positive semi-definite constraint $\boldsymbol{F}(\boldsymbol{\lambda}) \succeq \boldsymbol{0}$, we establish the following lemma.

Lemma 14.3 Let $\boldsymbol{v} \in \mathbb{C}^{N\times 1}$ be the eigenvector corresponding to the smallest eigenvalue of $\boldsymbol{F}(\boldsymbol{\lambda})$, i.e. $\boldsymbol{v} = \arg\min_{\|\xi\|=1} \xi^H \boldsymbol{F}(\boldsymbol{\lambda})\xi$. Then the constraint $\boldsymbol{F}(\boldsymbol{\lambda}) \succeq \boldsymbol{0}$ is equivalent to the constraint of $\boldsymbol{v}^H \boldsymbol{F}(\boldsymbol{\lambda})\boldsymbol{v} \geq 0$. In this case, the subgradient of $\boldsymbol{v}^H \boldsymbol{F}(\boldsymbol{\lambda})\boldsymbol{v}$ at the given $\boldsymbol{\lambda}$ and μ is

$$[\eta \boldsymbol{v}^H \boldsymbol{H}_1 \boldsymbol{v}, \dots, \eta \boldsymbol{v}^H \boldsymbol{H}_K \boldsymbol{v}, 0]^\dagger. \tag{14.20}$$

Furthermore, the subgradient of the ith constraint in C10 is given by the elementary vector $\boldsymbol{e}_i$ in $\mathbb{R}^{(K+1)\times 1}$ (i.e., $\boldsymbol{e}_i$ is of all zero entries except for the ith entry being one), while that of C11 is $\boldsymbol{e}_{K+1}$. With (14.19), (14.20), and the subgradients of C10 and C11, we can then apply the ellipsoid method to efficiently update $\boldsymbol{\lambda}$ and μ towards the optimal $\boldsymbol{\lambda}^{\text{opt}}$ and μ^{opt} for $(\mathcal{D}1.1)$.

Finding the Optimal Primary Solution to $(\mathcal{P}1)$. With $\boldsymbol{\lambda}^{\text{opt}}$ and μ^{opt} obtained, it remains to determine the optimal primary solution to $(\mathcal{P}1.1)$ (or equivalently to Problem 1). Specifically, by replacing $\boldsymbol{\lambda}$ and μ with $\boldsymbol{\lambda}^{\text{opt}}$ and μ^{opt} in Lemma 14.2, respectively, one obtains the optimal $(\boldsymbol{t}^{\text{opt}}, \boldsymbol{\ell}^{\text{opt}})$ for Problem 1 in a semi-closed form. Based on $(\boldsymbol{t}^{\text{opt}}, \boldsymbol{\ell}^{\text{opt}})$, one can then obtain the optimal local CPU frequencies $\{f_{i,n}^{\text{opt}}\}$ per user and the optimal transmit covariance matrix $\boldsymbol{Q}^{\text{opt}}$ for WPT. We then establish the following proposition.

Proposition 14.1 The optimal solution $(\{f_{i,n}^{\text{opt}}\}, \boldsymbol{Q}^{\text{opt}}, \boldsymbol{t}^{\text{opt}}, \boldsymbol{\ell}^{\text{opt}})$ for Problem 1 is given by

$$\ell_i^{\text{opt}} = \begin{cases} \left[R_i - \sqrt{\dfrac{T^2}{3\zeta_i C_i^3}\left(\dfrac{\alpha}{\lambda_i^{\text{opt}}} + \dfrac{\sigma^2 \ln 2}{B\tilde{g}_i} 2^{\frac{r_i^{\text{opt}}}{B}}\right)}\right]^+, & \lambda_i^{\text{opt}} > 0, \\ 0, & \lambda_i^{\text{opt}} = 0, \end{cases} \tag{14.21}$$

$$t_i^{\text{opt}} = \begin{cases} \ell_i^{\text{opt}} / r_i^{\text{opt}}, & \lambda_i^{\text{opt}} > 0, \\ 0, & \lambda_i^{\text{opt}} = 0, \end{cases} \tag{14.22}$$

$$f_{i,1}^{\text{opt}} = \dots = f_{i,C_i(R_i - \ell_i^{\text{opt}})}^{\text{opt}} = C_i(R_i - \ell_i^{\text{opt}})/T, \forall i \in \mathcal{K}, \tag{14.23}$$

and

$$\begin{aligned} \boldsymbol{Q}^{\text{opt}} &= \arg \underset{\boldsymbol{Q}\succeq \boldsymbol{0}}{\text{minimize}} \; T\text{tr}(\boldsymbol{Q}) \\ \text{subject to} \; & \frac{\zeta_i C_i^3 (R_i - \ell_i^{\text{opt}})^3}{T^2} + \frac{t_i^{\text{opt}}}{\tilde{g}_i}\beta\left(\frac{\ell_i^{\text{opt}}}{t_i^{\text{opt}}}\right) + p_{c,i} t_i^{\text{opt}} \leq T\eta \text{tr}(\boldsymbol{Q}\boldsymbol{H}_i), \; \forall i \in \mathcal{K}, \end{aligned} \tag{14.24}$$

where

$$r_i^{\text{opt}} \triangleq \frac{B}{\ln 2}\left(W_0\left(\frac{\tilde{g}_i}{\sigma^2 e}\left(\frac{\mu^{\text{opt}}}{\lambda_i^{\text{opt}}}+p_{c,i}\right)-\frac{1}{e}\right)+1\right) \tag{14.25}$$

corresponds to the offloading rate for user i, $\forall i \in \mathcal{K}$.

Proposition 14.1 can be verified by simply combining Lemmas 14.2 and 14.3. Note that the problem in (14.24) is a standard semi-definite program (SDP), which can thus be efficiently solved by off-the-shelf solvers, e.g., CVX [40].

Remark 14.1 Proposition 14.1 shows that the optimal joint computing and offloading design has the following interesting properties to minimize the energy consumption at the AP.

- First, if the energy harvesting constraint is not tight for user i (i.e., user i harvests sufficient wireless energy), then no computation offloading is required and user i should compute all the tasks locally (i.e., $\ell_i^{\text{opt}} = 0$). This can be explained based on the complementary slackness condition [38]. This property is intuitive: when the user has sufficient energy to accomplish the tasks locally, there is no need to employ computation offloading that incurs additional energy consumption for the MEC server's computation at the AP.
- Next, it is always beneficial to leave some bits for local computing at each user $i \in \mathcal{K}$, i.e. $\ell_i^{\text{opt}} < R_i$ always holds (see (14.21)). In other words, offloading all the bits to the AP is always suboptimal. This is because when $\ell_i^{\text{opt}} \rightarrow R_i$, the marginal energy consumption of local computing is almost zero, and thus it is beneficial to leave some bits for local computing in this case.
- Furthermore, it is observed that for each user i, the more stringent the energy harvesting constraint is (or the associated dual variable λ_i^{opt} is larger), the more bits should be offloaded to the AP with a smaller offloading rate r_i^{opt}. This property follows based on (14.21) and (14.25), in which a larger λ_i^{opt} admits a larger ℓ_i^{opt} and a smaller r_i^{opt}.
- Finally, the number of offloaded bits ℓ_i^{opt} and the offloading rate r_i^{opt} for each user i are affected by the channel gain $\tilde{g}_i$, the block duration T, the circuit power $p_{c,i}$, and the MEC energy consumption α per offloaded bit in the following way: (i) when the channel condition becomes better (i.e., $\tilde{g}_i$ becomes larger), both ℓ_i^{opt} and r_i^{opt} increase, and thus user i is likely to offload more bits with a higher offloading rate, (ii) a higher circuit power $p_{c,i}$ at the user leads to a higher offloading rate r_i^{opt}, and (iii) when T or α increases, ℓ_i^{opt} reduces and thus fewer bits are offloaded to the AP.

14.4 Numerical Results

In this section we provide numerical results to illustrate the performance of the proposed optimal joint MEC-WPT design. For comparison, we consider the following four baseline schemes:

- *Local computing only*: In this approach, each user $i \in \mathcal{K}$ accomplishes its computation task by only local computing, i.e. by setting $\ell_i = 0, \forall i \in \mathcal{K}$. The CPU frequencies at each user i are obtained as $f_i = C_i R_i / T_i$; the transmit energy covariance matrix $\boldsymbol{Q}^{\text{opt}}$ is obtained by solving problem (14.24) with $t_i^{\text{opt}} = 0$ and $\ell_i^{\text{opt}} = 0, \forall i \in \mathcal{K}$.
- *Computation offloading only*: In this approach, each user $i \in \mathcal{K}$ accomplishes its computation task by fully offloading to the AP, i.e. by setting $\ell_i = R_i$, $\forall i \in \mathcal{K}$. The CPU frequencies at all the users are set as zero; the allocated time slot t_i and the transmit covariance matrix $\boldsymbol{Q}^{\text{opt}}$ are determined by solving Problem 1 with $\ell_i^{\text{opt}} = R_i, \forall i \in \mathcal{K}$.
- *Joint design with isotropic WPT*: The N-antenna AP radiates the RF energy isotropically or omni-directionally over all directions by setting $\boldsymbol{Q} = p\boldsymbol{I}$, where $p \geq 0$ denotes the transmit power at each antenna. This scheme corresponds to solving Problem 1 by replacing $\boldsymbol{Q}$ as $p\boldsymbol{I}$ with p being another optimization variable.
- *Separate MEC-WPT design*: This scheme separately designs the computation offloading for MEC and the energy beamforming for WPT [19, 29]. First, the K users minimize their sum-energy consumption subject to the users' individual computation latency constraints [29]. Then, under the constraints of energy demand at the K users, the AP designs the transmit energy beamforming with minimum energy consumption [19].

In the simulations, the EH efficiency is set as $\eta = 0.3$. The system parameters are set as follows unless stated otherwise. The number of AP antennas is $N = 4$, $C_i = 10^3$ cycles/bit, $\zeta_i = 10^{-28}$, $\forall i \in \mathcal{K}$ [29], the circuit power is $p_{c,i} = 10^{-4}$ W, the energy consumption per offloaded bit by the MEC server is $\alpha = 10^{-4}$ J/bit, the receiver noise power is $\sigma^2 = 10^{-9}$ W, and the spectrum bandwidth for offloading is $B = 2$ MHz. The wireless channel from the AP to each user $i \in \mathcal{K}$ is modeled as an independent Rayleigh fading random vector with an average power loss of 5×10^{-6} (i.e., −53 dB), which corresponds to a distance of about 5 m from users to the AP in urban environment. The numerical results are obtained by averaging over 500 randomized channel realizations. Note that the simulation parameters are specifically chosen, but our approaches can be also applied to other system setups. Figures 14.2 and 14.3 show the average energy consumption at the AP under different system parameters. It is observed that the proposed joint design achieves the lowest average energy consumption at the AP among all the five schemes.

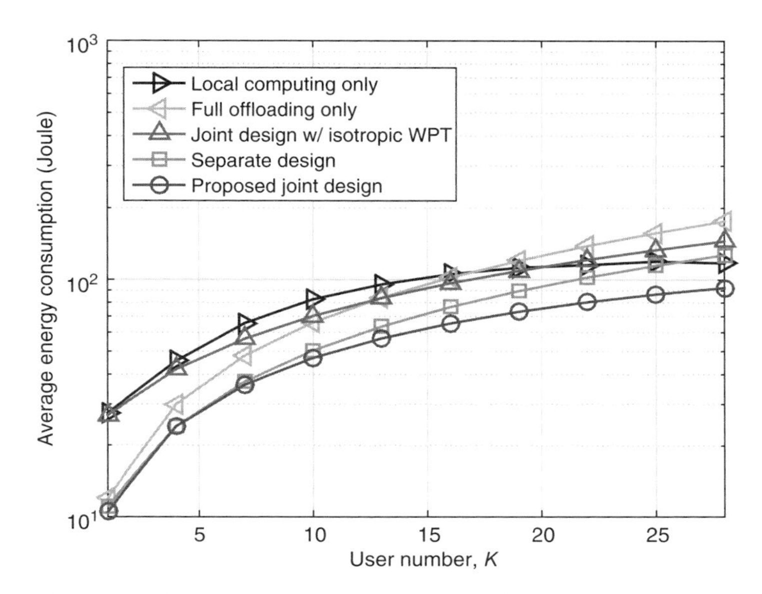

Figure 14.2 The average energy consumption at the AP versus user number K.

Figure 14.2 shows the average energy consumption versus the user number K, where $R = 10$ kbits and $T = 0.5$ s. It is shown that the gain achieved by the proposed joint design becomes more significant as the user number K becomes large. The full-offloading-only scheme outperforms the local-computing-only scheme, but with diminishing gain as K increases. This is because in the full-offloading-only scheme all users share the finite block and the offloading energy consumption would increase drastically when K becomes large.

Figure 14.3 shows the average energy consumption at the AP versus the block duration T, where $R = 10$ kbits and $K = 10$. First, with a small value of T (e.g., $T = 0.05$ s), the separate-design benchmark scheme is observed to achieve a near optimal performance obtained by the proposed joint design, while as T increases, the energy consumption with the local-computing-only scheme significantly decreases, approaching that with the proposed joint design. It is also observed that the energy consumption with the full-offloading-only scheme remains almost unchanged when $T \geq 0.1$ s. This is due to the fact that in this case the optimal offloading time for all users is fixed to be around 0.1 s for saving the circuit energy consumption, hence increasing T cannot further improve the energy efficiency. By contrast, the energy consumption with the local-computing-only scheme decreases monotonically as T increases. This is because as T increases, one can always lower the CPU frequency to save energy for local computing.

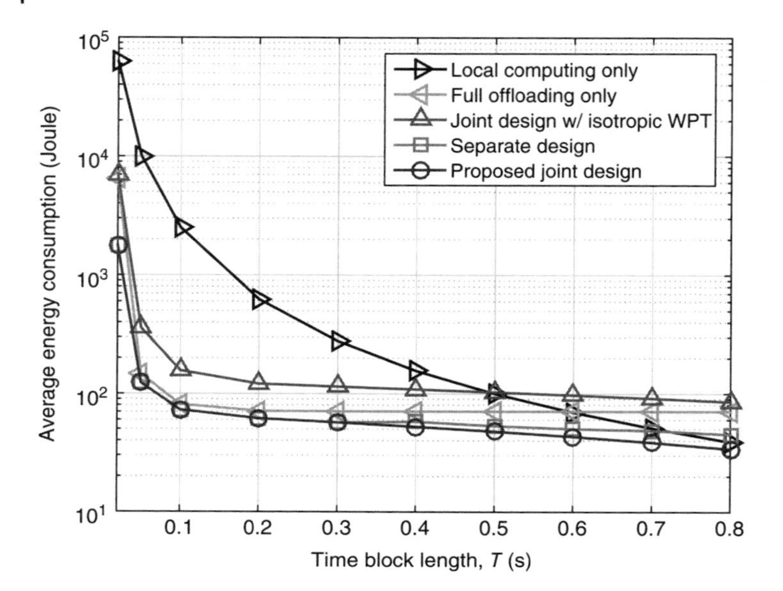

Figure 14.3 The average energy consumption at the AP versus block duration T.

14.5 Conclusion

We developed a unified MEC-WPT design framework with joint energy beamforming, offloading, and computing optimization in wireless-powered multi-user MEC systems. In particular, we pursue an energy-efficient design for the AP to minimize the total energy consumption subject to the users' individual computation latency constrains. Leveraging the Lagrange dual method, we obtained the optimal solution in a semi-closed form. Numerical results demonstrated the merits of the proposed joint design over alternative benchmark schemes. The proposed joint MEC-WPT design can pave the way to facilitate ubiquitous and self-sustainable computing for IoT devices.

Bibliography

1 F. Wang, J. Xu, X. Wang, and S. Cui (2017) Joint offloading and computing optimization in wireless powered mobile-edge computing systems. *IEEE Trans. Wireless Commun.*, accepted. [Online]. Available: https://arxiv.org/abs/1702.00606.

2 M. Chiang and T. Zhang (2016) Fog and IoT: An overview of research opportunities. *IEEE Internet Thing J.*, **3** (6): 854–864.

3 Y. Mao, C. You, J. Zhang, K. Huang, and K.B. Letaief (2017) A survey on mobile edge computing: The communication perspective. *IEEE Commun. Survey Tutorials*, **19** (4): 2322–2358.

4 S. Barbarossa, S. Sardellitti, and P.D. Lorenzo (2014) Communicating while computing: Distributed mobile cloud computing over 5G heterogenous networks. *IEEE Signal Process. Mag.*, **31** (6): 45–55.

5 E. Cuervo, A. Balasubramanian, D. Cho, A. Wolman, S. Saroiu, R. Chandra, and P. Bahl (2010) MAUI: Making smartphones last longer with code offload. In *Proceedings of the ACM MobiSys*, San Francisco, CA, pp. 49–62.

6 S. Kosta, A. Aucinas, P. Hui, R. Mortier, and X. Zhang (2012) ThinkAir: Dynamic resource allocation and parallel execution in the cloud for mobile code offloading. In *Proceedings of the IEEE INFOCOM*, Orlando, pp. 945–953.

7 R. Zhang and C.K. Ho (2013) MIMO broadcasting for simultaneous wireless information and power transfer. *IEEE Trans. Wireless Commun.*, **12** (5): 1989–2001.

8 D.W.K. Ng, E.S. Lo, and R. Schober (2014) Robust beamforming for secure communication in systems with wireless power transfer. *IEEE Trans. Wireless Commun.*, **13** (8): 4599–4615.

9 S. Timotheou, I. Krikidis, G. Zheng, and B. Ottersten (2014) Beamforming for MISO interference channels with QoS and RF energy transfer. *IEEE Trans. Wireless Commun.*, **13** (5): 2646–2658.

10 S. Bi, R. Zhang, and C. Ho (2015) Wireless powered communication: Opportunities and challenges. *IEEE Commun. Mag.*, **53** (4): 117–125.

11 H. Li, J. Xu, R. Zhang, and S. Cui (2015) A general utility optimization framework for energy harvesting based wireless communications. *IEEE Commun. Mag.*, **53** (4): 79–85.

12 J. Xu, L. Liu, and R. Zhang (2014) Multiuser MISO beamforming for simultaneous wireless information and power transfer. *IEEE Trans. Signal Process.*, **62** (18): 4798–4810.

13 F. Wang, T. Peng, Y. Huang, and X. Wang (2015) Robust transceiver optimization for power-splitting based downlink MISO SWIPT systems. *IEEE Signal Process. Lett.*, **22** (9): 1492–1496.

14 F. Wang, C. Xu, Y. Huang, X. Wang, and X.-Q. Gao (2017) REEL-BF design: Achieving the SDP bound for downlink beamforming with arbitrary shaping constraints. *IEEE Trans. Signal Process.*, **65** (10): 2672–2685.

15 Y. Zeng and R. Zhang (2015) Optimized training design for wireless energy transfer. *IEEE Trans. Commun.*, **63** (2): 536–550.

16 J. Xu and R. Zhang (2014) Energy beamforming with one-bit feedback. *IEEE Trans. Signal Process.*, **62** (20): 5370–5381.

17 J. Xu and R. Zhang (2016) A general design framework for MIMO wireless energy transfer with limited feedback. *IEEE Trans. Signal Process.*, **64** (10): 2475–2488.

18 C. Valenta and G. Durgin (2014) Harvesting wireless power: Survey of energy-harvester conversion efficiency in far-field, wireless power transfer systems. *IEEE Microwave Mag.*, **15** (4): 108–120.
19 Y. Zeng, B. Clerckx, and R. Zhang (2017) Communications and signals design for wireless power transmission. *IEEE Trans. Commun.*, **65** (5): 2264–2290.
20 S. Lee and R. Zhang (2017) Distributed wireless power transfer with energy feedback. *IEEE Trans. Signal Process.*, **65** (7): 1685–1699.
21 E. Boshkovska, D.W.K. Ng, N. Zlatanov, and R. Schober (2015) Practical non-linear energy harvesting model and resource allocation for SWIPT systems. *IEEE Commun. Lett.*, **19** (12): 2082–2085.
22 B. Clerckx and E. Bayguzina (2016) Waveform design for wireless power transfer. *IEEE Trans. Signal Process.*, **64** (23): 6313–6328.
23 F. Liu, P. Shu, H. Jin, L. Ding, J. Yu, D. Niu, and B. Li (2013) Gearing resource-poor mobile devices with powerful clouds: Architectures, challenges, and applications. *IEEE Wireless Commun.*, **20** (3): 14–22.
24 J. Liu, Y. Mao, J. Zhang, and K.B. Letaief (2016) Delay-optimal computation task scheduling for mobile-edge computing systems. In *Proceedings of the IEEE ISIT*, Barcelona, Spain, pp. 1451–1455.
25 D. Huang, P. Wang, and D. Niyato (2012) A dynamic offloading algorithm for mobile computing. *IEEE Trans. Wireless Commun.*, **11** (6): 1991–1995.
26 O. Muñoz, A. Pascual-Iserte, and J. Vidal (2015) Optimization of radio and computational resources for energy efficiency in latency-constrained application offloading. *IEEE Trans. Veh. Technol.*, **64** (10): 4738–4755.
27 M.-H. Chen, B. Liang, and M. Dong (2016) Joint offloading decision and resource allocation for multi-user multi-task mobile cloud. In *Proceedings of the IEEE ICC*, Kuala Lumpur, Malaysia, pp. 1–6.
28 X. Chen, L. Jiao, W. Li, and X. Fu (20160 Efficient multi-user computation offloading for mobile-edge cloud computing. IEEE/ACM Trans. Network., **24** (5): 2795–2808.
29 C. You, K. Huang, H. Chae, and B. Kim (2017) Energy-efficient resource allocation for mobile-edge computation offloading. *IEEE Trans. Wireless Commun.*, **16** (3): 1397–1411.
30 S. Sardellitti, G. Scutari, and S. Barbarossa (2015) Joint optimization of radio and computational resources for multicell mobile-edge computing. *IEEE Trans. Signal Inf. Process. Network.*, **1** (2): 89–103.
31 F. Wang, J. Xu, and Z. Ding (2017) Optimized multiuser computation offloading with multi-antenna NOMA. In *Proceedings of the IEEE GLOBECOM Workshop*, Singapore, pp. 1–7.
32 X. Cao, F. Wang, J. Xu, R. Zhang, and S. Cui (2017) Joint computation and communication cooperation for mobile edge computing. [Online]. Available: https://arxiv.org/abs/1704.06777.

33 C. You, K. Huang, and H. Chae (2016) Energy efficient mobile cloud computing powered by wireless energy transfer. *IEEE J. Selected Areas Commun.*, **34** (5): 1757–1770.
34 F. Wang (2017) Computation rate maximization for wirless powered mobile edge computing. In *Proceedigns of the APCC*, Perth, Australia, pp. 1–6.
35 S. Bi and Y. Zhang (2017) Computation rate maximization for wireless powered mobile-edge computing with binary computation offloading. [Online]. Available: https://arxiv.org/abs/1708.08810.
36 T.D. Burd and R.W. Brodersen (1996) Processor design for portable systems. *Kluwer J. VLSI Signal Process. Syst.*, **13** (2): 203–221.
37 R. Corless, G. Gonnet, D. Hare, D. Jeffrey, and D. Knuth (1996) On the Lambert *W* function. *Adv. Comput. Math.*, **5** (1): 329–359.
38 S. Boyd and L. Vandenberghe (2004) *Convex Optimization*. Cambridge University Press, Cambridge.
39 S. Boyd. Ellipsoid method notes, EE364b Lectures, Stanford University, California. [Online]. Available: http://stanford.edu/class/ee364b/lectures/ellipsoid_method_notes.pdf.
40 M. Grant, S. Boyd, and Y. Ye (2009) CVX: Matlab software for disciplined convex programming. [Online]. Available: http://cvxr.com/cvx/.

15

Wireless Power Transfer: A Macroscopic Approach

Constantinos Psomas and Ioannis Krikidis*

KIOS Research and Innovation Center of Excellence, Department of Electrical and Computer Engineering, University of Cyprus, Cyprus

Large-scale wireless networks are characterized by the existence of multi-user interference due to the concurrent transmission of the network's terminals. In conventional networks, i.e. networks where the main focus is information transfer, interference is a critical degrading factor of a terminal's performance. On the other hand, in networks where the main focus is wireless power transfer (WPT), interference is beneficial as it increases the harvesting efficiency. However, in both cases all wireless links are influenced by path-loss effects due to the distances between the terminals. As such, the terminals' deployment is another important element which affects their performance.

Therefore, a reasonable approach is to study radio frequency (RF) energy harvesting in large-scale networks by taking into account spatial randomness. The modeling of the geometrical characteristics of large-scale networks can be achieved with the employment of stochastic geometry, a suitable mathematical tool for studying spatial point processes [1]. Specifically, this chapter presents three sections, each dealing with a different wireless-powered communication scenario:

- Section 15.1 focuses on a wireless-powered cooperative network where the relays harvest energy from the source in order to convey its transmitted information
- Section 15.2 deals with a simultaneous wireless information and power transfer (SWIPT) scenario, where terminals employ successive interference cancellation (SIC) techniques to boost the harvesting efficiency
- Section 15.3 studies a wireless-powered communication network (WPCN) implementing a wireless-powered opportunistic beamforming (OBF) protocol.

*Corresponding author: Constantinos Psomas; psomas@ucy.ac.cy

Wireless Information and Power Transfer: Theory and Practice, First Edition.
Edited by Derrick Wing Kwan Ng, Trung Q. Duong, Caijun Zhong, and Robert Schober.

In each section closed-form mathematical expressions are provided for the considered performance metric of each scenario as well as numerical results with useful insights regarding the network's performance.

15.1 Wireless-Powered Cooperative Networks with Energy Storage

This section deals with relay selection in WPCNs, where spatially random relays are equipped with energy storage devices, e.g., batteries. In contrast to conventional techniques and to reduce complexity, the relays either harvest energy from the source signal (in the case of an uncharged battery) or attempt to decode and forward it (in the case of a charged battery). Several relay selection schemes that correspond to different state information requirements and implementation complexities are proposed. It is shown that energy storage significantly affects the system's performance and results in a zero diversity gain at high signal-to-noise ratios (SNRs). Moreover, the outage probability floors depend on the steady-state distribution of the battery.

15.1.1 System Model

Network and channel model. Consider a single cell where an access point (AP) communicates with a destination D via the help of a set of relay nodes R_i. The cell is modeled by a disc, denoted by $\mathcal{D}$, with radius ρ. The AP is located at its origin and the relay nodes R_i form a homogeneous Poisson point process (PPP) Φ of density λ inside $\mathcal{D}$; N denotes the number of relays. The distance between the AP and D is denoted by d_0 and no direct link between them exists, e.g., due to severe shadowing. Moreover, d_i denotes the distance between the AP and the ith relay R_i, while c_i denotes the distance between R_i and D, $1 \leq i \leq N$. All nodes are equipped with a single antenna and the relays have WPT capabilities. The AP transmits with power P and a spectral efficiency r_0 bits-per-channel-use (BPCU); the AP's transmitted signal is the only WPT source for the relays. Figure 15.1 illustrates the system model.

All wireless links suffer from both small-scale block fading and large-scale path-loss effects. The fading is Rayleigh distributed so the power of the channel fading is an exponential random variable with unit variance.[1] We denote by h_i and g_i the channel coefficients for the links between the AP and R_i, and R_i and D, respectively. The path-loss model assumes that the received power is proportional to $(1 + d^{\alpha})^{-1}$ where d is the Euclidean distance between the AP and a receiver, $\alpha > 2$ denotes the path-loss exponent, and we define

1 Rayleigh is used for simplicity throughout this chapter, but other channel models such as Rice or Nakagami could also be considered.

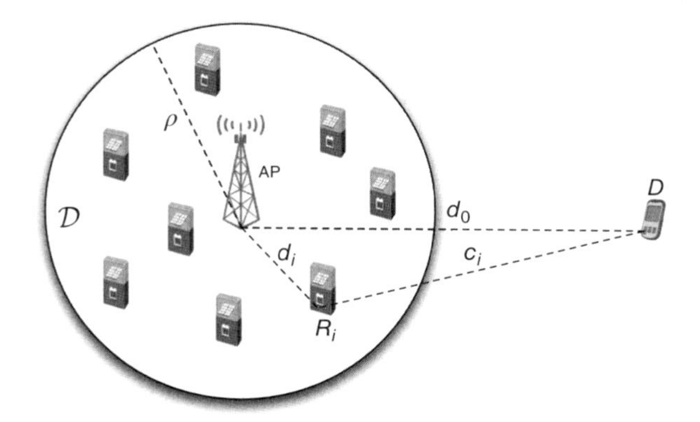

Figure 15.1 The considered system model. ©2015 IEEE. Reprinted, with permission, from [2].

$\delta \triangleq 2/\alpha$. The instantaneous fading channels are known only at the receivers, except if otherwise defined. In addition, all wireless links exhibit additive white Gaussian noise (AWGN) with variance σ^2; n_i denotes the AWGN at the ith node.

Cooperative protocol. The relay nodes have half-duplex capabilities and employ a decode-and-forward protocol. Time is slotted and communication is performed in two orthogonal time slots. In the first time slot (broadcast phase), the AP broadcasts the signal to the relays. In the second time slot, a relay (or a group of relays), which successfully decoded the source signal, forwards the signal to D according to the rules of the considered relay selection scheme. The selected relay transmits at a rate r_0 (same as rate of the source).

Battery model. Each relay node is equipped with a single energy storage device (e.g., battery, capacitor). At the beginning of the broadcast phase, the battery can be fully charged or empty [3]. A relay with a charged battery is active and participates in the relaying operation, while a relay with an empty battery is in harvesting mode and exploits the AP's signal for WPT purposes. An empty battery is fully charged if the input power is larger than the battery's size. Let $P_r = \Psi P$ be the battery's size/capacity with $\Psi < 1$. If the decoding is successful, the relay transmits the decoded message with power P_r. Otherwise, the available energy is used to operate the receiver's basic components (e.g., RF electronics, signal processing, etc.), transmit a negative-acknowledgment signal to indicate unsuccessful decoding, and for static maintenance operations (e.g., cooling system).

The charging/discharging behavior of the battery can be represented by a finite-state Markov chain (MC) with two states $\{s_0, s_1\}$, where the states s_0 and s_1 indicate that the battery is empty and charged, respectively. Then, the state-transition probability matrix is $\Pi = \begin{bmatrix} 1-\pi_0 & \pi_0 \\ \pi_1 & 1-\pi_1 \end{bmatrix}$, where π_0 denotes the probability that the input power is greater than the battery size of the relay, and π_1 is the probability that a charged relay node is selected for relaying. If $\eta = [\eta_0\ \eta_1]$ is the stationary steady-state probability vector of the MC, we have $\eta\Pi = \eta$. The solution of this system of linear equations gives the battery's steady-state distribution, which is $\eta_0 = \frac{\pi_1}{\pi_0+\pi_1}$ and $\eta_1 = \frac{\pi_0}{\pi_0+\pi_1}$ [2].

15.1.2 Relay Selection Schemes

Several relay selection schemes are now presented and the outage probability of each scheme is provided. Denote by $C(x) = \frac{1}{2}\log_2(1+x)$ the instantaneous capacity for a wireless link (one-hop transmission) with SNR x, and the associated outage probability is given by $\mathbb{P}\{C(x) < r_0\} = \mathbb{P}\{x < \epsilon\}$ where $\epsilon \triangleq 2^{2r_0} - 1$; to simplify the notation, we define $\Xi \triangleq \epsilon\sigma^2/P$.

Random relay selection. The random relay selection (RRS) scheme does not require any feedback regarding the battery status or the location of the relays and selects one at random. It corresponds to a low implementation complexity and is suitable for networks with strict power/bandwidth constraints. Without loss of generality, consider that the ith relay is selected to assist the source. If the ith relay is fully charged, it attempts to decode the source signal and, if successful, forwards it. If the relay's battery is empty, it switches to harvesting mode and uses the received signal for WPT purposes; in this case, the relay is inactive and an outage event occurs. On the other hand, the non-selected relays with empty batteries are in harvesting mode and use the source signal for potential charging. During the first time slot, the received signal at the ith relay can be written as $y_i = \sqrt{P}\frac{h_i}{\sqrt{1+d_i^\alpha}}s + n_i$, where s denotes the source signal with normalized power. If the ith relay is active, the received SNR is $\text{SNR}_i = \frac{P|h_i|^2}{(1+d_i^\alpha)\sigma^2}$. If the ith relay is inactive, the input at the WPT device is equal to $P_h = \zeta\frac{P|h_i|^2}{(1+d_i^\alpha)}$, where ζ is the WPT conversion efficiency; it is assumed that energy harvesting from AWGN is negligible and that $\zeta = 1$. In the second time slot, the signal at D can be written as $y_D = \sqrt{P_r}\frac{g_i}{\sqrt{1+c_i^\alpha}}s + n_D$ and the SNR is $\text{SNR}_D = \frac{P_r|g_i|^2}{(1+c_i^\alpha)\sigma^2}$.

The steady-state probability that a relay node is charged at the beginning of the broadcast phase, for the RRS scheme, is $\eta_1^{RRS} = \frac{\delta\exp(-\Psi)\frac{\gamma(\delta,\Psi\rho^\alpha)}{\Psi\delta}}{\delta\exp(-\Psi)\frac{\gamma(\delta,\Psi\rho^\alpha)}{\Psi\delta}+\frac{1}{\lambda\pi}}$ [2]. Hence, for highly dense networks, i.e. $\lambda \to \infty$, we have $\eta_1^{RRS} \to 1$. This was expected, since as the density of the network increases, the relay selection probability approaches zero ($1/N \to 0$) and so the relay nodes are most of the time in harvesting mode. On the other hand, η_1^{RRS} decreases as Ψ increases, since as the harvesting threshold increases (size of the battery), the probability to have an input power higher than the threshold decreases. The battery status is independent of the spectral efficiency, since uncharged relays observe only the energy content of the received signals, while charged relays are fully discharged in case of selection (independently of the decoding status (success or failure)).

An outage event for the RRS scheme occurs when (i) there are no available relays in $\mathcal{D}$ ($N = 0$), (ii) $\mathcal{D}$ contains at least one relay ($N \geq 1$) but the selected relay's battery is empty, (iii) $N \geq 1$, the selected relay is fully charged but cannot decode the source signal, and (iv) $N \geq 1$, the selected relay is fully charged, decodes the source signal but the destination D cannot support the targeted spectral efficiency. Therefore, the outage probability for the RRS scheme is [2]

$$\Pi_{\mathrm{RRS}} = \exp(-\lambda\pi\rho^2) + (1 - \exp(-\lambda\pi\rho^2))(1 - \eta_1^{\mathrm{RRS}} Q), \tag{15.1}$$

where

$$Q \triangleq \frac{\gamma(\delta, \Xi\rho^\alpha)\Xi^{-\delta}\delta}{\pi\rho^4 \exp\left(\Xi\left(1 + \frac{1}{\Psi}\right)\right)} \int_0^{2\pi}\int_0^{\rho} \frac{x dx d\theta}{\exp\left(\frac{\Xi}{\Psi}(x^2 + d_0^2 - 2xd_0\cos\theta)^{\frac{1}{\delta}}\right)}. \tag{15.2}$$

For the special case with P, $P_r \to \infty$, $\Psi = P_r/P$ (constant ratio), $\rho \ll d_0$, and $N \geq 1$, the outage probability of the RRS scheme is given by

$$\Pi_{RRS}^{\infty} \approx 1 - \eta_1^{RRS}\left(1 - \Xi\left(\frac{1 + d_0^\alpha}{\Psi} + 1\right)\right) \tag{15.3a}$$

$$\to 1 - \eta_1^{RRS}. \tag{15.3b}$$

Hence, the outage probability for the RRS scheme suffers from an outage floor at high SNRs, which depends on the steady-state distribution of the battery.

Relay selection based on the closest distance. The relay selection based on the closest distance (RCS) requires an a priori knowledge of the location of the relay nodes. We assume that the AP monitors the relay locations via a low-rate feedback channel or a global positioning system (GPS) mechanism, and selects its closest relay. The RCS scheme does not take into account battery status and/or instantaneous fading and so corresponds to a low implementation complexity, specifically for scenarios with low mobility. The mathematical description of the RCS scheme follows the one for the RRS scheme. The RRS and RCS schemes select a relay without taking into account the battery status and the selected relay is discharged at the end of the relaying slot independently of its decoding efficiency. As both handle the selected relay in the same way and the probability to select a relay in the RCS scheme is also $\mathbb{E}\{1/N\} \approx 1/\lambda\pi\rho^2$ (a relay can be the closest with the same probability), the steady-state distribution of the RCS scheme is equivalent to the RRS scheme, that is, $\eta_1^{RCS} = \eta_1^{RRS}$.

The scenarios for an outage event follow the discussion for the RRS scheme. Therefore, the outage probability achieved by the RCS scheme is given by [2]

$$\Pi_{RCS} = \exp(-\lambda\pi\rho^2) + (1 - \exp(-\lambda\pi\rho^2))(1 - \eta_1^{\mathrm{RCS}} Q'(\lambda)), \tag{15.4}$$

where

$$Q'(\lambda) \triangleq \frac{2\pi\lambda^2 \exp(-\Xi - \frac{\Xi}{\Psi})}{(1-\exp(-\pi\lambda\rho^2))^2} \int_0^{2\pi}\int_0^{\rho}\int_0^{\rho} \frac{\exp(-\lambda\pi r^2 - \Xi x^\alpha - \lambda\pi x^2) rx dr dx d\theta}{\exp\left(\frac{\Xi}{\Psi}(r^2 + d_0^2 - 2rd_0\cos\theta)^{\frac{1}{\delta}}\right)}. \tag{15.5}$$

For the special case with $P, P_r \to \infty$, $\Psi = P_r/P$ (constant ratio), $\rho \ll d_0$, $N \geq 1$, and $\alpha = 2$, the outage probability of the RCS scheme is given by

$$\Pi_{RCS}^{\infty} \approx 1 - \frac{\eta_1^{RCS}\lambda\pi}{\lambda\pi + \Xi}\left(1 + \frac{\Xi\rho^2\exp(-\lambda\pi\rho^2)}{1-\exp(-\lambda\pi\rho^2)}\right)\left(1 - \Xi\left(\frac{1+d_0^2}{\Psi} + 1\right)\right) \tag{15.6a}$$

$$\to 1 - \eta_1^{RCS}. \tag{15.6b}$$

Therefore, the outage probability of the RCS scheme converges to an outage floor at high SNRs, which depends on the battery's charging behavior. By comparing expressions (15.3a) and (15.6a), it is clear that both schemes converge to the same outage floor and thus become asymptotically equivalent (the convergence floor is independent of α and d_0). However, it can be seen that the RCS scheme converges to the outage floor faster than the RRS scheme.

Random relay selection with battery information. The random relay selection with battery information (RRSB) scheme randomly selects a relay node among the charged relays (if any). The RRSB scheme is based on a priori knowledge of the battery status and requires relays to feed their battery status (1-bit feedback) at the beginning of each broadcast phase. The steady-state probability that a relay node is charged at the beginning of the broadcast phase, for the RRSB scheme, is given by $\eta_1^{RRSB} = 1 - \frac{\Psi^\delta}{\delta\lambda\pi\exp(-\Psi)\gamma(\delta,\Psi\rho^\alpha)}$ [2].

In this case, an outage event occurs when (i) there is no charged relay in the system or (ii) the first or the second hop of the relay transmission is in outage. As such, the outage probability achieved by the RRSB scheme is given by [2]

$$\Pi_{RRSB} = \exp(-\lambda\eta_1^{RRSB}\pi\rho^2) + (1 - \exp(-\lambda\eta_1^{RRSB}\pi\rho^2))(1-Q), \tag{15.7}$$

where Q is the success probability for the relaying link given by (15.2). When $P, P_r \to \infty$, $\Psi = P_r/P$ (constant ratio), we have $Q \to 1$ and so the outage probability is dominated by the event where no relay is fully charged. Therefore, the outage probability asymptotically converges to $\Pi_{\mathrm{RRSB}}^{\infty} \to \exp(-\lambda\eta_1^{\mathrm{RRSB}}\pi\rho^2)$.

Relay selection based on the closest distance with battery information. The relay selection based on the closest distance with battery information (RCSB) scheme follows the principles of the RCS scheme, but considers the battery status of the relay nodes. Specifically, the RCSB scheme selects the closest charged relay according to $R^* = \arg_{R_i \in \Omega} \min_{i=1,\ldots,N'} d_i$. Regarding the steady-state probability of the battery, it follows the discussion of the RRSB scheme. Therefore, the steady-state probability that a relay node is charged at the beginning of the broadcast phase, for the RCSB scheme, is given by $\eta_1^{RCSB} = \eta_1^{RRSB}$. The outage probability achieved by the RCSB scheme is [2]

$$\Pi_{RCSB} = \exp(-\lambda_{\Omega}\pi\rho^2) + (1 - \exp(-\lambda_{\Omega}\pi\rho^2))(1 - Q'(\lambda_{\Omega})), \tag{15.8}$$

where $\lambda_{\Omega} = \lambda\eta_1^{\mathrm{RCSB}}$ and $Q'(\cdot)$ is given by (15.5). When $P, P_r \to \infty$ with $\Psi = P/P_r$ (constant ratio), by using the same arguments as with the RRSB scheme, expression (15.8) asymptotically converges to $\Pi_{\mathrm{RCSB}}^{\infty} \to \exp(-\lambda\eta_1^{\mathrm{RCSB}}\pi\rho^2)$.

Distributed beamforming. The distributed beamforming (DB) scheme selects the charged relays at the beginning of the broadcast phase; it is a useful performance benchmark for the single-relay selection schemes. Specifically, all relays with charged batteries become active and attempt to decode the AP's signal. The ones that succeed form a virtual multiple antenna array and coherently transmit to the destination D. The DB scheme requires perfect time synchronization and signaling between the relays as well as channel state information at the relays. The broadcast phase follows the description of the RRS scheme. The received signal at D during the second phase is $y_D = \sqrt{P_r}\sum_{i\in C}\frac{w_i g_i}{\sqrt{1+c_i^{\alpha}}}s + n_D$, where $w_i = g_i^* / \sqrt{\sum_{i\in C}|g_i|^2}$ is the precoding coefficient at the ith relay that ensures coherent combination of the signals at D and C is the set of relays which participate in the relaying transmission [4]. Therefore, the SNR at D is $\mathrm{SNR}_D = P_r\sum_{i\in C}\frac{|g_i|^2}{(1+c_i^{\alpha})\sigma^2}$. In this case, the steady-state probability that a relay is charged at the beginning of the broadcast phase is $\eta_1^{DB} = \frac{\delta\exp(-\Psi)\frac{\gamma(\delta,\Psi\rho^{\alpha})}{\Psi^{\delta}}}{\delta\exp(-\Psi)\frac{\gamma(\delta,\Psi\rho^{\alpha})}{\Psi^{\delta}}+\rho^2}$ [2], which is independent of the density λ. The charged relays are fully discharged at the end of the relaying time slot and so the network's size does not affect the battery status distribution.

For the DB scheme, an outage event occurs when (i) the relay set C is empty or (ii) when the coherent relaying transmission is in outage, i.e. the destination is not able to decode the relaying signal. For the outage probability performance of the DB scheme, we state the following expression when $\rho \ll d_0$ [2],

$$\Pi_{DB} = \sum_{k=0}^{\infty} \frac{\gamma\left(k, \frac{\Xi(1+d_0^{\alpha})}{\Psi}\right)}{\Gamma(k)} \exp(-\lambda'\pi\rho^2)\frac{(\lambda'\pi\rho^2)^k}{k!}, \tag{15.9}$$

where $\lambda' = \lambda\eta_1^{DB}\frac{\delta}{\rho^2}\exp(-\Xi)\frac{\gamma(\delta,\Xi\rho^{\alpha})}{\Xi^{\delta}}$; the sum in (15.9) quickly converges to the outage probability and only a small number of terms is required (less than 10). For $P, P_r \to \infty$, $\Psi = P_r/P$ (constant ratio), the outage probability is

$$\Pi_{DB}^{\infty} \approx \exp(-\lambda' \pi \rho^2) I_0 \left(2\rho \sqrt{\frac{\Xi(1 + d_0^{\alpha})\lambda' \pi}{\Psi}} \right) \rightarrow \exp(-\lambda \eta_1^{DB} \pi \rho^2). \quad (15.10)$$

Therefore, for high SNRs, the outage probability of the DB scheme is equal to the probability that the set C is empty. This probability is an exponential function of η_1^{DB} and approaches zero as $\lambda \eta_1^{\mathrm{DB}} \rho^2$ increases. However, the DB scheme corresponds to a higher system complexity and signaling overhead, since it requires a continuous feedback to enable coherent combination of the relaying signals at the destination.

15.1.3 Numerical Results

Unless otherwise stated, the simulations use the following parameters: $\rho = 3$ m, $\lambda = 1$, $\Psi = 0.1$, $\alpha = 3$, $\sigma^2 = 1$, $r_0 = 0.01$ BPCU; the dashed lines represent the theoretical results.

Figure 15.2 plots the outage probability performance of the proposed relay selection schemes versus the transmitted power P. The RRS and the RCS schemes converge to the same outage floor at high SNRs but the RCS scheme

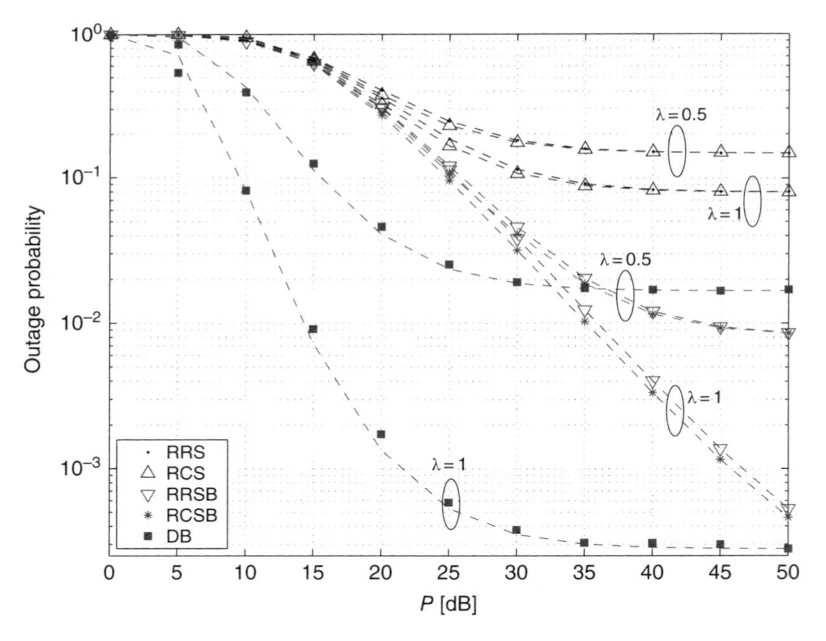

Figure 15.2 Outage probability versus P, $d_0 = 2\rho$. ©2015 IEEE. Reprinted, with permission, from [2].

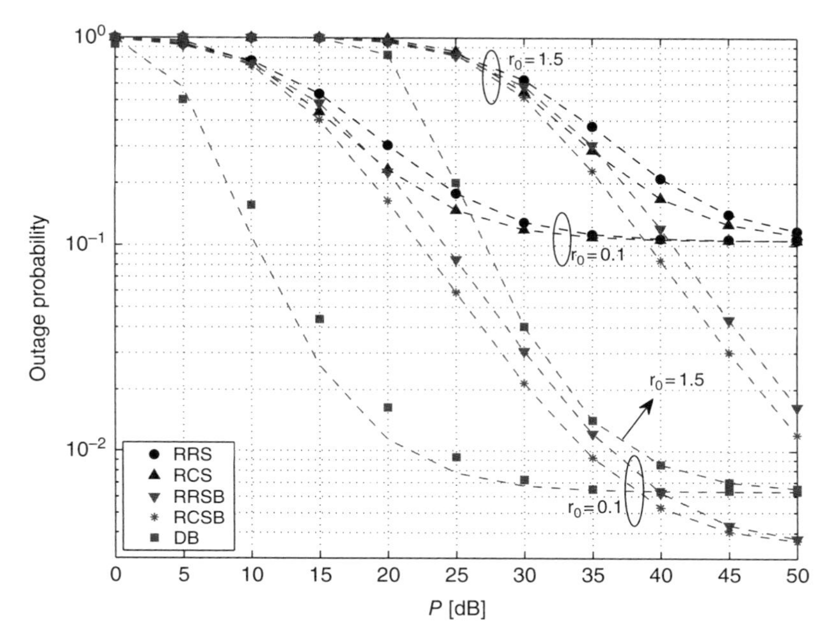

Figure 15.3 Outage probability versus P, $d_0 = \rho$. ©2015 IEEE. Reprinted, with permission, from [2].

slightly outperforms the RRS scheme at moderate SNRs and thus converges to the outage floor much faster. The RRSB and RCSB schemes, which take into account the battery status and avoid selection of uncharged relays, significantly improve the achieved performance and converge to the lowest outage floor; both schemes converge to the same outage floor at high SNRs. On the other hand, the DB scheme outperforms RRSB/RCSB schemes at low and moderate P. For these values, the transmission by multiple relays through beamforming boosts the SNR at the destination and improves the outage probability. In addition, as the density λ increases, more relays participate in the relaying operation and therefore the gap between DB and single-relay selection schemes increases.

In Figure 15.3 we plot the outage probability performance for different spectral efficiencies r_0. As it can be seen, the convergence outage floor of the relay selection schemes is independent of the spectral efficiency. Increasing the spectral efficiency affects only the convergence rate of the selection schemes (slower convergence). This observation has been expected, since according to our analysis the convergence floor only depends on the steady-state distribution of the battery, which is not a function of the spectral efficiency.

15.2 Wireless-Powered Ad Hoc Networks with SIC and SWIPT

In this section the employment of the SIC technique is studied in bipolar ad hoc networks, where the receivers employ SWIPT with the power splitting (PS) technique [5]. It is shown how each receiver can utilize SIC in order to boost the WPT without affecting the information decoding. Analytical and numerical results are presented for coverage probability and average harvested energy before and after the employment of SIC. The provided results demonstrate that SIC is significantly beneficial for SWIPT systems and it is shown that in certain scenarios the harvested energy converges to its upper bound, i.e. the case where all the power from the received signal is used for harvesting.

15.2.1 System Model

Network and channel model. Consider a large-scale bipolar ad hoc wireless network consisting of a random number of transmitter–receiver pairs [1]. The bipolar model is ideal for modeling ultra-dense ad hoc networks, where devices such as smartphones, tablets, sensors, etc. will be pairwise connected. The transmitters form a homogeneous PPP $\Phi = \{x_i \in \mathbb{R}^2\}$, $i \geq 1$, of density λ, where x_i denotes the location of the ith transmitter. Each transmitter x_i has a unique receiver at a distance d_0 in some random direction. Figure 15.4

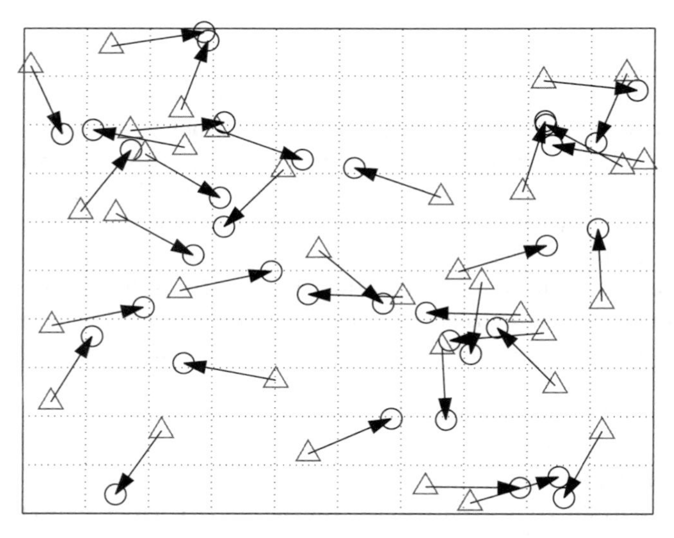

Figure 15.4 A snapshot of a bipolar ad hoc network; the receivers and transmitters are depicted by circles and triangles, respectively. ©2016 IEEE. Reprinted, with permission, from [6].

schematically shows the network model. All nodes are equipped with a single antenna. Time is slotted and in each time slot all transmitters are active without any coordination or scheduling. All wireless links suffer from both small-scale block fading and large-scale path-loss effects. The fading is Rayleigh distributed so the power of the channel fading is an exponential random variable with unit variance. We denote by h_i the channel coefficient for the link between the ith transmitter and the typical receiver. Moreover, all wireless links exhibit AWGN with variance σ^2. The path-loss model assumes the received power is proportional to $(1 + d_i^\alpha)^{-1}$ where d_i is the Euclidean distance from the origin to the ith transmitter and $\alpha > 2$ is the path-loss exponent.

Joint wireless information and power transfer model. The transmitters have a continuous power supply, such as a battery or the power grid, and transmit with the same power P_t. Each receiver has SWIPT capabilities and employs the PS method such that the received signal is split into two parts: one is converted to a baseband signal for information decoding and the other is directed to the rectenna for energy harvesting. Let $0 < \nu \leq 1$ denote the PS parameter, i.e. $100\nu\%$ of the received power is used for decoding. The additional circuit noise incurred during the RF to baseband conversion phase is modeled as an AWGN with zero mean and variance σ_C^2. Therefore, the signal-to-interference-plus-noise ratio (SINR) at the typical receiver is $\text{SINR}_0 = \frac{\nu P_t h_0 \tau^{-1}}{\nu(\sigma^2 + P_t I_0) + \sigma_C^2}$, where $\tau = 1 + d_0^\alpha$ and $I_0 = \sum_{x_i \in \Phi} h_i (1 + d_i^\alpha)^{-1}$, $i > 0$ is the aggregate interference at the typical receiver. The RF energy harvesting is a long-term operation and is expressed in terms of average harvested energy [5]. As $100(1 - \nu)\%$ of the received energy is used for rectification, the average energy harvested at the typical receiver is $E(\nu) = \zeta\, \mathbb{E}\{(1 - \nu) P_t (h_0 \tau^{-1} + I_0)\}$, where $0 < \zeta \leq 1$ denotes the conversion efficiency from RF to direct current (DC) voltage. Any RF energy harvesting from the AWGN noise is considered to be negligible.

Successive interference cancellation model. The receivers are assumed to apply SIC to cancel the n strongest interfering signals, $n \in \mathbb{N}$. Specifically, the receiver tries to decode the useful signal from its associated transmitter; if successful, no SIC is employed. Otherwise, the receiver attempts to decode the strongest interfering signal and remove it from the received signal. The SINR is then re-evaluated and the receiver re-attempts to decode the useful signal. If the decoding of the useful signal is still unsuccessful, the receiver proceeds to decode and remove the next strongest interfering signal. This procedure repeats up to n times, during which the receiver will either manage to decode the useful signal or after the nth attempt it will be in outage.

15.2.2 SWIPT with SIC

The performance of SIC is evaluated in terms of the coverage probability, that is, the probability the SINR is above a threshold θ, i.e. $\mathbb{P}\{\text{SINR} > \theta\}$. The coverage probability of a receiver which has not applied SIC is given by [6]

$$\Pi_{\text{NC}}(\nu) = \exp\left(-\frac{\theta\tau}{P_t}\left(\sigma^2 + \frac{\sigma_C^2}{\nu}\right) - \frac{2\pi^2\lambda\theta\tau}{\alpha(1+\theta\tau)^{\left(1-\frac{2}{\alpha}\right)}} \csc\left(\frac{2\pi}{\alpha}\right)\right). \quad (15.11)$$

Also, the coverage probability of a receiver attempting to decode the nth interferer is [6]

$$\Pi_{\text{D}}(\nu, n) = \int_0^{\infty} \frac{f(r,n)\exp\left(-\frac{2\pi\lambda\xi r^{2-\alpha}}{\alpha-2}\,{}_2F_1\left(1, 1-\frac{2}{\alpha}; 2-\frac{2}{\alpha}; -\frac{1+\xi}{r^{\alpha}}\right)\right)}{(1+\xi\tau^{-1})\exp\left(\frac{\xi}{P_t}\left(\sigma^2 + \frac{\sigma_C^2}{\nu}\right)\right)} dr, \quad (15.12)$$

where $\xi = \theta(1 + r^{\alpha})$ and $f(r, n) = \frac{2(\pi\lambda)^n}{\Gamma(n)} r^{2n-1} \exp(-\pi\lambda r^2)$. The coverage probability of a receiver which has successfully canceled n interferers is [6]

$$\Pi_{\text{C}}(\nu, n) = \int_0^{\infty} \frac{\exp\left(-\frac{\theta\tau}{P_t}\left(\sigma^2 + \frac{\sigma_C^2}{\nu}\right)\right) f(r,k)}{\exp\left(\frac{2\pi\lambda\theta\tau r^{2-\alpha}}{\alpha-2}\,{}_2F_1\left(1, 1-\frac{2}{\alpha}; 2-\frac{2}{\alpha}; -\frac{1+\theta\tau}{r^{\alpha}}\right)\right)} dr, \quad (15.13)$$

where $f(r, k) = \frac{2(\pi\lambda)^n}{\Gamma(n)} r^{2n-1} \exp(-\pi\lambda r^2)$.

Note that when no interferers have been canceled, $\Pi_{\text{C}}(\nu, 0) = \Pi_{\text{NC}}(\nu)$. The coverage probability of a receiver attempting to cancel up to n interferers is [6]

$$\Pi_{\text{SIC}}(\nu, n) = \Pi_{\text{NC}}(\nu) + \sum_{i=1}^{n} \left(\prod_{j=0}^{i-1} (1 - \Pi_{\text{C}}(\nu, j)) \right) \prod_{j=1}^{i} \Pi_{\text{D}}(\nu, j)\Pi_{\text{C}}(\nu, i) \quad (15.14)$$

and the average harvested energy by the typical receiver is [6]

$$E(\nu) = \zeta(1-\nu)P_t\left(\tau^{-1} + \frac{2}{\alpha}\pi^2\lambda\csc\left(\frac{2\pi}{\alpha}\right)\right). \quad (15.15)$$

We now look at how SIC can be exploited in order to increase the average harvested energy. As the employment of SIC provides an improvement to the coverage probability, the PS parameter can be reduced in such a way that the achieved performance is still as good as the case where SIC is not applied. The benefit from this method is that as the PS parameter decreases, more power is provided to the harvesting operation thus increasing the average harvested energy. In other words, we would like to compute the following

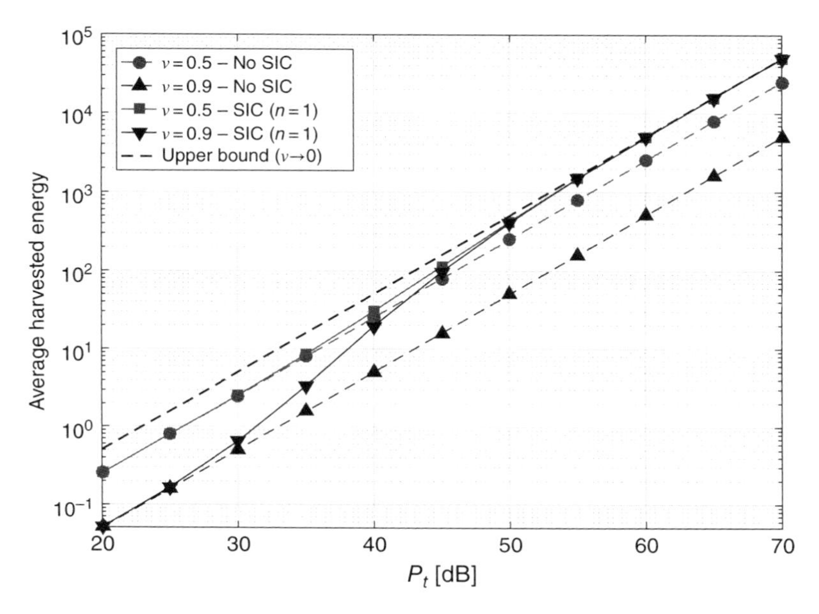

Figure 15.5 Average harvested energy versus P_t. ©2016 IEEE. Reprinted, with permission, from [6].

$$
\begin{aligned}
&\underset{\nu}{\text{maximize}}\ E(\nu) \\
&\text{subject to} \quad 0 < \nu \leq 1, \Pi_{\text{SIC}}(\nu, n) \geq \eta,
\end{aligned} \tag{15.16}
$$

where $\eta \geq \Pi_{\text{NC}}(\nu)$ is the target outage probability. In order to maximize the average harvested energy at the receiver, it suffices to minimize ν subject to the above constraints.

15.2.3 Numerical Results

Unless otherwise stated, we use $\lambda = 10^{-3}$, $P_t = 50$ dB, $d_0 = 10$ m, $\theta = -5$ dB, $\sigma^2 = 1$, $\sigma_C^2 = 1$, $\nu = 0.5$, $\zeta = 1$, and $\alpha = 4$. Furthermore, we set $\eta = \Pi_{\text{NC}}(\nu)$; solid and dashed lines represent the theoretical results.

Figure 15.5 shows the impact of the transmit power P_t and SIC on the average harvested energy. Clearly, the employment of our method provides significant energy harvesting gains. This is because an increase in P_t provides better quality signal, which results in more power for harvesting. For $\nu = 0.9$, the harvesting gains are noticeable from $P_t = 25$ dB whereas for $\nu = 0.5$ they are noticeable from $P_t = 40$ dB. This is again due to the fact that there is more than enough power at the receiver to satisfy the performance threshold θ so the adjustment of ν starts from smaller P_t. Moreover, for large values of P_t the energy harvested with the proposed method converges to the upper bound ($\nu \to 0$).

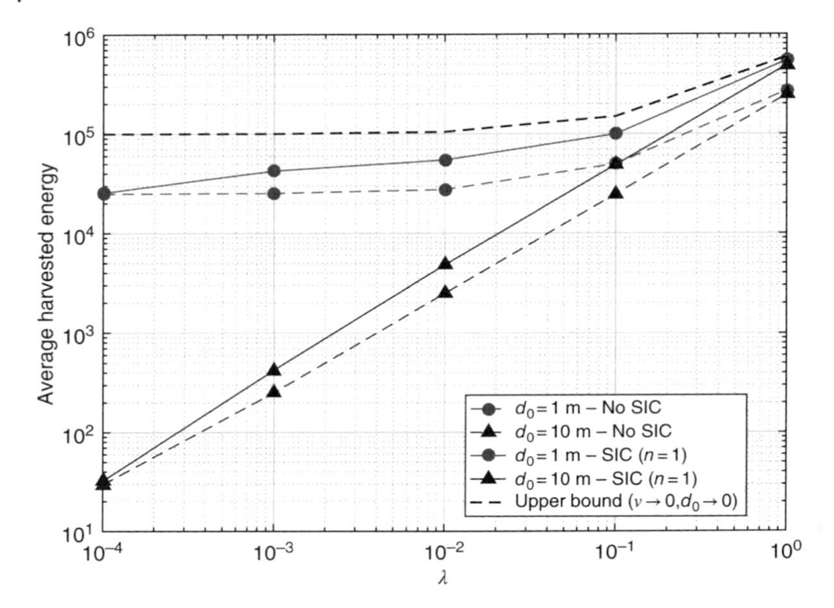

Figure 15.6 Average harvested energy versus λ. ©2016 IEEE. Reprinted, with permission, from [6].

Figure 15.6 illustrates how the density of the network λ and the distance d_0 affect the average harvested energy. As expected, the smaller the distance d_0 and the greater the density λ, the larger the average harvested energy. Also, our proposed method provides energy harvesting gains which increase with λ since the probability of decoding the strongest interfering signal is higher in a denser network. Thus SIC increases the coverage probability, which in turn provides energy harvesting benefits. For large values of λ the harvested energy converges to the same value regardless of d_0 since in this case the interfering signals dominate the network and so the energy harvested from the direct link is insignificant. In Figure 15.6, we also plot the upper bound of the average harvested energy for $\nu \to 0$ and $d_0 \to 0$; it is clear that our proposed method converges to the upper bound for large values of λ.

15.3 A Wireless-Powered Opportunistic Feedback Protocol

This section presents an OBF protocol with limited feedback in the context of WPCNs. The terminals are randomly deployed around an AP and are equipped with multiple antennas as well as a rectenna array to harvest energy. In the OBF scheme, the AP acquires channel-related feedback from the network's terminals and allocates the orthonormal beams to the users with the best link. We consider the case where the feedback is solely powered by energy harvested from electromagnetic radiation. Specifically, each terminal adjusts the length of its feedback based on the amount of harvested energy; when the energy is not sufficient to feed back at least one bit, the terminal is considered to be in outage. Two fundamental rectenna architectures are studied, the DC combiner and the RF combiner, as well as a hybrid architecture of these two. The beam outage probability is provided for all considered architectures and scenarios.

15.3.1 System Model

Network and channel model. Consider a WPCN with multiple randomly deployed terminals in a single cell. The terminals are spatially distributed according to a homogeneous PPP Φ in the Euclidean plane, with density λ [1]. The coverage area, denoted by $\mathcal{B}$, is modeled as a disc of radius ρ. An AP and a power beacon (PB) are co-located at the origin of $\mathcal{B}$ with a circular exclusion zone of radius ξ around them. The exclusion zone acts as a prohibited area for safety reasons and guarantees that all terminals are in the far-field of the PB. The PB operates at a different frequency band to the AP to avoid interfering with the communication links. Figure 15.7 schematically presents the considered system topology. The AP is equipped with M antennas, while the PB has a single transmit antenna. The terminals employ K antennas connected to one of two different combiner circuits: one for data decoding with N antenna elements and one for energy harvesting with L antenna elements, where $K = N + L$. All antennas are considered to be omnidirectional. Each terminal harvests energy from the PB's transmitted signals through its rectenna array configuration of L antenna elements [7]. Depending on the energy harvesting performance of the rectenna array, a feedback of b bits is returned to the AP; b is a discrete random variable that takes an integer value between 1 and M. We consider a batteryless architecture so any energy harvested is not stored but is used immediately to potentially operate the terminal's feedback mechanism.

All downlink wireless links suffer from both small-scale block fading and large-scale path-loss effects. The fading is Rayleigh distributed so the power of the channel fading is an exponential random variable with unit variance. We denote by $h_{k,i,j}$ the channel coefficient for the link between the AP's kth transmit antenna and the jth receive antenna of the ith terminal, and by $g_{i,j} \equiv |g_{i,j}| \exp(\jmath\theta_{i,j})$ the channel coefficient for the link between the PB and the jth rectenna of the ith terminal. All wireless links exhibit AWGN with variance σ^2; $n_{i,j}$ denotes the AWGN at the jth receive antenna of the ith terminal. The path-loss model assumes the received power is proportional to $d_i^{-\alpha}$ where d_i is the Euclidean distance from the origin to the ith terminal and $\alpha > 2$ is the path-loss exponent. The feedback channel at the uplink is assumed to suffer

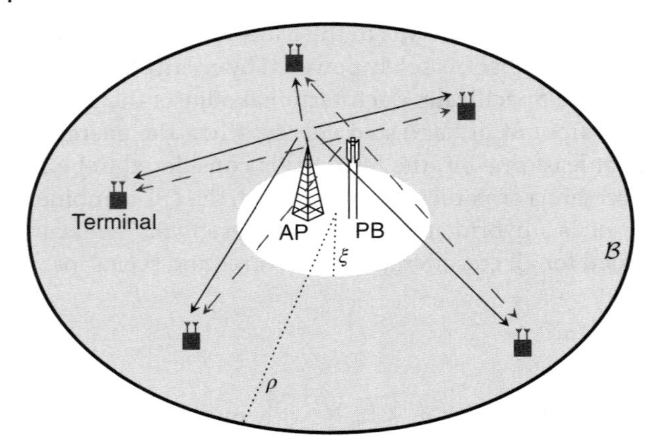

Figure 15.7 The considered system model; solid and dashed lines depict communication and power links, respectively.

only from path-loss effects, so a single antenna is randomly selected for the transmission and reception of the feedback channel.

Information and power transfer. The AP employs the OBF scheme to serve the terminals. It transmits M information streams, one for each beam, by generating M isotropic distributed random orthonormal vectors $\{\boldsymbol{u}_1, \dots, \boldsymbol{u}_M\}$ with $\boldsymbol{u}_m \in \mathbb{C}^{M\times 1}$, $1 \le m \le M$. By omitting time index and carriers, the baseband-equivalent transmitted signal is given by $\boldsymbol{v} = \sum_{m=1}^{M} \boldsymbol{u}_m s_m$, where $\mathbb{E}\{||\boldsymbol{v}||^2\} = M$ and s_m is the mth transmitted symbol with $\mathbb{E}\{|s_m|^2\} = 1$ for all $m = 1, \dots, M$. Then, the signal received at the jth antenna of the ith terminal is given by $r_{i,j} = \sqrt{P_t d_i^{-\alpha}} \boldsymbol{h}_{i,j}^T \boldsymbol{v} + n_{i,j}$, where $\boldsymbol{h}_{i,j}^T = [h_{1,i,j}, \dots, h_{M,i,j}]$ and P_t denotes the AP's transmit power. Therefore, the SINR for the lth beam at the jth receive antenna of the ith terminal is $\gamma_{i,j,l} = \frac{P_t |\boldsymbol{h}_{i,j}^T \boldsymbol{u}_l|^2}{\sigma^2 d_i^{\alpha} + P_t \sum_{m\neq l}^{M} |\boldsymbol{h}_{i,j}^T \boldsymbol{u}_m|^2}$. In order to increase the energy efficiency and decrease the implementation complexity, each terminal employs a selection combiner scheme. Hence, the achieved SINR for the lth beam at the ith terminal is $\gamma_{i,l} = \max_{1\le j\le N} \gamma_{i,j,l}$.

Each terminal harvests energy from the PB's transmitted signal to power its feedback mechanism. The transmitted RF signal from the PB is $s(t) = \sqrt{2P_h}\Re\{\exp(\jmath[2\pi f_c t + \arg x(t)])\}$, where $P_h = \mathbb{E}\{s^2(t)\}$ is the PB's transmit power, f_c is the carrier frequency, and $x(t)$ is a modulated energy signal with $|x(t)|^2 = 1$. Thus, the received signal at the jth antenna of the ith terminal is $y_{i,j}(t) = \sqrt{2P_h d_i^{-\alpha}} |g_{i,j}(t)| \cos(2\pi f_c t + \arg x(t) + \theta_{i,j}(t))$, where $|g_{i,j}(t)|$ is a Rayleigh random variable with unit scale parameter. The received signal is converted to DC with the use of a rectifier; energy harvesting from the AWGN is considered negligible and it is ignored. The output current of the diode for the jth antenna of the ith terminal is given by $I_{i,j}(t) = I_s \left(\exp\left(\frac{y_{i,j}(t)}{\mu V_T}\right) - 1\right) = I_s \sum_{k=1}^{\infty} \frac{1}{k!} \left(\frac{y_{i,j}(t)}{\mu V_T}\right)^k$, where I_s denotes the reverse saturation current of the diode, $\mu \in [1\ 2]$ is an ideality factor which is a function of the operating conditions and physical contractions, and V_T is the thermal voltage.

Rectenna array architectures. Each terminal is equipped with a rectenna array of L elements to boost the rectification process and increase its efficiency. The interconnection of these elements is performed in the DC domain, the RF domain or a hybrid combination of the two [7, 8]. Consider a hybrid DC/RF combiner, implementing $L_r \leq L$ elements over ℓ subarrays of the RF combiner such that $L_r = \sum_{k=1}^{\ell} L_{r_k}$, where L_{r_k} is the number of antenna elements of the kth RF combiner, and a subarray of $L_d = L - L_r$ elements directly connected to the DC combiner. Each RF combiner merges its antenna inputs in the RF domain and all $\ell + L_d$ outputs are then combined in the DC domain. Each element of the DC combiner has its own rectification circuit and is not affected by the remaining elements; the signal at the jth antenna of the ith terminal is as given above. On the other hand, the RF combiner merges the inputs in the RF domain and requires a single rectification circuit. In this case, the signals can be combined in two ways: coherently and non-coherently.

In the coherent case, the rectenna array aligns the phases of the incoming signals so that they are combined co-phased. A phase shifter circuit can be employed to form a beam by continuously shifting the phase of the incoming signal at each antenna; this can be implemented with passive elements [9]. Therefore, the combined signal at the kth RF combiner of the ith terminal is $y_{i,k}^{c}(t) = \sum_{j=1}^{L_{r_k}} y_{i,j,k}(t) = \sqrt{2P_h d_i^{-\alpha}} \cos(2\pi f_c t + \arg x(t) + \theta_i(t)) \sum_{j=1}^{L_{r_k}} |g_{i,j,k}(t)|$, where $y_{i,j,k}(t)$ is the received signal at the kth RF combiner's jth antenna of the ith terminal and $|g_{i,j,k}(t)|$ is a Rayleigh random variable with unit scale parameter. For the non-coherent case, no RF alignment is required as the signals are combined non-coherently. Here, the combined signal is $y_{i,k}^{c}(t) = \sum_{j=1}^{L_{r_k}} y_{i,j,k}(t) = \sqrt{2P_h d_i^{-\alpha}} |c_{i,k}(t)| \cos(2\pi f_c t + \arg x(t) + \theta_{i,k}(t))$, where $|c_{i,k}(t)| \exp(\jmath\theta_{i,k}(t))$ is a circularly symmetric complex Gaussian random variable with zero mean and variance L_{r_k}. The output current from each diode is processed by a LPF which produces a relatively smooth DC current, given by $I_i^{\mathrm{DC}}(t) = \frac{I_s P_h d_i^{-\alpha}}{(\mu V_T)^2} \left(\sum_{j=1}^{L_d} |g_{i,j}(t)|^2 + \sum_{k=1}^{\ell} \omega_k(t) \right)$, where $|g_{i,j}(t)|^2$ is the output of the jth rectenna connected to the DC combiner, and $\omega_k(t)$ is the output of the kth RF combiner given by $|c_{i,k}(t)|^2$ for the non-coherent case, and $\left(\sum_{j=1}^{L_{r_k}} |g_{i,j,k}(t)| \right)^2$ for the coherent case where $|g_{i,j,k}(t)|$ is the input from the jth antenna element to the kth coherent RF combiner. Then, the total harvested DC power is a linear function of $I_i^{\mathrm{DC}}(t)$, that is, $P_i(t) = \zeta I_i^{\mathrm{DC}}(t)$, where ζ is the the combiner's conversion efficiency. Note that the total harvested DC power for a strictly DC combiner is given by simply setting $\ell = 0$ to $I_i^{\mathrm{DC}}(t)$ or when $\ell > 0$ and $L_{r_i} = 1$, for all i; similarly, for an RF combiner, it is given by setting $L_d = 0$ to $I_i^{\mathrm{DC}}(t)$.

15.3.2 Wireless-Powered OBF Protocol

The OBF scheme aims to maximize the sum-rate by assigning the M beams to the M "best" terminals. In the conventional scheme, each terminal returns a feedback to the AP for the achieved SINR from each beam. Then, the AP assigns each beam to the terminal with the highest SINR for that specific beam. However, this has significant demands in terms of system resources, such as bandwidth and energy. The proposed scheme considers terminals that harvest energy from RF signals to power their feedback transmission. The transmitted feedback is of finite length and it is entirely determined by the harvested energy. A detailed step-by-step description of the proposed protocol is given below:

1) A random ordered pre-assignment of the M beams for each terminal is performed. All assignments are known to the AP and each terminal has knowledge of its own ordered assignment.
2) The AP broadcasts M beamforming vectors and the PB transmits RF signals to the terminals over different frequency bands.
3) Each terminal harvests energy through its rectenna array. If it is sufficient to return $b \geq 1$ bits of feedback, the terminal measures the SINR for the first b beams in its ordered assignment. Otherwise, the terminal remains idle.
4) Each active terminal transmits b bits of feedback back to the AP, $1 \leq b \leq M$. The feedback consists of one bit per beam where the ith bit describes whether or not the ith measured SINR is above a pre-assigned threshold τ.
5) The AP evaluates the received feedback and randomly assigns each beam to a terminal which returned a positive feedback for that beam. If all feedback for a specific beam is negative, the assignment is done randomly.

The one-bit-per-beam approach is considered since a certain pre-assignment of the beams can achieve the optimal scaling law of the sum-rate with just one bit of feedback [10]. Moreover, the small WPT efficiency and the associated doubly near-far problem makes the one-bit feedback channel ideal [11].

15.3.3 Beam Outage Probability

A terminal harvests energy from the PB's transmitted signals and attempts to return a feedback of length up to M bits. When the harvested energy is sufficient to return at least one bit, the terminal becomes active, otherwise it remains idle. For the ith terminal to transmit a feedback of random number of b bits, $1 \leq b \leq M$, the Shannon capacity of the uplink between the terminal and the AP should be greater or equal to b bits per channel use, that is, $C_Q = \log_2\left(1 + \frac{P_i^Q(t)}{d_i^\alpha \sigma^2}\right) \geq b \Rightarrow P_i^Q(t) \geq \phi(b) d_i^\alpha \sigma^2$, where $\phi(b) \triangleq 2^b - 1$, $P_i^Q(t)$ is the total harvested DC power by the $\mathcal{Q}$ combiner, and $\mathcal{Q} \in \{\mathrm{H}, \mathrm{D}, \mathrm{R}_n, \mathrm{R}_c\}$ refers to the hybrid, DC, non-coherent RF, and coherent RF combiner, respectively. We now provide the probability of a terminal returning less than b bits of feedback, i.e. $\Pi^{\mathcal{Q}}(b) = \mathbb{P}\{P_i^{\mathcal{Q}} < \phi(b) d_i^{\alpha} \sigma^2\}$; note that $\Pi^{\mathcal{Q}}(1)$ is the idle probability. We first consider the hybrid combiner and, for the sake of tractability, we consider the one implementing the non-coherent RF combiner; the number of antenna elements are assumed to be equal for all ℓ subarrays, which we denote by L_{r_0}.

The probability of returning fewer than b bits of feedback for a terminal employing the hybrid combiner is [8]

$$\Pi^{\mathrm{H}}(b) = 1 - \frac{(\phi(b)\chi)^{-\frac{1}{\alpha}}}{\alpha(\rho^2 - \xi^2) L_{r_0}^{\ell}} \left[\sum_{k=1}^{L_d} \Gamma(k) c_k \Delta_1(k) + \sum_{k=1}^{\ell} \Gamma(k) L_{r_0}^{k+\frac{1}{\alpha}} c'_k \Delta_2(k) \right], \tag{15.17}$$

where $\chi \triangleq \frac{(\mu V_T)^2 \sigma^2}{\zeta I_s P_h}$, $L_{r_0} > 1$, $\Delta_i(k) \triangleq \sum_{m=0}^{k-1} \frac{\Gamma\left(m+\frac{1}{\alpha}, \mathcal{L}_i \phi(b) \chi \xi^{2\alpha}\right) - \Gamma\left(m+\frac{1}{\alpha}, \mathcal{L}_i \phi(b) \chi \rho^{2\alpha}\right)}{m!}$, $\mathcal{L}_i \in \{1, 1/L_{r_0}\}$, $c_{s_1} = \frac{(\mathcal{L}_2 - \mathcal{L}_1)^{-s_2}}{(s_1-1)!}$, $c_{s_1-n_1} = \frac{s_2}{n_1} \sum_{j=1}^{n_1} \frac{(s_1-n_1+j-1)! c_{s_1-(n_1-j)}}{(s_1-n_1-1)!(\mathcal{L}_1-\mathcal{L}_2)^j}$, $n_1 = 1, \ldots, s_1 - 1$, $c'_{s_2} = \frac{(\mathcal{L}_1 - \mathcal{L}_2)^{-s_1}}{(s_2-1)!}$, $c'_{s_2-n_2} = \frac{s_1}{n_2} \sum_{j=1}^{n_2} \frac{(s_2-n_2+j-1)! c'_{s_2-(n_2-j)}}{(s_2-n-1)!(\mathcal{L}_2-\mathcal{L}_1)^j}$, $n_2 = 1, \ldots, s_2 - 1$, and $s_i \in \{L_d, \ell\}$.

By setting $\ell = 0$ or $L_{r_0} = 1$ in (15.17) we get the probability of returning fewer than b bits of feedback for a terminal employing the DC combiner [8]

$$\Pi^{\mathrm{D}}(b) = 1 - \frac{(\phi(b)\chi)^{-\frac{1}{\alpha}}}{\alpha(\rho^2 - \xi^2)} \sum_{m=0}^{L-1} \frac{\Gamma\left(m + \frac{1}{\alpha}, \phi(b)\chi\xi^{2\alpha}\right) - \Gamma\left(m + \frac{1}{\alpha}, \phi(b)\chi\rho^{2\alpha}\right)}{m!}. \tag{15.18}$$

Similarly, by setting $L_d = 0$ and $\ell = 1$, the probability of returning fewer than b bits of feedback for the non-coherent RF combiner with one subarray is [8]

$$\Pi^{\mathrm{R}_n}(b) = 1 - \frac{\left(\frac{L}{\phi(b)\psi}\right)^{\frac{1}{\alpha}}}{\alpha(\rho^2 - \xi^2)} \left[\Gamma\left(\frac{1}{\alpha}, \phi(b)\frac{\psi}{L}\xi^{2\alpha}\right) - \Gamma\left(\frac{1}{\alpha}, \phi(b)\frac{\psi}{L}\rho^{2\alpha}\right) \right], \tag{15.19}$$

where $\psi \triangleq \frac{(\mu V_T)^2 \sigma^2}{\zeta I_s P_h}$. Finally, the probability of returning fewer than b bits of feedback for the coherent RF combiner with one subarray ($\ell = 1$) is [8]

$$\Pi^{\mathrm{R}_c}(b) = 1 - \frac{\left(\frac{2\beta}{\phi(b)\psi}\right)^{\frac{1}{\alpha}}}{\alpha(\rho^2 - \xi^2)} \sum_{k=0}^{L-1} \frac{\Gamma\left(k + \frac{1}{\alpha}, \frac{\phi(b)\psi\xi^{2\alpha}}{2\beta}\right) - \Gamma\left(k + \frac{1}{\alpha}, \frac{\phi(b)\psi\rho^{2\alpha}}{2\beta}\right)}{k!}, \tag{15.20}$$

where $\beta = ((2L - 1)!!)^{1/L}$.

A beam is in outage if the achieved SINR is less than a target SINR τ. The cumulative distribution function of the observed beam SINR γ is given by [8]

$$F_\gamma(\tau) = 1 + \frac{2\left(\frac{P_t}{\sigma^2\tau}\right)^{\frac{2}{\alpha}}}{\alpha(\rho^2 - \xi^2)} \sum_{m=1}^{N} \frac{(-1)^m \binom{N}{m} \left[\Gamma\left(\frac{2}{\alpha}, \frac{\sigma^2\tau m\xi^\alpha}{P_t}\right) - \Gamma\left(\frac{2}{\alpha}, \frac{\sigma^2\tau m\rho^\alpha}{P_t}\right)\right]}{m^{\frac{2}{\alpha}}(\tau+1)^{m(M-1)}}. \tag{15.21}$$

An outage event occurs when no terminal returns a positive feedback and the link (if any) between the AP and the randomly selected terminal is in outage. So, the beam outage probability for the lth beam and the Q combiner is [8]

$$P_{\text{out}}^{Q} = \exp(-\lambda^Q|\mathcal{B}|(1 - F_\gamma(\tau)))[\exp(-\overline{\lambda^Q}|\mathcal{B}|)(1 - F_\gamma(\tau)) + F_\gamma(\tau)], \tag{15.22}$$

where $Q \in \{\text{H}, \text{D}, \text{R}_n, \text{R}_c\}$, $F_\gamma(\tau)$ is defined by (15.21), $|\mathcal{B}| = \pi(\rho^2 - \xi^2)$, λ^Q is the density of the terminals that return feedback for the lth beam given by $\lambda^Q = \left[1 - \Pi^Q(M) + \sum_{b=1}^{M-1} \frac{b}{M}(\Pi^Q(b+1) - \Pi^Q(b))\right] \lambda$, and $\overline{\lambda^Q} = \lambda - \lambda^Q$ is the density of the ones that are idle. For $P_t, P_h \to \infty$, the beam outage probability for all rectenna architectures converges to

$$P_{\text{out}}^{\infty} = \exp\left(-\lambda\pi(\rho^2 - \xi^2)\left(1 - \left[1 - \frac{1}{(\tau+1)^{M-1}}\right]^N\right)\right). \tag{15.23}$$

The error floor is independent of both the size of the rectenna array and the implemented combiner. This is expected since in this case all terminals harvest enough energy to feed back M bits, irrespective of the rectenna architecture.

15.3.4 Numerical Results

Unless otherwise stated, the following parameters are used: $M = 2$, $\lambda = 0.5$, $\xi = 2$ m, $\rho = 10$ m, $\alpha = 3$, $\tau = 20$ dB, $I_s = 1$ mA, $V_T = 28.85$ mV, $\mu = 2$, $\zeta = 0.9$, $\sigma^2 = -50$ dBm, $P_t = P_h$, $L_d = L_r = L/2$, and $\ell = 1$; solid and dashed lines represent the theoretical results.

Figure 15.8 illustrates the probability $\Pi^Q(b)$ of a terminal returning fewer than b bits of feedback. As expected, $\Pi^Q(b)$ decreases as the transmit power increases. Also, when the terminals are required to feed back $b = 3$ bits, their performance is much lower than when $b = 2$ bits. Comparing the architectures, the best performance is provided by the coherent RF combiner since combining the signals co-phased produces significant gains. The worst performance is provided by the non-coherent RF combiner whereas the DC combiner comes second best. These observations are true when all combiners have the same parameter ζ, that is, a more efficient non-coherent RF combiner could outperform a DC combiner. Finally, as expected, the hybrid combiner's performance lies between the performance of the DC and the non-coherent RF combiner.

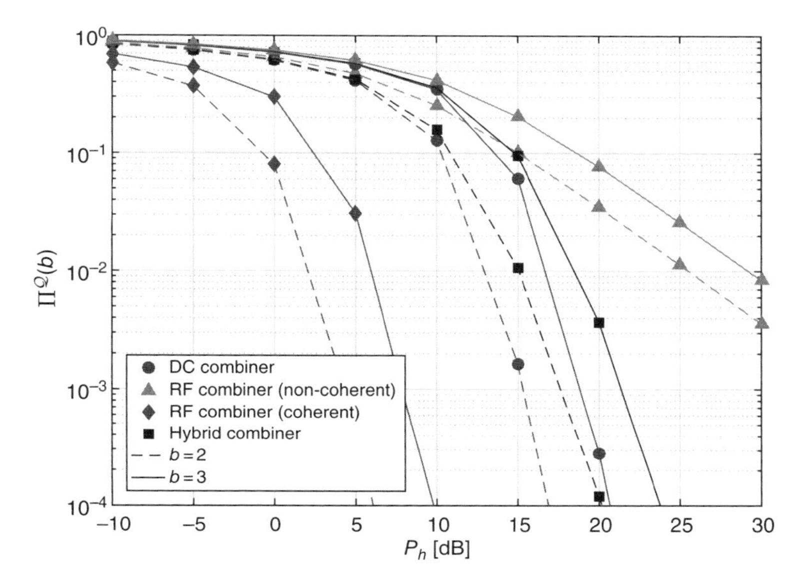

Figure 15.8 $\Pi^{\mathcal{Q}}(b)$ versus P_h, $L = 8$.

Figure 15.9 depicts the beam outage probability versus the transmit power with $N = 4$ and $L = 6$. The proposed scheme is compared with the random beamforming (no feedback) and the full-feedback beamforming, i.e. when the terminals return SINR information for all beams. The first main observation is that the proposed scheme significantly outperforms the random beamforming. However, the proposed scheme does not perform as well as full-feedback beamforming since full-feedback means that the AP has full knowledge of the terminals' SINRs. However, the performances converge to the same outage probability floor for $P_t, P_h \to \infty$. It is obvious that the coherent RF combiner achieves a lower outage probability for low to moderate P_t, P_h values, whereas for $P_t, P_h \to \infty$ all architectures converge to the same outage probability floor.

15.4 Conclusion

This chapter dealt with a macroscopic approach of WPT. In particular, three different wireless-powered communication scenarios were presented from a large-scale point of view with the use of stochastic geometry, a useful mathematical tool. Stochastic geometry is ideal for studying large-scale wireless-powered networks since it captures the main issues of such networks, e.g., the doubly near-far problem, and can provide closed-form expressions as well as study the impact of the considered system's main parameters.

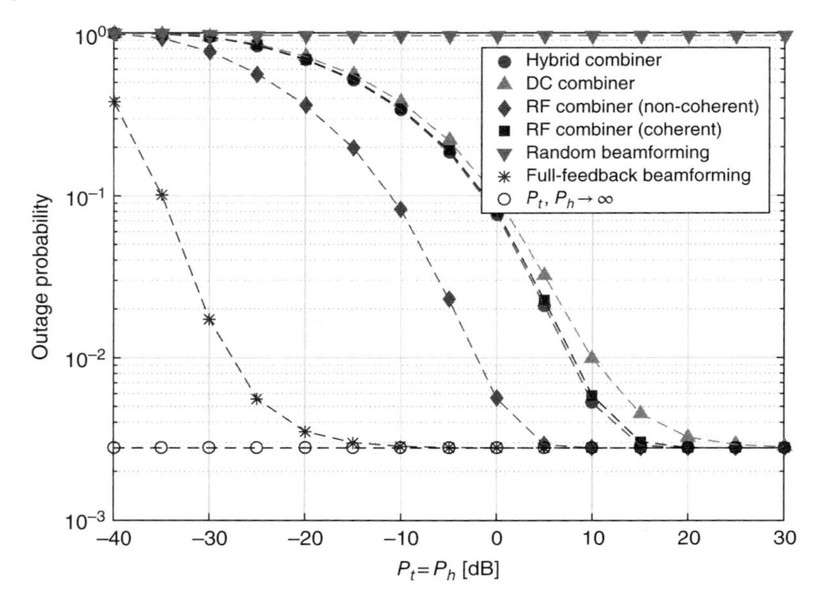

Figure 15.9 Outage probability versus $P_t = P_h$, $N = 4$, $L = 6$.

Bibliography

1 M. Haenggi (2013) *Stochastic geometry for wireless networks.* Cambridge University Press.

2 I. Krikidis (2015) Relay selection in wireless powered cooperative networks with energy storage. *IEEE J. Selected Areas Commun.*, **33**: 2596–2610.

3 S. Lee, R. Zhang, and K. Huang (2013) Opportunistic wireless energy harvesting in cognitive radio networks. *IEEE Trans. Wireless Commun.*, **12**: 4788–4799.

4 Z. Ding, I. Krikidis, B. Sharif, and H.V. Poor (2014) Wireless information and power transfer in cooperative networks with spatially random relays. *IEEE Trans. Wireless Commun.*, **13**: 4440–4453.

5 I. Krikidis (2014) Simultaneous information and energy transfer in large-scale networks with/without relaying. *IEEE Trans. Commun.*, **62**: 900–912.

6 C. Psomas and I. Krikidis (2016) Successive interference cancellation in bipolar ad hoc networks with SWIPT. *IEEE Wireless Commun. Lett.*, **5**: 364–367.

7 U. Olgun, C.-C. Chen, and J.L. Volakis (2011) Investigation of rectenna array configurations for enhanced RF power harvesting. *IEEE Ant. Wireless Prop. Lett.*, **10**: 262–265.

8 C. Psomas and I. Krikidis (2018) A wireless powered feedback protocol for opportunistic beamforming using rectenna arrays. *IEEE Trans. Green Commun. Network.*, **2** (1): 100–113.

9 F. Meng, K. Ma, K. S. Yeo, and S. Xu (2016) A 57-to-64-GHz 0.094-mm^2 5-bit passive phase shifter in 65-nm CMOS. *IEEE Trans. Very Large Scale Integr. (VLSI) Syst.*, **24** (5): 1917–1925.

10 J. Diaz, O. Simeone, and Y. Bar-Ness (2006) How many bits of feedback is multiuser diversity worth in MIMO downlink? In *Proceedings of the IEEE Symposium on Spread Spectral Technical Applications*, Manaus, Brazil, pp. 505–509.

11 I. Krikidis (2015) Opportunistic beamforming with wireless powered 1-bit feedback through rectenna array. *IEEE Signal Proc. Lett.*, **22** (11): 2054–2058.

Index

Wireless Information and Power Transfer: Theory and Practice, First Edition.
Edited by Derrick Wing Kwan Ng, Trung Q. Duong, Caijun Zhong, and Robert Schober.